Hydraulic Design of Energy Dissipators for Culverts and Channels

Hydraulic Engineering Circular 14 – "Energy Dissipators"
Listing of Updates and Corrections (errata & corrigenda)

DATE	ACTION	BY
13 Aug 2012	Page 4-2, first paragraph: text should have parentheses around the '3Fr' quantity: "… if the $\tan\theta$ is greater than 1/**(3Fr)**, …"	JSK
1 Oct 2012	Page 4-2 Figure 4.2 Blaisdell's (1/3Fr) should be Blaisdell's (1/(3Fr))	CLN
13 Aug 2012	Page 4-4, Equation 4.3 should be: "$\theta = \tan^{-1}(1/$**(3Fr)**$)$"	JSK
13 Aug 2012	Page 4-4, third paragraph: text should have parentheses around the '3Fr' quantity: "… flaring the wingwall more than 1/**(3Fr)** (for example 45°) …"	JSK
13 Aug 2012	Page 4-5, Step 3: equation should depict: "$\tan\theta = 1/$**(3Fr)** $= 1/$**(3(1.52))** $= 0.22$"	JSK
1 Oct 2012	Page 4-6, Alternative 2 … $\tan\theta = 1/3Fr$ should be $\tan\theta = 1/(3Fr)$	CLN
13 Aug 2012	Page 4-7, Step 3: equation should depict: "$\tan\theta = 1/$**(3Fr)** $= 1/$**(3(1.52))** $= 0.22$"	JSK
1 Oct 2012	Page 4-8, Alternative 2 … $\tan\theta = 1/3Fr$ should be $\tan\theta = 1/(3Fr)$	CLN

This list includes all known items as of Monday 1 October 2012

Notes:
No update to the publication is planned at this time.
FHWA does not have any printed copies of this document. NHI allows purchase of some FHWA documents.
See the FHWA Hydraulics website to report any additional errata and corrigenda.

Technical Report Documentation Page

1. Report No. FHWA-NHI-06-086 HEC 14	2. Government Accession No.	3. Recipient's Catalog No.
4. Title and Subtitle Hydraulic Design of Energy Dissipators for Culverts and Channels Hydraulic Engineering Circular Number 14, Third Edition		5. Report Date July 2006
		6. Performing Organization Code
7. Author(s) Philip L. Thompson and Roger T. Kilgore		8. Performing Organization Report No.
9. Performing Organization Name and Address Kilgore Consulting and Management 2963 Ash Street Denver, CO 80207		10. Work Unit No. (TRAIS)
		11. Contract or Grant No. DTFH61-02-D-63009/T-63047
12. Sponsoring Agency Name and Address Federal Highway Administration National Highway Institute Office of Bridge Technology 4600 North Fairfax Drive 400 Seventh Street, SW Suite 800 Room 3203 Arlington, Virginia 22203 Washington D.C. 20590		13. Type of Report and Period Covered Final Report (3rd Edition) July 2004 – July 2006
		14. Sponsoring Agency Code

15. Supplementary Notes
Project Manager: Cynthia Nurmi – FHWA Resource Center Technical Assistance: Jorge Pagan, Bart Bergendahl, Sterling Jones (FHWA); Rollin Hotchkiss (consultant)

16. Abstract
The purpose of this circular is to provide design information for analyzing and mitigating energy dissipation problems at culvert outlets and in open channels. The first three chapters provide general information on the overall design process (Chapter 1), erosion hazards (Chapter 2), and culvert outlet velocity and velocity modification (Chapter 3). These provide a background and framework for anticipating dissipation problems. In addition to describing the overall design process, Chapter 1 provides design examples to compare selected energy dissipators. The next three chapters provide assessment tools for considering flow transitions (Chapter 4), scour (Chapter 5), and hydraulic jumps (Chapter 6). For situations where the tools in the first six chapters are insufficient to fully mitigate a dissipation problem, the remaining chapters address the design of six types of constructed energy dissipators. Although any classification system for dissipators is limited, this circular uses the following breakdown: internal (integrated) dissipators (Chapter 7), stilling basins (Chapter 8), streambed level dissipators (Chapter 9), riprap basins and aprons (Chapter 10), drop structures (Chapter 11), and stilling wells (Chapter 12). Much of the information presented has been taken from the literature and adapted, where necessary, to fit highway needs. Research results from the Turner Fairbank Highway Research Center and other facilities have also been incorporated. A survey of state practices and experience was also conducted to identify needs for this circular.

17. Key Word energy dissipator, culvert, channel, erosion, outlet velocity, hydraulic jump, internal dissipator, stilling basin, impact basin, riprap basin, riprap apron, drop structure, stilling well	18. Distribution Statement This document is available to the public from the National Technical Information Service, Springfield, Virginia, 22151

19. Security Classif. (of this report) Unclassified	20. Security Classif. (of this page) Unclassified	21. No. of Pages 287	22. Price

Form DOT F 1700.7 (8-72)   Reproduction of completed page authorized

ACKNOWLEDGMENTS

First Edition

The first edition of this Circular was prepared in 1975 as an integral part of Demonstration Project No. 31, "Hydraulic Design of Energy Dissipators for Culverts and Channels," sponsored by Region 15. Mr. Philip L. Thompson of Region 15 and Mr. Murray L. Corry of the Hydraulics Branch wrote sections, coordinated, and edited the Circular. Dr. F. J. Watts of the University of Idaho (on a year assignment with Hydraulics Branch), Mr. Dennis L. Richards of the Hydraulics Branch, Mr. J. Sterling Jones of the Office of Research, and Mr. Joseph N. Bradley, Consultant to the Hydraulics Branch, contributed substantially by writing sections of the Circular. Mr. Frank L. Johnson, Chief, Hydraulics Branch, and Mr. Gene Fiala, Region 10 Hydraulics Engineer, supported the authors by reviewing and discussing the drafts of the Circular. Mr. John Morris, Region 4 Hydraulics Engineer, collected research results and assembled a preliminary manual that was used as an outline for the first draft. Mrs. Linda L. Gregory and Mrs. Silvia M. Rodriguez of the Region 15 Word Processing center and Mrs. Willy Rudolph of the Hydraulics Branch aided in manual preparation. The authors wish to express their gratitude to the many individuals and organizations whose research and designs are incorporated into this Circular.

Second Edition

Mr. Philip Thompson and Mr. Dennis Richards updated the first edition in 1983 so that HEC 14 could be reprinted and distributed as a part of Demonstration Project 73. The 1983 edition did not add any new dissipators, but did correct all the identified errors in the first edition. A substantial revision for Chapter 5, Estimating Erosion at Culvert Outlets, was accomplished using material that was published by Dr. Steven Abt, Dr. James Ruff, and Dr. A Shaikh in 1980. The second edition was prepared in U.S. customary units.

Third Edition

Mr. Philip Thompson and Mr. Roger Kilgore prepared this third edition of the Circular with the assistance of Dr. Rollin Hotchkiss. This edition retains all of the dissipators featured in the second edition, except the Forest Service (metal), USBR Type II stilling basin, and the Manifold stilling basin. The following dissipators have been added: USBR Type IX baffled apron, riprap aprons, broken-back culverts, outlet weir, and outlet drop followed by a weir. This edition is in both U.S. customary and System International (SI) units. A previous SI unit version of HEC 14 was published in 2000 as a part of the FHWA Hydraulics Library on CDROM, FHWA-IF-00-022.

TABLE OF CONTENTS

LIST OF TABLES

LIST OF FIGURES

LIST OF SYMBOLS

a	=	Acceleration, m/s^2 (ft/s^2)
A	=	Area of flow, m^2 (ft^2)
A_o	=	Area of flow at culvert outlet, m^2 (ft^2)
B	=	Width of rectangular culvert barrel, m (ft)
D	=	Diameter or height of culvert barrel, m (ft)
D_{50}	=	Particle size of gradation, of which 50 percent, of the mixture is finer by weight, m (ft)
E	=	Energy, m (ft)
f	=	Darcy-Weisbach resistance coefficient
F	=	Force, N (lb)
Fr	=	Froude number, ratio of inertial forces to gravitational force in a system
g	=	gravitational acceleration, m/s^2 (ft/s^2)
H_L	=	Head loss (total), m (ft)
H_f	=	Friction head loss, m (ft)
n	=	Manning's flow roughness coefficient
P	=	Wetted perimeter of flow prism, m (ft)
q	=	Discharge per unit width, m^2/s (ft^2/s)
Q	=	Discharge, m^3/s (ft^3/s)
r	=	Radius
R	=	Hydraulic radius, A/P, m (ft)
R_e	=	Reynolds number
S	=	Slope, m/m (ft/ft)
S_f	=	Slope of the energy grade line, m/m (ft/ft)
S_o	=	Slope of the bed, m/m (ft/ft)
S_w	=	Slope of the water surface, m/m (ft/ft)
T	=	Top width of water surface, m (ft)
TW	=	Tailwater depth, m (ft)
V	=	Mean Velocity, m/s (ft/s)
V_n	=	Velocity at normal depth, m/s (ft/s)
y	=	Depth of flow, m (ft)
y_e	=	Equivalent depth $(A/2)^{1/2}$, m (ft)
y_m	=	Hydraulic depth (A/T), m (ft)
y_n	=	Normal depth, m (ft)
y_c	=	Critical depth, m (ft)
y_o	=	Outlet depth, m (ft)
Z	=	Side slope, sometimes expressed as 1:Z (Vertical:Horizontal)
α	=	Unit conversion coefficient (varies with application)
α	=	Kinetic energy coefficient; inclination angle
β	=	Velocity (momentum) coefficient; wave front angle
γ	=	Unit Weight of water, N/m^3 (lb/ft^3)

θ   = Angle: inclination, contraction, central

μ   = Dynamic viscosity, $N \cdot s/m^2$ ($lb \cdot s/ft^2$)

ν   = Kinematic viscosity, m^2/s (ft^2/s)

ρ   = Mass density of fluid, kg/m^3 ($slugs/ft^3$)

τ   = Shear stress, N/m^2 (lb/ft^2)

GLOSSARY

Basin: Depressed or partially enclosed space.

Customary Units (CU): Foot-pound system of units also referred to as English units.

Depth of Flow: Vertical distance from the bed of a channel to the water surface.

Design Discharge: Peak flow at a specific location defined by an appropriate return period to be used for design purposes.

Freeboard: Vertical distance from the water surface to the top of the channel at design condition.

Hydraulic Radius: Flow area divided by wetted perimeter.

Hydraulic Roughness: Channel boundary characteristic contributing to energy losses, commonly described by Manning's n.

Normal Depth: Depth of uniform flow in a channel or culvert.

Riprap: Broken rock, cobbles, or boulders placed on side slopes or in channels for protection against the action of water.

System International (SI): Meter-kilogram-second system of units often referred to as metric units.

Uniform flow: Hydraulic condition in a prismatic channel where both the energy (friction) slope and the water surface slope are equal to the bed slope.

Velocity, Mean: Discharge divided by the area of flow.

CHAPTER 1: ENERGY DISSIPATOR DESIGN

Under many circumstances, discharges from culverts and channels may cause erosion problems. To mitigate this erosion, discharge energy can be dissipated prior to release downstream. The purpose of this circular is to provide design procedures for energy dissipator designs for highway applications. The first six chapters of this circular provide general information that is used to support the remaining design chapters. Chapter 1 (this chapter) discusses the overall analysis framework that is recommended and provides a matrix of available dissipators and their constraints. Chapter 2 provides an overview of erosion hazards that exist at both inlets and outlets. Chapter 3 provides a more precise approach for analyzing outlet velocity than is found in HDS 5. Chapter 4 provides procedures for calculating the depth and velocity through transitions. Chapter 5 provides design procedures for calculating the size of scour holes at culvert outlets. Chapter 6 provides an overview of hydraulic jumps, which are an integral part of many dissipators.

For some sites, appropriate energy dissipation may be achieved by design of a flow transition (Chapter 4), anticipating an acceptable scour hole (Chapter 5), and/or allowing for a hydraulic jump given sufficient tailwater (Chapter 6). However, at many other sites more involved dissipator designs may be required. These are grouped as follows:

- Internal Dissipators (Chapter 7)

- Stilling Basins (Chapter 8)

- Streambed Level Dissipators (Chapter 9)

- Riprap Basins and Aprons (Chapter 10)

- Drop Structures (Chapter 11)

- Stilling Wells (Chapter 12)

The designs included are listed in Table 1.1. Experienced designers can use Table 1.1 to determine the dissipator type to use and go directly to the appropriate chapter. First time designers should become familiar with the recommended energy dissipator design procedure that is discussed in this chapter.

Most of the information presented has been taken from the literature and adapted, where necessary, to fit highway needs. Recent research results have been incorporated, wherever possible, and a field survey was conducted to determine States' present practice and experience.

1.1   ENERGY DISSIPATOR DESIGN PROCEDURE

The designer should treat the culvert, energy dissipator, and channel protection designs as an integrated system. Energy dissipators can change culvert performance and channel protection requirements. Some debris-control structures represent losses not normally considered in the culvert design procedure. Velocity can be increased or reduced by changes in the culvert design. Downstream channel conditions (velocity, depth, and channel stability) are important considerations in energy dissipator design. A combination of dissipator and channel protection might be used to solve specific problems.

Table 1.1. Energy Dissipators and Limitations

| Chapter | Dissipator Type | Froude Number[7] (Fr) | Allowable Debris [1] | | | Tailwater (TW) |
			Silt/ Sand	Boulders	Floating	
4	Flow transitions	na	H	H	H	Desirable
5	Scour hole	na	H	H	H	Desirable
6	Hydraulic jump	> 1	H	H	H	Required
7	Tumbling flow[2]	> 1	M	L	L	Not needed
7	Increased resistance[3]	na	M	L	L	Not needed
7	USBR Type IX baffled apron	< 1	M	L	L	Not needed
7	Broken-back culvert	> 1	M	L	L	Desirable
7	Outlet weir	2 to 7	M	L	M	Not needed
7	Outlet drop/weir	3.5 to 6	M	L	M	Not needed
8	USBR Type III stilling basin	4.5 to 17	M	L	M	Required
8	USBR Type IV stilling basin	2.5 to 4.5	M	L	M	Required
8	SAF stilling basin	1.7 to 17	M	L	M	Required
9	CSU rigid boundary basin	< 3	M	L	M	Not needed
9	Contra Costa basin	< 3	H	M	M	< 0.5D
9	Hook basin	1.8 to 3	H	M	M	Not needed
9	USBR Type VI impact basin[4]	na	M	L	L	Desirable
10	Riprap basin	< 3	H	H	H	Not needed
10	Riprap apron[8]	na	H	H	H	Not needed
11	Straight drop structure[5]	< 1	H	L	M	Required
11	Box inlet drop structure[6]	< 1	H	L	M	Required
12	USACE stilling well	na	M	L	N	Desirable

[1]Debris notes: N = none, L = low, M = moderate, H = heavy
[2]Bed slope must be in the range $4\% < S_o < 25\%$
[3]Check headwater for outlet control
[4]Discharge, $Q < 11$ m³/s (400 ft³/s) and Velocity, $V < 15$ m/s (50 ft/s)
[5]Drop < 4.6 m (15 ft)
[6]Drop < 3.7 m (12 ft)
[7]At release point from culvert or channel
[8]Culvert rise less than or equal to 1500 mm (60 in)
na = not applicable.

The energy dissipator design procedure, illustrated in Figure 1.1, shows the recommended design steps. The designer should apply the following design procedure to one drainage channel/culvert and its associated structure at a time.

Step 1. Identify and Collect Design Data. Energy dissipators should be considered part of a larger design system that includes a culvert or a chute, channel protection requirements (both upstream and downstream), and may include a debris control structure. Much of the input data will be available to the energy dissipator design phase from previous design efforts.

a. Culvert Data: The culvert design should provide: type (RCB, RCP, CMP, etc); height, D; width, B; length, L; roughness, n; slope, S_o; design discharge, Q; tailwater, TW; type of control (inlet or outlet); outlet depth, y_o; outlet velocity, V_o; and outlet Froude number, Fr_o. Culvert outlet velocity, V_o, is discussed in Chapter 3. HDS 5 (Normann, et al., 2001) provides design procedures for culverts.

b. Transition Data: Flow transitions are discussed in Chapter 4. For most culvert designs, the designer will have to determine the flow depth, y, and velocity, V, at the exit of standard wingwall/apron combinations.

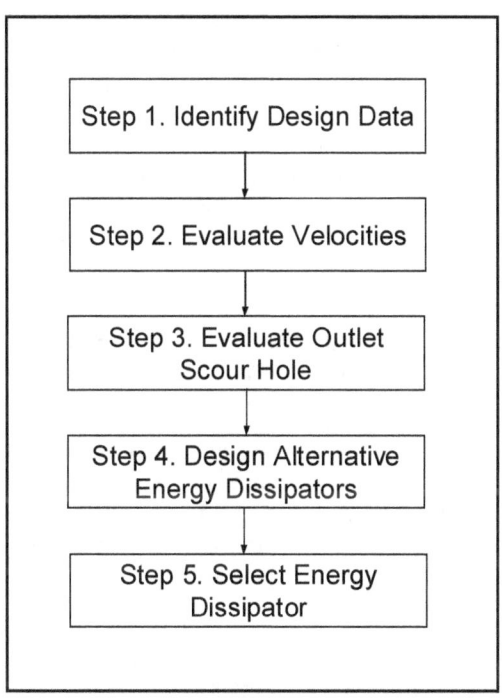

Figure 1.1. Energy Dissipator Design Procedure

c. Channel Data: The following channel data is used to determine the TW for the culvert design: design discharge, Q; slope, S_o; cross section geometry; bank and bed roughness, n; normal depth, y_n = TW; and normal velocity, V_n. If the cross section is a trapezoid, it is defined by the bottom width, B, and side slope, Z, which is expressed as 1 unit vertical to Z units horizontal (1V:ZH). HDS 4 (Schall, et al., 2001) provides examples of how to compute normal depth in channels. The size and amount of debris should be estimated using HEC 9 (Bradley, J.B., et al., 2005). The size and amount of bedload should be estimated.

d. Allowable Scour Estimate: In the field, the designer should determine if the bed material at the planned exit of the culvert is erodible. If it is, the potential extent of scour should be estimated: depth, h_s; width, W_s; and length, L_s. These estimates should be based on the physical limits to scour at the site. For example, the length, L_s, can be limited by a rock ledge or vegetation. The following soils parameters in the vicinity of planned culvert outlets should be provided. For non-cohesive soil, a grain size distribution including D_{16} and D_{84} is needed. For cohesive soil, the values needed are saturated shear strength, S_v, and plasticity index, PI.

    e.   Stability Assessment: The channel, culvert, and related structures should be evaluated for stability considering potential erosion, as well as buoyancy, shear, and other forces on the structure (see Chapter 2). If the channel, culvert, and related structures are assessed as unstable, the depth of degradation or height of aggradation that will occur over the design life of the structure should be estimated.

Step 2.   Evaluate Velocities. Compute culvert or chute exit velocity, V_o, and compare with downstream channel velocity, V_n. (See Chapter 3.) If the exit velocity and flow depth approximates the natural flow condition in the downstream channel, the culvert design is acceptable. If the velocity is moderately higher, the designer can evaluate reducing velocity within the barrel or chute (see Chapter 3) or reducing the velocity with a scour hole (step 3). Another option is to modify the culvert or chute (channel) design such that the outlet conditions are mitigated. If the velocity is substantially higher and/or the scour hole from step 3 is unacceptable, the designer should evaluate energy dissipators (step 4). Definition of the terms "approximately equal," "moderately higher," and "substantially higher" is relative to site-specific concerns such as sensitivity of the site and the consequences of failure. However, as rough guidelines that should be re-evaluated on a site-specific basis, the ranges of less than 10 percent, between 10 and 30 percent, and greater than 30 percent, respectively, may be used.

Step 3.   Evaluate Outlet Scour Hole. Compute the outlet scour hole dimensions using the procedures in Chapter 5. If the size of the scour hole is acceptable, the designer should document the size of the expected scour hole for maintenance and note the monitoring requirements. If the size of the scour hole is excessive, the designer should evaluate energy dissipators (step 4).

Step 4.   Design Alternative Energy Dissipators. Compare the design data identified in step 1 to the attributes of the various energy dissipators in Table 1.1. Design one or more of the energy dissipators that substantially satisfy the design criteria. The dissipators fall into two general groups based on Fr:

1. Fr < 3, most designs are in this group

2. Fr > 3, tumbling flow, USBR Type III stilling basin, USBR Type IV stilling basin, SAF stilling basin, and USBR Type VI impact basin

Debris, tailwater channel conditions, site conditions, and cost must also be considered in selecting alternative designs.

Step 5.   Select Energy Dissipator. Compare the design alternatives and select the dissipator that has the best combination of cost and velocity reduction. Each situation is unique and the exercise of engineering judgment will always be necessary. The designer should document the alternatives considered.

1.2 DESIGN EXAMPLES

The energy dissipator design procedure is best illustrated by applying it and the material presented in the energy dissipator design chapters to a series of design problems. These examples are intended to provide an overview of the design process. Pertinent chapters should be consulted for design details. The two design examples illustrate the process for cases where the Froude number is greater than 3 with a defined channel (tailwater) and less than 3 without a defined channel (no tailwater), respectively.

Design Example: RCB (Fr > 3) with Defined Downstream Channel (SI)

Evaluate the outlet velocity from a 3048 mm x 1829 mm RCB and determine the need for an energy dissipator.

Solution

Step 1. Identify Design Data.

 a. Culvert Data: Type, D, B, L, n, S_o, Q, TW, Control, y_o, V_o, Fr_o

  RCB, D = 1.829 m, B = 3.048 m, L = 91.44 m, n = 0.012

  S_o = 6.5%, Q = 11.8 m^3/s, TW = 0.579 m, inlet control

  Elevation of outlet invert = 30.48 m

  y_o = 0.457 m, V_o = 8.473 m/s, Fr_o = 4

 b. Transition Data: y and V at end of apron, Chapter 4

  The standard outlet with 45° wingwalls is an abrupt expansion. Since the culvert is in inlet control, the flow at the end of the apron will be supercritical: y = y_o = 0.457 m and V = V_o = 8.473 m/s

 c. Channel Data: Q, S_o, geometry, n, z, b, y_n, V_n, debris, bedload

  Q = 11.8 m^3/s, S_o = 6.5%, trapezoidal, 1:2 (V:H), b = 3.048 m, n = 0.03

  y_n = 0.579 m, V_n = 4.846 m/s

  Graded gravel bed with no boulders, little floating debris

 d. Allowable Scour Estimate: h_s, W_s, L_s, D_{16}, D_{84}, σ, S_v, PI

  Scour hole should be contained within channel W_s = L_s = 3.048 m and should be no deeper than 1.524 m. This allowable estimate can be obtained by observing scour holes in the vicinity.

 e. Stability Assessment:

  The channel, culvert, and related structures are evaluated for stability considering potential erosion, as well as buoyancy, shear, and other forces on the structure. If the channel, culvert, and related structures are assessed as unstable, the depth of degradation or height of aggradation that will occur over the design life of the structure should be estimated. In this case, the channel appears to be stable. No long-term degradation or head cutting was observed in the field.

Step 2. Evaluate Velocities.

 Since V_o = 8.473 m/s is much larger than V_n = 4.846 m/s, increasing culvert n is not practical. Determine if a scour hole is acceptable (Step 3) or design an energy dissipator (Step 4).

Step 3. Evaluate Outlet Scour Hole.

h_s, W_s, L_s, V_s from Chapter 5. If these values exceed allowable values in step 1, protection is required.

y_e = 0.835 m, h_s = 2.530 m, W_s = 15.850 m, L_s = 21.640 m, V_s = 737 m^3

Scour appears to be a problem and consideration should be given to reducing the V_o = 8.473 m/s to the 4.846 m/s in the channel.

Step 4. Design Alternative Energy Dissipators.

The following dissipators were determined from Table 1.1 by comparing the limitations shown against the site conditions. Since Fr > 3, tumbling flow, increased resistance, as well as, USBR Type IV, SAF stilling basin, and USBR Type VI streambed level dissipators will be designed. The outlet weir and outlet drop/weir were also assessed, but were not feasible without increasing the size of the culvert. Furthermore, a broken-back culvert was not considered and the culvert is too large for a riprap apron.

    a.   Tumbling flow (Chapter 7): Five elements 0.59 m in height spaced 5.02 m apart are required to reduce the velocity to V_c = 3.36 m/s. In order to accomplish this reduction, the last 25.1 m of culvert is used for the elements (4 spacing lengths between elements plus one-half spacing length before the first element and after the last element). In addition, this portion of the culvert must be increased in height to 2.1 m to accommodate the elements.

    b.   Increased resistance (Chapter 7): For a roughness height, h = 0.12 m, the internal resistance, n_{LOW} = 0.039 for velocity check and n_{HIGH} = 0.052 for Q check. The velocity at the outlet is 4.4 m/s. The elements are 1.2 m apart for 28 rows. Therefore, the modified culvert length required to accommodate the roughness elements is 33.6 m (27 spacing lengths between elements plus one-half spacing length before the first element and after the last element).

    c.   USBR Type IV stilling basin (Chapter 8): The dissipator length, L_B = 21.6 m, is located below the streambed at elevation 25.0 m. The total length of the stilling basin including transitions is 38.6 m. The exit velocity, V_2, is 4.85 m/s, which matches the channel velocity, V_n, of 4.846 m/s.

    d.   SAF stilling basin (Chapter 8): The dissipator length, L_B = 3.353 m, is located below the streambed at elevation 27.889 m. The total length of the stilling basin including transitions is 12.192 m. The exit velocity, V_2, is 4.877 m/s, which is close to channel velocity, V_n, of 4.846 m/s.

    e.   USBR Type VI impact basin (Chapter 9): The dissipator width, W_B, is 3.5 m. The height, h_1, equals 2.68 m and length, L, equals 4.65 m. The exit velocity, V_B, equals 3.7 m/s, which is calculated knowing the energy loss is 61 percent.

Step 5. Select Energy Dissipator.

The dissipator selected should be governed by comparing the efficiency, cost, natural channel compatibility, and anticipated scour for all the alternatives.

In this example, all the structures highlighted fit the channel, meet the velocity criteria, and produce significant energy losses. However, the costs of the USBR Type VI are lower than the other dissipators, so becomes the dissipator of choice.

Design Example: RCB (Fr > 3) with Defined Downstream Channel (CU)

Evaluate the outlet velocity from a 10 ft x 6 ft reinforced concrete box (RCB) culvert and determine the need for an energy dissipator.

Solution

Step 1. Identify Design Data:

    a. Culvert Data: Type, D, B, L, n, S_o, Q, TW, Control, y_o, V_o, Fr_o

        RCB, D = 6 ft, B = 10 ft, L = 300 ft, n = 0.012

        S_o = 6.5%, Q = 417 ft³/s, TW = 1.9 ft, inlet control

        Elevation of outlet invert = 100 ft

        y_o = 1.5 ft, V_o = 27.8 ft/s, Fr_o = 4

    b. Transition Data: y and V at end of apron, Chapter 4

        The standard outlet with 45° wingwalls is an abrupt expansion. Since the culvert is in inlet control, the flow at the end of the apron will be supercritical: y = y_o = 1.5 ft and V = V_o = 27.8 ft/s

    c. Channel Data: Q, S_o, geometry, n, z, b, y_n, V_n, debris, bedload

        Q = 417 ft³/s., S_o = 6.5%, trapezoidal, 1:2 (V:H), b = 10 ft, n = 0.03

        y_n = 1.9 ft, V_n = 15.9 ft/s

        Graded gravel bed with no boulders, little floating debris

    d. Allowable Scour Estimate: h_s, W_s, L_s, D_{16}, D_{84}, σ, S_v, PI

        Scour hole should be contained within channel W_s = L_s = 10 ft and should be no deeper than 5 ft. This allowable estimate can be obtained by observing scour holes in the vicinity.

    e. Stability Assessment:

        The channel, culvert, and related structures are evaluated for stability considering potential erosion, as well as buoyancy, shear, and other forces on the structure. If the channel, culvert, and related structures are assessed as unstable, the depth of degradation or height of aggradation that will occur over the design life of the structure should be estimated. In this case, the channel appears to be stable. No long-term degradation or head cutting was observed in the field.

Step 2. Evaluate Velocities.

    Since V_o = 27.8 ft/s is much larger than V_n = 15.9 ft/s, increasing culvert n is not practical. Determine if a scour hole is acceptable (step 3) or design an energy dissipator (step 4).

Step 3. Evaluate Outlet Scour Hole.

    h_s, W_s, L_s, V_s from Chapter 5. If these values exceed allowable values in step 1, protection is required.

    y_e = 2.74 ft, h_s = 8.3 ft, W_s = 52 ft, L_s = 71 ft, V_s = 963 ft³

Scour appears to be a problem and consideration should be given to reducing the V_o = 27.8 ft/s to the 15.9 ft/s in the channel.

Step 4. Design Alternative Energy Dissipators.

The following dissipators were determined from Table 1.1 by comparing the limitations shown against the site conditions. Since Fr > 3, tumbling flow, increased resistance, as well as the USBR Type IV, SAF stilling basin, and USBR Type VI streambed level dissipators will be designed. The outlet weir and outlet drop/weir were also assessed, but were not feasible without increasing the size of the culvert. Furthermore, a broken-back culvert was not considered and the culvert is too large for a riprap apron.

a. Tumbling flow (Chapter 7): Five elements 1.92 ft in height spaced 16.3 ft apart are required to reduce the velocity to V_c = 11.0 ft/s. In order to accomplish this reduction, the last 81.5 ft of culvert is used for the elements (4 spacing lengths between elements plus one-half spacing length before the first element and after the last element). In addition, this portion of the culvert must be increased in height to 6.7 ft to accommodate the elements.

b. Increased resistance (Chapter 7): For a roughness height, h = 0.4 ft, the internal resistance, n_{Low}, equals 0.039 for velocity check and n_{HIGH} equals 0.052 for Q check. The velocity at the outlet is 14.5 ft/s. The elements are 4.0 ft apart for 28 rows. Therefore, the modified culvert length required to accommodate the roughness elements is 112 ft (27 spacing lengths between elements plus one-half spacing length before the first element and after the last element).

c. USBR Type IV stilling basin (Chapter 8): The dissipator length, L_B = 70.9 ft, is located below the streambed at elevation 82.0 ft. The total length of the stilling basin including transitions is 126.5 ft. The exit velocity, V_2, is 16 ft/s, which is close to channel velocity, V_n, of 15.9 ft/s.

d. SAF stilling basin (Chapter 8): The dissipator length, L_B = 11 ft is located below the streambed at elevation 91.5 ft. The total length of the stilling basin including transitions is 40 ft. The exit velocity, V_2, is 16 ft/s, which is close to channel velocity, V_n, of 15.9 ft/s.

e. USBR Type VI impact basin (Chapter 9): The dissipator width, W_B, is 12 ft. The height, h_1 = 9.17 ft and length, L = 16 ft. The exit velocity, V_B, equals 12.9 ft/s, which is calculated knowing the energy loss is 61 percent.

Step 5. Select Energy Dissipator.

The dissipator selected should be governed by comparing the efficiency, cost, natural channel compatibility, and anticipated scour for all the alternatives.

In this example, all the structures highlighted fit the channel, meet the velocity criteria, and produce significant energy losses. However, the costs of the USBR Type VI are lower than the other dissipators, so becomes the dissipator of choice.

Design Example: RCB (Fr < 3) with Undefined Downstream Channel (SI)

Evaluate the outlet velocity from a 3048 mm x 1829 mm reinforced concrete box (RCB) and determine the need for an energy dissipator.

Solution

Step 1. Identify Design Data.

    a.   Culvert Data: Type, D, B, L, n, S_o, Q, TW, Control, y_o, V_o, Fr_o

        RCB, D = 1.524 m, B = 1.524 m, L = 64.922 m, n = 0.012

        S_o = 3.0%, Q = 5.66 m^3/s, TW = 0.0 m, inlet control

        Elevation of outlet invert = 30.480 m

        y_o = 0.655 m, V_o = 5.791 m/s, Fr_o = 2.3

    b.   Transition Data: y and V at end of apron, Chapter 4

        The standard Outlet with 90° headwall is an abrupt expansion. Since the culvert is in inlet control, the flow at the end of the apron will be supercritical: y = y_o = 0.655 m and V = V_o = 5.791 m/s

    c.   Channel Data: Q, S_o, geometry, n, z, b, y_n, V_n, debris, bedload

        The downstream channel is undefined. The water will spread and decrease in depth as it leaves the culvert making tailwater essentially zero. The channel is graded sand with no boulders and has moderate to high amounts of floating debris.

    d.   Allowable Scour Estimate: h_s, W_s, L_s, D_{16}, D_{84}, σ, S_v, PI

        A scour basin not more than 0.914 meters deep is allowable at this site. Allowable outlet velocity should be about 3 m/s.

    e.   Stability Assessment:

        The channel, culvert, and related structures are evaluated for stability considering potential erosion, as well as buoyancy, shear, and other forces on the structure. If the channel, culvert, and related structures are assessed as unstable, the depth of degradation or height of aggradation that will occur over the design life of the structure should be estimated. In this case, the channel appears to be stable. No long-term degradation or head cutting was observed in the field.

Step 2. Evaluate Velocities.

    Since V_o = 5.791 m/s is much larger than V_{allow} = 3.0 m/s, increasing culvert n is not practical. Determine if a scour hole is acceptable (step 3) or design an energy dissipator (step 4).

Step 3. Evaluate Outlet Scour Hole.

    h_s, W_s, L_s, V_s from Chapter 5. If these values exceed allowable values in step 1, protection is required.

    y_e = 0.707 m, h_s = 1.707 m, W_S = 9.449 m, L_S = 14.935 m, V_S = 62 m^3

    Since 1.707 m is greater than the 0.914 m allowable, an energy dissipator will be necessary.

Step 4. Design Alternative Energy Dissipators.

The following dissipators were determined from Table 1.1 by comparing the limitations shown against the site conditions. For comparison purposes all the Fr < 3 dissipators will be designed (even those that cannot handle a moderate amount of debris). Dissipators meeting the Froude number requirement, but not designed are as follows (reason for exclusion in parentheses): SAF stilling basin (requires tailwater), Contra Costa basin (no defined channel); Broken-back culvert (mild site slope); outlet weir (infeasible without increasing culvert size); and riprap apron (culvert too large).

a. Tumbling flow (Chapter 7): The S_o = 3% is less than the 4% required, but the design is included for comparison. Five elements 0.55 m in height spaced 4.68 m apart are required to reduce the velocity to V_c = 3.32 m/s. In order to accomplish this reduction, the last 23.4 m of the culvert is used for the elements (4 spacing lengths between elements plus one-half spacing length before the first element and after the last element). In addition, this portion of the culvert must be increased in height to 2.0 m to accommodate the elements.

b. Increased resistance (Chapter 7): For a roughness height, h = 0.09 m, the internal resistance, n_{LOW} = 0.032 for velocity check and n_{HIGH} = 0.043 for Q check. The discharge check indicates that the culvert height has to be increased to 1.7 m. The velocity at the outlet is 3.2 m/s. The elements are 0.9 m apart for 34 rows. Therefore, the modified culvert length required to accommodate the roughness elements is 30.6 m (33 spacing lengths between elements plus one-half spacing length before the first element and after the last element).

c. CSU rigid boundary basin (Chapter 9): Width of basin, W_B = 9.144 m, length of basin, L_B = 8.534 m, number of roughness rows, N_r = 4, number of elements, N = 17, divergence, U_e = 1.9:1, width of elements, W_1 = 0.914 m, height of elements, h = 0.229 m, velocity at basin outlet, V_B = 2.896 m/s, depth at basin outlet, y_B = 0.213 m.

d. USBR Type VI impact basin (Chapter 9): The dissipator width, W_B, is 4.0 m. The height, h_1 = 3.12 m, and length, L = 5.33 m. The exit velocity, V_B, equals 4.2 m/s, which is calculated knowing the energy loss is 47 percent.

e. Hook basin (Chapter 9): Assuming the downstream velocity, V_n, equals the allowable, 3.0 m/s, V_o/V_n = 5.791/3.0 = 1.93. The dimensions for a straight trapezoidal basin are: length, L_B = 4.572 m, width, W_6 = $2W_o$ = 3.048 m, side slope = 2:1, length to first hook, L_1 = 1.905 m, length to second hooks, L_2 = 3.179 m, height of hook, h_3 = 0.716 m, target exit velocity, V_B = V_n = 3.0 m/s. From Figure 9.12, V_o/V_B = 2.0; actual V_B = 5.8/2.0 = 2.896 m/s, which is less than the target.

f. Riprap basin (Chapter 10): Assuming a diameter of rock, D_{50} = 0.38 m, the depth of pool, h_s = 0.78 m, length of pool = 7.8 m, length of apron = 3.9 m, length of basin = 11.7 m, thickness of riprap on approach, $3D_{50}$ = 1.14 m, and thickness of riprap for the remainder of basin, $2D_{50}$ = 0.76 m.

Step 5. Select Energy Dissipator.

The dissipator selected should be governed by comparing the efficiency, cost, natural channel compatibility, and anticipated scour for all the alternatives.

Right-of-way (ROW), debris, and dissipator cost are all constraints at this site. ROW is expensive making the longer dissipators more costly. Debris will affect the operation of the impact basin and may be a problem with the CSU roughness elements and tumbling flow designs. In the final analysis, the riprap basin was selected based on cost and anticipated maintenance.

Design Example: RCB (Fr < 3) with undefined Downstream Channel (CU)

Evaluate the outlet velocity from a 5 ft by 5 ft reinforced concrete box (RCB) and determine the need for an energy dissipator.

Solution

Step 1. Identify Design Data.

a. Culvert Data: Type, D, B, L, n, S_o, Q, TW, Control, y_o, V_o, Fr_o

RCB, D = 5 ft, B = 5 ft, L = 213 ft, n = 0.012

S_o = 3.0%, Q = 200 ft^3/s, TW = 0.0 ft, inlet control

Elevation of outlet invert = 100 ft

y_o = 2.15 ft, V_o = 19 ft/s, Fr_o = 2.3

b. Transition Data: y and V at end of apron, Chapter 4

The standard Outlet with 90° headwall is an abrupt expansion. Since the culvert is in inlet control, the flow at the end of the apron will be supercritical: y = y_o = 2.15 ft and V = V_o = 19 ft/s

c. Channel Data: Q, S_o, geometry, n, z, b, y_n, V_n, debris, bedload

The downstream channel is undefined. The water will spread and decrease in depth as it leaves the culvert making tailwater essentially zero. The channel is graded sand with no boulders and has moderate to high amounts of floating debris.

d. Allowable Scour Estimate: h_s, W_s, L_s, D_{16}, D_{84}, σ, S_v, PI

A scour basin not more than 0.914 meters deep is allowable at this site. Allowable outlet velocity should be about 10 ft/s.

e. Stability Assessment:

The channel, culvert, and related structures are evaluated for stability considering potential erosion, as well as buoyancy, shear, and other forces on the structure. If the channel, culvert, and related structures are assessed as unstable, the depth of degradation or height of aggradation that will occur over the design life of the structure should be estimated. In this case, the channel appears to be stable. No long-term degradation or head cutting was observed in the field.

Step 2.   Evaluate Velocities.

Since V_o = 19 ft/s is much larger than V_{allow} = 10 ft/s, increasing culvert n is not practical.   Determine if a scour hole is acceptable (step 3) or design an energy dissipator (step 4).

Step 3.   Evaluate Outlet Scour Hole.

h_s, W_s, L_s, V_s from Chapter 5. If these values exceed allowable values in step 1, protection is required.

y_e = 2.32 ft, h_s = 5.6 ft, W_S = 32 ft, L_S = 49 ft, V_S = 81 yd^3

Since 5.6 ft is greater than the 3.0 ft allowable, an energy dissipator will be necessary.

Step 4.   Design Alternative Energy Dissipators.

The following dissipators were determined from Table 1.1 by comparing the limitations shown against the site conditions. For comparison purposes all the Fr < 3 dissipators will be designed (even those that cannot handle a moderate amount of debris). Dissipators meeting the Froude number requirement, but not designed are as follows (reason for exclusion in parentheses): SAF stilling basin (requires tailwater), Contra Costa basin (no defined channel); Broken-back culvert (mild site slope); outlet weir (infeasible without increasing culvert size); and riprap apron (culvert too large).

a.   Tumbling Flow (Chapter 7): The S_o = 3% is less the 4% required, but the design is included for comparison. Five elements 1.8 ft in height spaced 15.4 ft apart are required to reduce the velocity to V_c = 10.9 ft/s. In order to accomplish this reduction, the last 77.0 ft of the culvert is used for the elements (4 spacing lengths between elements plus one-half spacing length before the first element and after the last element). In addition, this portion of the culvert must be increased in height to 6.5 ft to accommodate the elements.

b.   Increased resistance (Chapter 7): For a roughness height, h = 0.3 ft, the internal resistance, n_{LOW} = 0.032 for velocity check and n_{HIGH} = 0.043 for Q check. The discharge check indicates that the culvert height has to be increased to 5.6 ft. The velocity at the outlet is 10.6 ft/s. The elements are 3.0 ft apart for 34 rows. Therefore, the modified culvert length required to accommodate the roughness elements is 102 ft (33 spacing lengths between elements plus one-half spacing length before the first element and after the last element)

c.   CSU Rigid Boundary basin (Chapter 9): Width of basin, W_B = 30 ft, length of basin, L_B = 28 ft, number of roughness rows, N_r = 4, number of elements, N = 17, divergence, U_e = 1.9:1, width of elements, W_1 = 3.0 ft, height of elements, h = 0.75 ft, velocity at basin outlet, V_B = 9.5 ft/s, depth at basin outlet, y_B = 0.70 ft.

d.   USBR Type VI (Chapter 9): The dissipator width, W_B, is 13 ft. The height, h_1 = 10.17 ft, and length, L = 17.33 ft. The exit velocity, V_B, equals 13.9 ft/s, which is calculated knowing the energy loss is 47 percent.

e. Hook (Chapter 9): Assuming the downstream velocity, V_n, equals the allowable, 10 ft/s, V_o/V_n = 19/10 = 1.9. The dimensions for a straight trapezoidal basin are: length, L_B = 15 ft, width, W_6 = $2W_o$ = 10 ft, side slope = 2:1, length to first hook, L_1 = 6.25 ft, length to second hooks, L_2 = 10.43 ft, height of hook, h_3 = 2.35 ft, target exit velocity, V_B = V_n = 10 ft/s. From Figure 9.12, V_o/V_B = 2.0; actual V_B = 19/2.0 = 9.5 ft/s which is less than the target.

f. Riprap basin (Chapter 10): Assuming a diameter of rock, D_{50} = 1.2 ft, the depth of pool, h_s = 2.7 ft, length of pool = 27 ft, length of apron = 13.5 ft, length of basin = 40.5 ft, thickness of riprap on approach, $3D_{50}$ = 3.6 ft, and thickness of riprap for the remainder of basin, $2D_{50}$ = 2.4 ft.

Step 5. Select Energy Dissipator.

The dissipator selected should be governed by comparing the efficiency, cost, natural channel compatibility, and anticipated scour for all the alternatives.

Right-of-way (ROW), debris, and dissipator cost are all constraints at this site. ROW is expensive making the longer dissipators more costly. Debris will affect the operation of the impact basin and may be a problem with the CSU roughness elements and tumbling flow designs. In the final analysis, the riprap basin was selected based on cost and anticipated maintenance.

This page intentionally left blank.

CHAPTER 2: EROSION HAZARDS

This chapter discusses potential erosion hazards at culverts and countermeasures for these hazards. Section 2.1 presents the hazards associated with culvert inlets: channel alignment and approach velocity, depressed inlets, headwalls and wingwalls, and inlet and barrel failures. Section 2.2 presents the hazards associated with culvert outlets: local scour, channel degradation, and standard culvert end treatments.

2.1 EROSION HAZARDS AT CULVERT INLETS

The erosion hazard at culvert inlets from vortices, flow over wingwalls, and fill sloughing is generally minor and can be addressed by maintenance if it occurs. Designers should focus their attention on the following concerns and associated mitigation measures.

2.1.1 Channel Alignment and Approach Velocity

An erosion hazard may exist if a defined approach channel is not aligned with the culvert axis. Aligning the culvert with the approach channel axis will minimize erosion at the culvert inlet. When the culvert cannot be aligned with the channel and the channel is modified to bend into the culvert, erosion can occur at the bend in the channel. Riprap or other revetment may be needed (see Lagasse, et al., 2001).

At design discharge, water will normally pond at the culvert inlet and flow from this pool will accelerate over a relatively short distance. Significant increases in velocity only extend upstream from the culvert inlet at a distance equal to the height of the culvert. Velocity near the inlet may be approximated by dividing the flow rate by the area of the culvert opening. The risk of channel erosion should be judged on the basis of this average approach velocity. The protection provided should be adequate for flow rates that are less than the maximum design rate. Since depth of ponding at the inlet is less for smaller discharges, greater velocities may occur. This is especially true in channels with steep slopes where high velocity flow prevails.

2.1.2 Depressed Inlets

Culvert inverts are sometimes placed below existing channel grades to increase culvert capacity or to meet minimum cover requirements. Hydraulic Design Series No. 5 (HDS 5) (Normann, et al., 2001) discusses the advantages of providing a depression or fall at the culvert entrance to increase culvert capacity. However, the depression may result in progressive degradation of the upstream channel unless resistant natural materials or channel protection is provided.

Culvert invert depressions of 0.30 or 0.61 m (1 to 2 ft) are usually adequate to obtain minimum cover and may be readily provided by modification of the concrete apron. The drop may be provided in two ways. A vertical wall may be constructed at the upstream edge of the apron, from wingwall to wingwall. Where a drop is undesirable, the apron slab may be constructed on a slope to reduce or eliminate the vertical face.

Caution must be exercised in attempting to gain the advantages of a lowered inlet where placement of the outlet flow line below the channel would also be required. Locating the entire culvert flow line below channel grade may result in deposition problems.

2.1.3   Headwalls and Wingwalls

Recessing the culvert into the fill slope and retaining the fill by either a headwall parallel to the roadway or by a short headwall and wingwalls does not produce significant erosion problems. This type of design decreases the culvert length and enhances the appearance of the highway by providing culvert ends that conform to the embankment slopes. A vertical headwall parallel to the embankment shoulder line and without wingwalls should have sufficient length so that the embankment at the headwall ends remain clear of the culvert opening.   Normally riprap protection of this location is not necessary if the slopes are sufficiently flat to remain stable when wet.

The inlet headwall (with or without wingwalls) does not have to extend to the maximum design headwater elevation. With the inlet and the slope above the headwall submerged, velocity of flow along the slope is low. Even with easily erodible soils, a vegetative cover is usually adequate protection in this area.

Wingwalls flared with respect to the culvert axis are commonly used and are more efficient than parallel wingwalls. The effects of various wingwall placements upon culvert capacity are discussed in HDS 5 (Normann, et al., 2001). Use of a minimum practical wingwall flare has the advantage of reducing the inlet area requiring protection against erosion. The flare angle for the given type of culvert should be consistent with recommendations of HDS 5.

If the flow velocity near the inlet indicates a possibility of scour threatening the stability of wingwall footings, erosion protection should be provided. A concrete apron between wingwalls is the most satisfactory means for providing this protection. The slab has the further advantage that it may be reinforced and used to support the wingwalls as cantilevers.

2.1.4   Inlet and Barrel Failures

Most inlet failures reported have occurred on large, flexible-type pipe culverts with projected or mitered entrances without headwalls or other entrance protection. The mitered or skewed ends of corrugated metal pipes, cut to conform to the embankment slopes, offer little resistance to bending or buckling. When soils adjacent to the inlet are eroded or become saturated, pipe inlets can be subjected to buoyant forces. Lodged drift and constricted flow conditions at culvert entrances cause buoyant and hydrostatic pressures on the culvert inlet edges that, while difficult to predict, have significant effect on the stability of culvert entrances.

To aid in preventing inlet failures of this type, protective features generally should include full or partial concrete headwalls and/or slope paving. Riprap can serve as protection for the embankment, but concrete inlet structures anchored to the pipe are the only protection against buoyant failure. Manufactured concrete or metal sections may be used in lieu of the inlet structures shown. Metal end sections for culvert pipes larger than 1350 mm (54 in) in height must be anchored to increase their resistance to failure.

Failures of inlets are of primary concern, but other types of failures have occurred. Seepage of water along the culvert barrel has caused piping or the washing out of supporting material. Hydrostatic pressure from seepage water or from flow under the culvert barrel has buckled the bottoms of large corrugated metal arch pipes. Good compaction of backfill material is essential to reduce the possibility of these types of failures. Where soils are quite erosive, special impervious bedding and backfill materials should be placed for a short distance at the culvert entrance. Further protection may be provided by cutoff collars placed at intervals along the culvert barrel or by a special subdrainage system.

2.2 EROSION HAZARDS AT CULVERT OUTLETS

Erosion at culvert outlets is a common condition. Determination of the local scour potential and channel erodibility should be standard procedure in the design of all highway culverts. Chapter 3 provides procedures for determining culvert outlet velocity, which will be the primary indicator of erosion potential.

2.2.1 Local Scour

Local scour is the result of high-velocity flow at the culvert outlet, but its effect extends only a limited distance downstream as the velocity transitions to outlet channel conditions. Natural channel velocities are almost always less than culvert outlet velocities because the channel cross-section, including its flood plain, is generally larger than the culvert flow area. Thus, the flow rapidly adjusts to a pattern controlled by the channel characteristics.

Long, smooth-barrel culverts on steep slopes will produce the highest velocities. These cases will no doubt require protection of the outlet channel at most sites. However, protection is also often required for culverts on mild slopes. For these culverts flowing full, the outlet velocity will be critical velocity with low tail-water and the full barrel velocity for high tail-water. Where the discharge leaves the barrel at critical depth, the velocity will usually be in the range of 3 to 6 m/s (10 to 20 ft/s). Estimating local scour at culvert outlets is an important topic discussed in more detail in Chapter 5.

A common mitigation measure for small culverts is to provide at least minimum protection (see Riprap Aprons in Chapter 10), and then inspect the outlet channel after major storms to determine if the protection must be increased or extended. Under this procedure, the initial protection against channel erosion should be sufficient to provide some assurance that extensive damage could not result from one runoff event. For larger culverts, the designer should consider estimating the size of the scour hole using the procedures in Chapter 5.

2.2.2 Channel Degradation

Culverts are generally constructed at crossings of small streams, many of which are eroding to reduce their slopes. This channel erosion or degradation may proceed in a fairly uniform manner over a long length of stream or it may occur abruptly with drops progressing upstream with every runoff event. The latter type, referred to as headcutting, can be detected by location surveys or by periodic maintenance inspections following construction. Information regarding the degree of instability of the outlet channel is an essential part of the culvert site investigation. If substantial doubt exists as to the long-term stability of the channel, measures for protection should be included in the initial construction. HEC 20 "Stream Stability at Highway Structures" (Lagasse, et al., 2001) provides procedures for evaluating horizontal and vertical channel stability.

2.2.3 Standard Culvert End Treatments

Standard practice is to use the same end treatment at the culvert entrance and exit. However, the inlet may be designed to improve culvert capacity or reduce head loss while the outlet structure should provide a smooth flow transition back to the natural channel or into an energy dissipator. Outlet transitions should provide uniform redistribution or spreading of the flow without excessive separation and turbulence. Therefore, it may not be possible to satisfy both inlet and outlet requirements with the same end treatment or design. As will be illustrated in Chapter 4, properly designed outlet transitions are essential for efficient energy dissipator design. In some cases, they may substantially reduce or eliminate the need for other end treatments.

CHAPTER 3: CULVERT OUTLET VELOCITY AND VELOCITY MODIFICATION

This chapter provides an overview of outlet velocity computation. The purpose of this discussion is to identify culvert configurations that are candidates for velocity reduction within the barrel or for more detailed velocity computation. Outlet velocities can range from 3 m/s (10 ft/s) for culverts on mild slopes up to 9 m/s (30 ft/s) for culverts on steep slopes. The discussion in this chapter is limited to changing culvert material or increasing culvert size to modify or reduce the velocity within the culvert. The discussion of energy dissipator designs for reducing velocity within the barrel is found in Chapter 7.

The continuity equation, which states that discharge is equal to flow area times average velocity (Q = AV), is used to compute culvert velocities within the barrel and at the outlet. The discharge, Q, is determined during culvert design. The flow area, A, for determining outlet velocity is calculated using the culvert outlet depth that is consistent with the culvert flow type. The culvert flow types and recommended outlet depths from HDS 5 (Normann, et al., 2001) are summarized in the following sections.

3.1 CULVERTS ON MILD SLOPES

Figure 3.1 (Normann, et al., 2001) shows the types of flow for culverts on mild slopes, that is, culverts flowing with outlet control. Culverts A and B have unsubmerged inlets. Culverts C and D have submerged inlets. Culverts A, B and C have unsubmerged outlets. The higher of critical depth or tailwater depth at the outlet is used for calculating outlet velocity. Since the barrel for Culvert D flows full to the exit, the full barrel area is used for calculating outlet velocity. Each of these cases as well as refinements is discussed in the following sections.

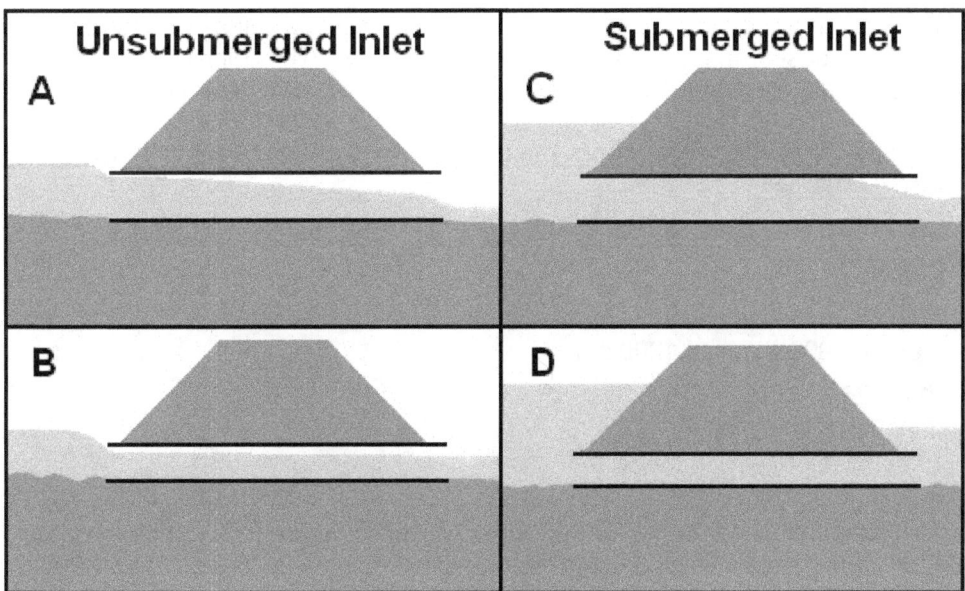

Figure 3.1. Outlet Control Flow Types

3.1.1 Submerged Outlets

In Figure 3.1D, the tailwater controls the culvert outlet velocity. Outlet velocity is determined using the full barrel area. As long as the tailwater is above the culvert, the outlet velocity can be reduced by increasing the culvert size. The degree of reduction is proportional to the reciprocal of the culvert area. Table 3.1 illustrates the amount of reduction that can be achieved.

Table 3.1. Example Velocity Reductions by Increasing Culvert Diameter

Culvert Diameter Change (SI) mm	914 to 1219	1219 to 1524	1524 to1829
Culvert Diameter Change (CU) ft	3 to 4	4 to 5	5 to 6
Percent Reduction in Outlet Velocity (V=Q/A)	44%	35%	31%

For high tailwater conditions, erosion may not be a serious problem. The designer should determine if the tailwater will always control or if the outlet will be unsubmerged under some circumstances. Full flow can also exist when the discharge is high enough to produce critical depth equal to or higher than the crown of the culvert barrel. As long as critical depth is higher than the crown, outlet velocity reduction can be achieved by increasing the barrel size as illustrated above.

3.1.2 Unsubmerged Outlets (Critical Depth) and Tailwater

In Figures 3.1A, B, and C, the tailwater is below the crown of the culvert. Outlet velocity is determined using the flow area at the outlet that is calculated using the higher of the tailwater or critical depth. For Figure 3.1B, the tailwater controls; for Figures 3.1A and 3.1C, critical depth controls. (Appendix B includes useful figures for estimating critical depth for a variety of culvert shapes.) If critical depth is above the culvert, the culvert will flow full and the outlet velocity can be reduced by increasing the culvert size as shown above. The following example illustrates critical depth and velocity computation for full and partial full flow at the outlet.

Design Example: Velocity Reduction by Increasing Culvert Size When Critical Depth Occurs at the Outlet (SI)

Evaluate the reduction in velocity by replacing a 914 mm diameter culvert with a 1219 mm diameter culvert. Given:

CMP Culvert
Diameter, D = 900 mm and 1200 mm
Q = 2.83 m^3/s
Tailwater, TW = 0.610 m

Solution

Step 1. Read critical depth, y_c, for 900 mm CMP from Figure B.2. Since y_c exceeds 0.900 m, the barrel is flowing full to the end even though TW is less than 0.900 m.

Step 2. Calculate flow area, A, and velocity, V, with the pipe flowing full.

$A = \pi D^2/4 = 3.14(0.900)^2/4 = 0.636$ m^2

$V = Q/A = 2.83/0.656 = 4.4$ m/s.

Step 3. Read critical depth, y_c, for 1200 mm CMP from Figure B.2. The new y_c = 0.95 m which is less than D so y_c controls outlet velocity.

Step 4. Calculate flow area, A, using Table B.2. With y/D = 0.95/1.2 = 0.79, A/D^2 = 0.6655, and V = Q/A = 2.832/(0.6655 $(1.2)^2$) = 2.95 m/s.

This is a reduction of about 32 percent. The reduction is less than shown in Table 3.1 because the 1.2 m pipe is not flowing full at the exit.

Design Example: Velocity Reduction by Increasing Culvert Size When Critical Depth Occurs at the Outlet (CU)

Evaluate the reduction in velocity by replacing a 3-ft-diameter culvert with a 4-ft-diameter culvert. Given:

CMP Culvert
Diameter, D = 3 ft and 4 ft
Q = 100 ft^3/s
Tailwater, TW = 2.0 ft

Solution

Step 1. Read critical depth, y_c, for 3 ft CMP from Figure B.2. Since y_c exceeds 3 ft, the barrel is flowing full to the end even though TW is less than 3 ft.

Step 2. Calculate flow area, A, and velocity, V, with the pipe flowing full.

$A = \pi D^2/4 = 3.14(3)^2/4 = 7.065$ ft^2

$V = Q/A = 100/7.065 = 14.2$ ft/s.

Step 3. Read critical depth, y_c, for 4 ft CMP from Figure B.2. The new y_c = 3.1 ft which is less than 4 ft so y_c controls outlet velocity.

Step 4. Calculate flow area, A, using Table B.2. With y/D = 3.1/4 = 0.78, A/D^2 = 0.6573, and V = Q/A = 100/0.6573$(4)^2$ = 9.5 ft/s.

This is a reduction of about 33 percent. The reduction is less than shown in Table 3.1 because the 4 ft pipe is not flowing full at the exit.

3.1.3 Unsubmerged Outlets (Brink Depth)

Brink depth, y_o, which is shown in Figure 3.2, is the depth that occurs at the exit of the culvert. The flow goes through critical depth upstream of the outlet when the tailwater elevation is below the critical depth elevation in the culvert. Figures 3.3 and 3.4 may be used to determine outlet brink depths for rectangular and circular sections. These figures are dimensionless rating curves that indicate the effect on brink depth of tailwater for culverts on mild or horizontal slopes. In order to use these curves, the designer must determine normal depth or tailwater (TW) in the outlet channel and $Q/(BD^{3/2})$ or $Q/D^{5/2}$ for the culvert. Table B.1 (Appendix B) can be used to estimate TW if the downstream channel can be approximated with a trapezoidal channel.

For culvert shapes other than rectangular and circular, the brink depth for low tailwater can be approximated from the critical depth curves found in Appendix B. Since critical depth is larger than brink depth, determining brink depth in this manner is not conservative, but is acceptable.

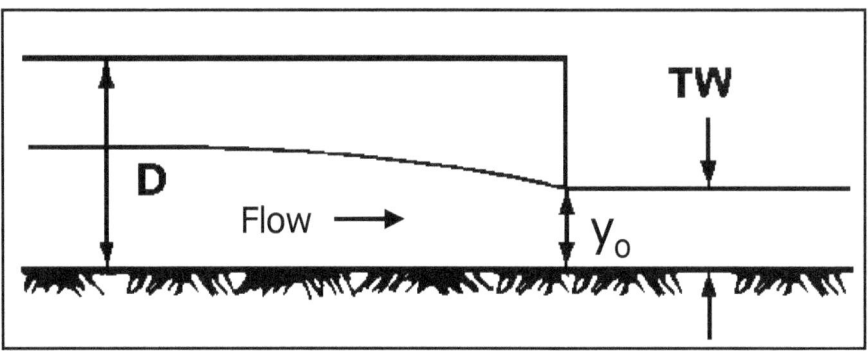

Figure 3.2. Definition Sketch for Brink Depth.

When the tailwater depth is low, culverts on mild or horizontal slopes will flow with critical depth near the outlet. This is indicated on the ordinate of Figures 3.3 and 3.4. As the tailwater increases, the depth at the brink increases at a variable rate along the $Q/(BD^{3/2})$ or $Q/D^{5/2}$ curve, until a point where the tailwater and brink depth vary linearly at the $45°$ line on the figures. The following example illustrates the use of these figures and the effect of changing culvert size for a constant Q and TW.

Design Example: Velocity Reduction by Increasing Culvert Size for Brink Depth Conditions (SI)

Evaluate the reduction in velocity by replacing a 1.050 m pipe culvert with a larger pipe culvert. Given:

$$Q \quad = \quad 1.7 \text{ m}^3/\text{s}$$
$$TW \quad = \quad 0.610 \text{ m, constant}$$

Solution

Step 1. Calculate the quantity $K_u Q/D^{5/2}$ and TW/D. From Figure 3.4 determine y_o/D. (See following table for calculations.)

Step 2. Calculate y_c from Figure B.2 or other appropriate method. Note that critical depth is greater than brink depth.

Step 3. Determine flow area based on y_o/D using Table B.2 and outlet velocity.

D (m)	$1.811Q/D^{5/2}$	TW/D	y_o/D	y_o (m)	y_c (m)	A/D^2	A (m²)	V=Q/A (m/s)
1.050	2.73	0.58	0.64	0.67	0.73	0.5308	0.585	2.90
1.200	1.95	0.51	0.55	0.66	0.70	0.4426	0.637	2.67
1.350	1.45	0.45	0.47	0.63	0.70	0.3627	0.661	2.57
1.500	1.12	0.41	0.42	0.63	0.67	0.3130	0.704	2.41

Changing culvert diameter from 1.050 to 1.500 m, a 43 percent increase, results in a decrease of only 17 percent in the outlet velocity.

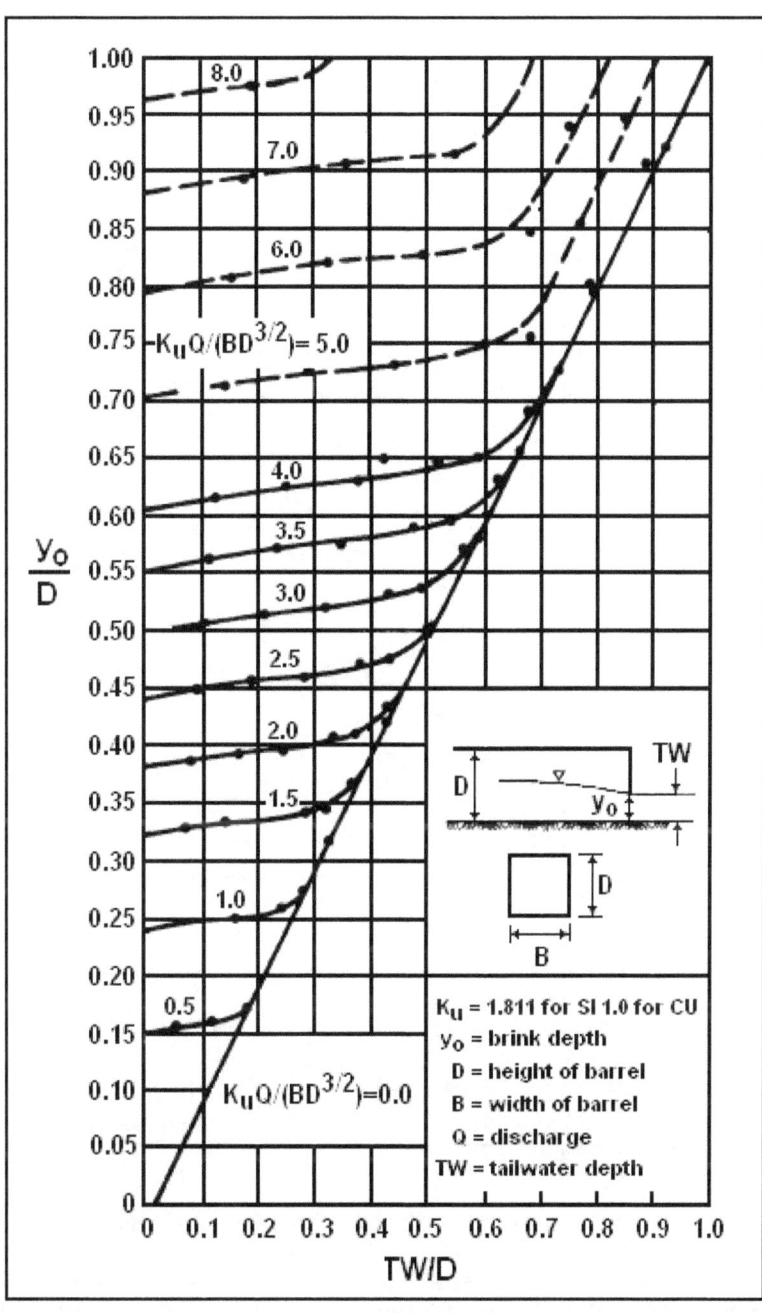

Figure 3.3. Dimensionless Rating Curves for the Outlets of Rectangular Culverts on Horizontal and Mild Slopes (Simons, 1970)

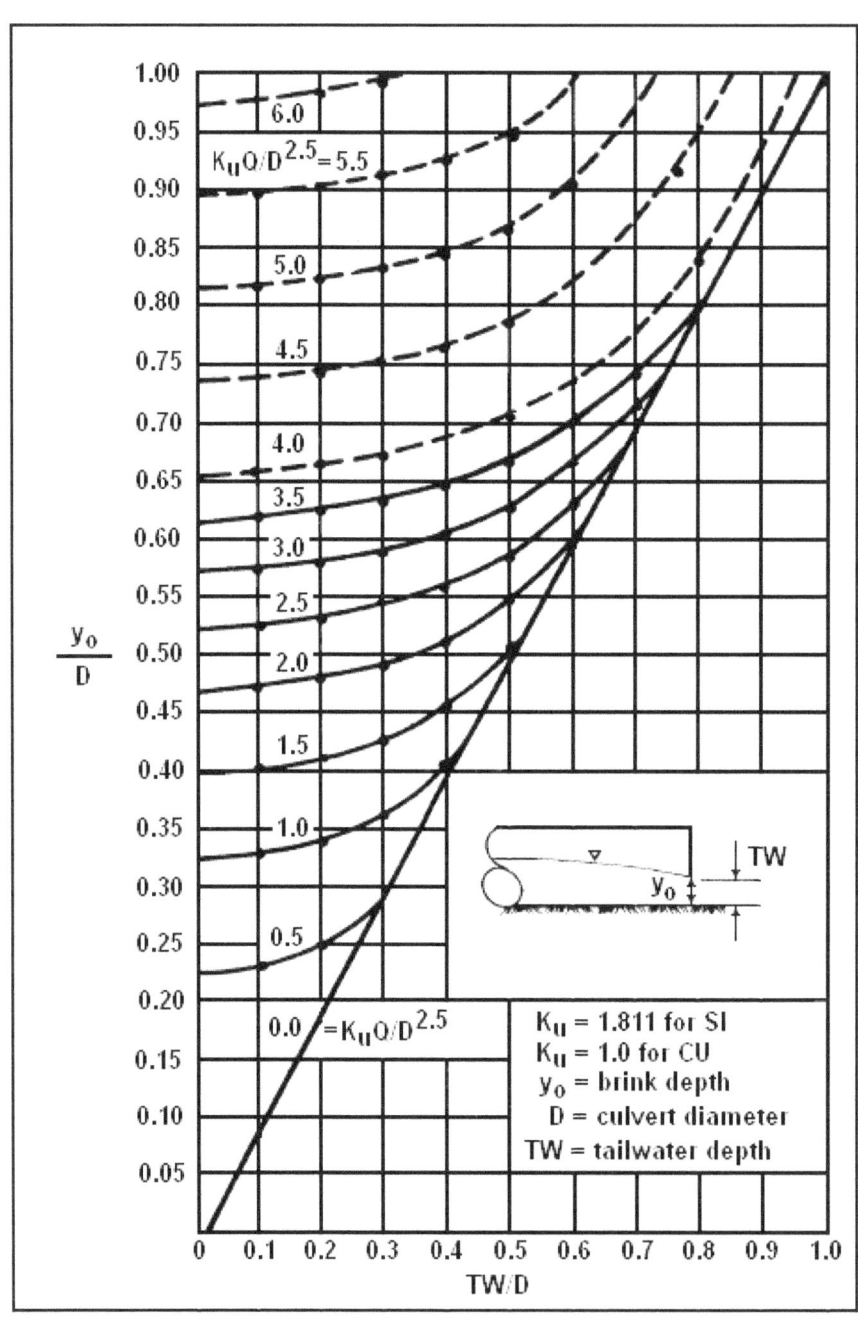

Figure 3.4. Dimensionless Rating Curves for the Outlets of Circular Culverts on Horizontal and Mild Slopes (Simons, 1970)

Design Example: Velocity Reduction by Increasing Culvert Size for Brink Depth Conditions CU)

Evaluate the reduction in velocity by replacing a 3.5 ft pipe culvert with a larger pipe culvert. Given:

$$Q = 60 \text{ ft}^3/\text{s}$$
$$TW = 2 \text{ ft, constant}$$

Solution

Step 1. Calculate the quantity $K_u Q/D^{5/2}$ and TW/D. From Figure 3.4 determine y_o/D. (See following table for calculations.)

Step 2. Calculate y_c from Figure B.2 or other appropriate method. Note that critical depth is greater than brink depth.

Step 3. Determine flow area based on y_o/D using Table B.2 and outlet velocity.

D (ft)	$Q/D^{5/2}$	TW/D	y_o/D	y_o (ft)	y_c (ft)	A/D^2	A (ft^2)	V=Q/A (ft/s)
3.5	2.62	0.57	0.63	2.20	2.4	0.52	6.4	9.4
4.0	1.88	0.50	0.54	2.16	2.3	0.43	6.9	8.7
4.5	1.40	0.44	0.46	2.10	2.3	0.35	7.1	8.5
5.0	1.07	0.40	0.41	2.05	2.2	0.30	7.5	8.0

Changing culvert diameter from 3.5 to 5 ft, a 43 percent increase, results in a decrease of only 15 percent in the outlet velocity.

3.2   CULVERTS ON STEEP SLOPES

Figure 3.5 (Normann, et al., 2001) shows the types of flow for culverts on steep slopes, i.e., culverts flowing with inlet control.

3.2.1   Submerged Outlets (Full Flow)

For culvert flow types shown in Figure 3.5B and D, full flow is assumed at the outlet. The outlet velocity is calculated using the full barrel area. See Section 3.1.1 for a discussion on the effect of increasing culvert diameter to decrease outlet velocity.

3.2.2   Unsubmerged Outlets (Normal Depth)

For culvert flow types shown in Figure 3.5A and C, normal flow is assumed at the culvert outlet and the outlet velocity is computed using Manning's Equation. Hydraulic Design Series No. 3 (FHWA, 1961) provides charts for a direct solution of Manning's Equation for circular and rectangular culverts. Tables B.1 and B.2 (Appendix B) can also be used to determine normal depth for circular and rectangular culverts. The following example illustrates how to compute normal depth and the effect on outlet velocity of increasing the roughness of the culvert.

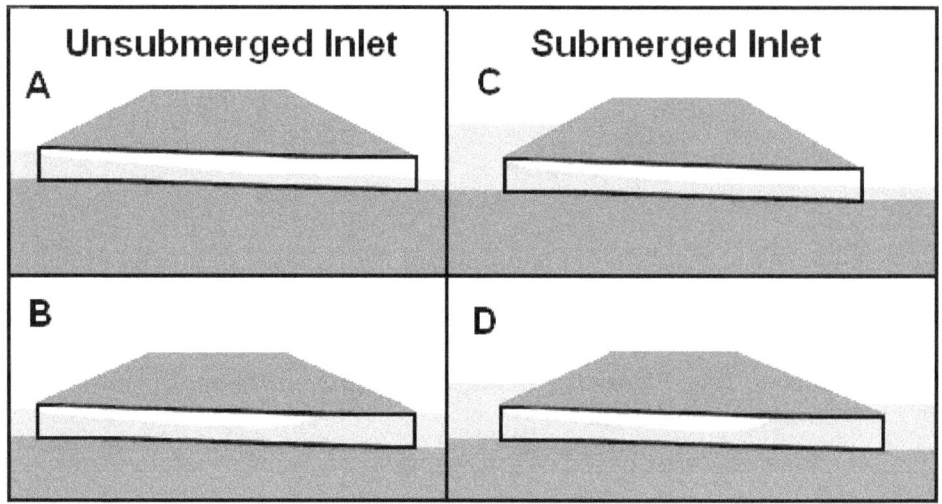

Figure 3.5. Inlet Control Flow Types

<u>**Design Example: Increasing Roughness to Reduce Velocity (SI)**</u>

Evaluate increasing roughness for reducing velocity. Given:

    Culvert Diameter, D, = 1.524 m

    Q   = 2.832 m^3/s

    n   = 0.012 for concrete and 0.024 for corrugated metal

    S_o = 0.01 m/m (1 percent slope)

<u>**Solution**</u>

For a smooth pipe (concrete):

    Step 1.   Calculate the quantity $\alpha Qn/(D^{8/3}S^{1/2})$ = 1.49(2.832)(0.012)/((1.524)$^{8/3}$(0.01)$^{1/2}$) = 0.1646

    Step 2.   Calculate depth, y, from Table B.2. y/D = 0.41, y = 0.41(1.524) = 0.625 m

    Step 3.   Calculate area, A, from Table B.2. A/D^2 = 0.3032, A = 0.3032(1.524)^2 = 0.704 m^2

    Step 4.   Calculate velocity, V_o, = Q/A = 2.832/0.704 = 4.02 m/s.

    Step 5.   Read critical depth, y_c, from Figure B.2. y_c = 0.9 m.

                Since y_c > y, the flow is supercritical and exit depth is normal depth.

For a rough pipe (corrugated metal):

    Step 1.   Calculate $\alpha Qn/(D^{8/3}S^{1/2})$ = 1.49(2.832)(0.024)/((1.524)$^{8/3}$(0.01)$^{1/2}$) = 0.3293

    Step 2.   Calculate depth, y, from Table B.2. y/D = 0.62, y = 0.62(1.524) = 0.945 m

    Step 3.   Calculate area, A, from Table B.2. A/D^2 = 0.5115, A = 0.5115(1.524)^2 = 1.19 m^2

    Step 4.   Calculate velocity, V_o, = Q/A = 2.832/1.19 = 2.38 m/s.

Step 5. Read critical depth, y_c, from Figure B.2. $y_c = 0.9$ m.

Since $y_c < y$, the flow is subcritical. The exit depth will be critical depth of 0.9 m and the exit velocity will be critical velocity of 2.41 m/s.

Design Example: Increasing Roughness to Reduce Velocity (CU)

Evaluate increasing roughness for reducing velocity. Given:

Culvert Diameter, D = 5 ft

Q = 100 ft³/s

n = 0.012 for concrete and 0.024 for corrugated metal

S_o = 0.01 ft/ft (1 percent slope)

Solution

For a smooth pipe (concrete):

Step 1. Calculate the quantity $\alpha Qn/(D^{8/3}S^{1/2}) = 1(100)(0.012)/((5)^{8/3}(0.01)^{1/2}) = 0.1642$

Step 2. Calculate depth, y, from Table B.2. $y/D = 0.41$, $y = 0.41(5) = 2.05$ ft

Step 3. Calculate area, A, from Table B.2. $A/D^2 = 0.3032$, $A = 0.3032(5)^2 = 7.58$ ft²

Step 4. Calculate velocity, V_o, = Q/A = 100/7.58 = 13.2 ft/s.

Step 5. Read critical depth, y_c, from Figure B.2. $y_c = 2.9$ ft. Since $y_c > y$, the flow is supercritical and exit depth is normal depth.

For a rough pipe (corrugated metal):

Step 1. Calculate $\alpha Qn/(D^{8/3}S^{1/2}) = 1(100)(0.024)/((5)^{8/3}(0.01)^{1/2}) = 0.3283$

Step 2. Calculate depth, y, from Table B.2. $y/D = 0.62$, $y = 0.62(5) = 3.1$ ft

Step 3. Calculate area, A, from Table B.2. $A/D^2 = 0.5115$, $A = 0.5115(5)^2 = 12.78$ ft²

Step 4. Calculate velocity, V_o, = Q/A = 100/12.78 = 7.82 ft/s.

Step 5. Read critical depth, y_c, from Figure B.2. $y_c = 2.9$ ft. Since $y_c < y$, the flow is subcritical. The exit depth will be critical depth of 2.9 ft and the exit velocity will be critical velocity of 7.9 ft/s.

For culverts on steep slopes, increasing the barrel size for a given discharge and slope has little effect on velocity. For example, using the 1.524 m (5 ft) diameter concrete pipe in the previous example, a $V_o = 4.02$ m/s (13.2 ft/s) was calculated. If a 2.438 m (8 ft) pipe is put at the same location, the velocity in the larger pipe will be 3.84 m/s (12.6 ft/s). The pipe diameter was more than doubled, but the velocity was only decreased by 4 percent.

Some reduction in outlet velocity can be obtained by increasing the number of barrels carrying the total discharge. Reducing the flow rate per barrel reduces velocity at normal depth if the flow line slopes are the same. Substituting two smaller pipes with the same depth to diameter ratio for a large one reduces Q per barrel to one-half the original rate and the outlet velocity to approximately 87 percent of that in the single-barrel design. However, this 13 percent reduction must be considered in light of the increased cost of the culverts. In addition, the percentage reduction decreases as the number of barrels is increased. For example, using four pipes instead of three provides only an additional 5 percent reduction in outlet velocity. A design using more barrels may still result in velocities requiring protection, with a large increase in the area to be protected.

For culverts on slopes greater than critical, rougher material will cause greater depth of flow and less velocity in equal size pipes. Velocity varies inversely with resistance; therefore, using a corrugated metal pipe instead of a concrete pipe will reduce velocity approximately 40 percent, and substitution of a structural plate corrugated metal pipe for concrete will result in about 50 percent reduction in velocity. Barrel resistance is obviously an important factor in reducing velocity at the outlets of culverts on steep slopes. Chapter 7 contains detailed discussion and specific design information for increasing barrel resistance.

3.2.3 Broken-back Culvert

Substituting a "broken-slope" flow line for a steep, continuous slope can be used for controlling outlet velocity. Chapter 7 contains detailed discussion and specific design information for designing broken-back culverts.

CHAPTER 4: FLOW TRANSITIONS

A flow transition is a change of open channel flow cross section designed to be accomplished in a short distance with a minimum amount of flow disturbance. Five types of transitions are shown in Figure 4.1: cylindrical quadrant, straight line, square end, warped, and wedge. Expansion transitions are illustrated, but contraction transitions would have similar geometry.

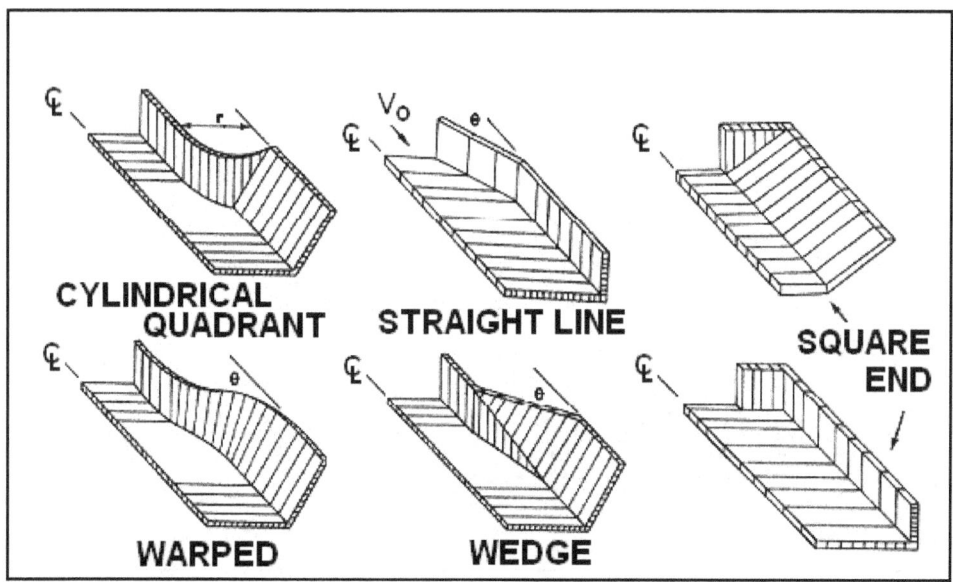

Figure 4.1. Transition Types

The most common flow transitions are the square end expansion (headwall) and the straight-line (wingwall) transitions. Both of these transitions are considered abrupt expansions and are discussed in Section 4.1. Procedures are provided for determining the velocity and depth exiting these standard headwall and wingwall configurations. An apron, which is an integral part of these transitions, protects the channel bottom at the culvert outlet from erosion.

Specially designed open channel flow inlet transitions (contractions) are normally not required for highway culverts. The economical culvert is designed to operate with an upstream headwater pool that dissipates the channel approach velocity and, therefore, negates the need for an approach flow transition. Side and slope tapered culvert inlets are designed as submerged transitions and do not fall within the intended limits of open channel transitions discussed in this chapter (see Normann, et al., 2001). Special inlet transitions are useful when the conservation of energy is essential because of allowable headwater considerations such as an irrigation structure in subcritical flow (see Section 4.2) or where it is desirable to maintain a small cross section with supercritical flow in a steep channel (see Section 4.3). Section 4.4 addresses supercritical flow expansions.

Expansions/transitions upstream of stilling basins are designed to decrease depth, increase velocity, and, therefore, increase Froude number. These supercritical expansions include design of a chute and determination of the needed depression below the streambed to force an efficient hydraulic jump. This topic is addressed in detail in Section 8.1.

4.1 ABRUPT EXPANSION

As a jet of water, which is not laterally constrained, leaves a culvert flowing in outlet control, the water surface plunges or drops very rapidly (see Figure 4.2). As the water surface drops and the flow spreads out, the potential energy stored as depth is converted to kinetic energy or velocity. Therefore, the velocity leaving the wingwall apron can be higher than the culvert outlet velocity and must be considered in determining outlet protection. The straight-line transition may also be considered an abrupt transition if the $\tan\theta$ is greater than 1/3Fr, where θ is the angle between the wingwall and culvert axis.

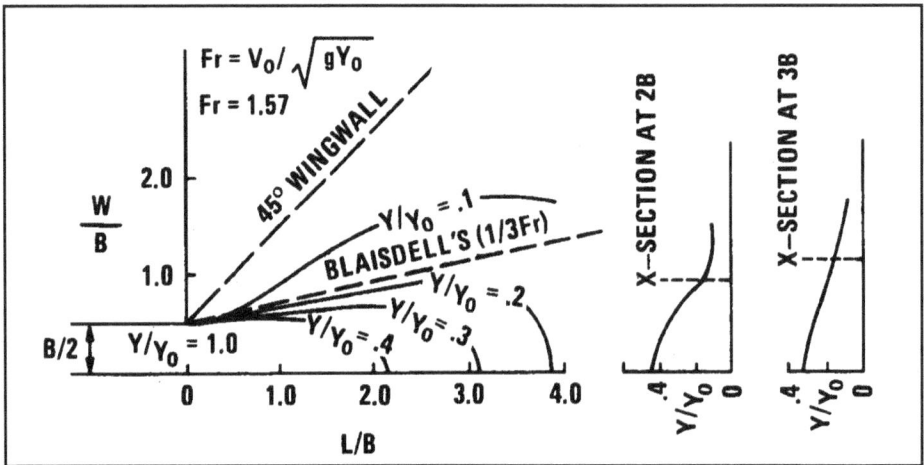

Figure 4.2. Dimensionless Water Surface Contours (Watts, 1968)

A reasonable estimate of transition exit velocity can be obtained by using the energy equation and assuming the losses to be negligible. By neglecting friction losses, a higher velocity than actually occurs is predicted making the error on the conservative side.

A more accurate way to determine transition exit flow conditions was developed by Watts (1968). Watts' experimental data has been converted to Equation 4.1 (for boxes) and Equation 4.2 (for circular pipes) for determining V_A/V_o.

$$\frac{V_A}{V_o} = 1.65 - 0.3Fr \tag{4.1}$$

$$\frac{V_A}{V_o} = 1.65 - 0.45\frac{Q}{\sqrt{gD^5}} \tag{4.2}$$

where,

V_A =   average velocity on the apron, m/s (ft/s)

V_o =   velocity at the culvert outlet, m/s (ft/s)

Also based on Watts' work, Figures 4.3 and 4.4 relate Froude number (Fr) or $Q/(gD^5)^{0.5}$ to the average depth/brink depth ratio (y_A/y_o). These equations and curves were developed for Fr from 1 to 3, which are applicable for most abrupt culvert outlet transitions. Normally, low tailwater is encountered at the culvert outlet and flow is supercritical on the outlet apron.

4-2

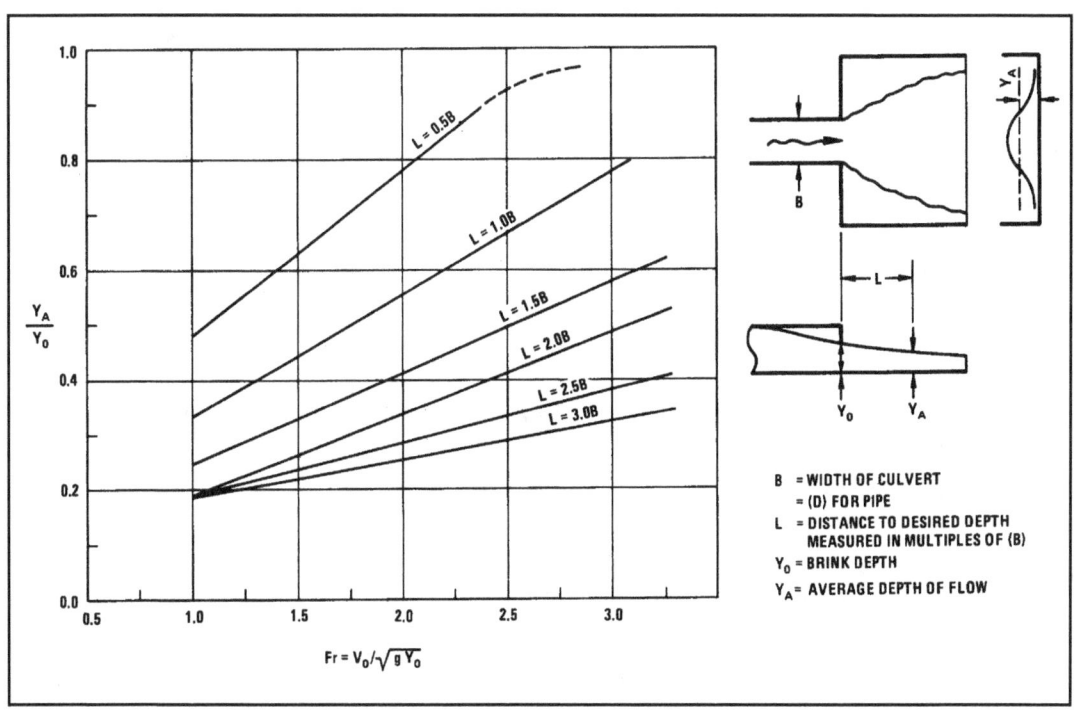

Figure 4.3. Average Depth for Abrupt Expansion Below Rectangular Culvert Outlet

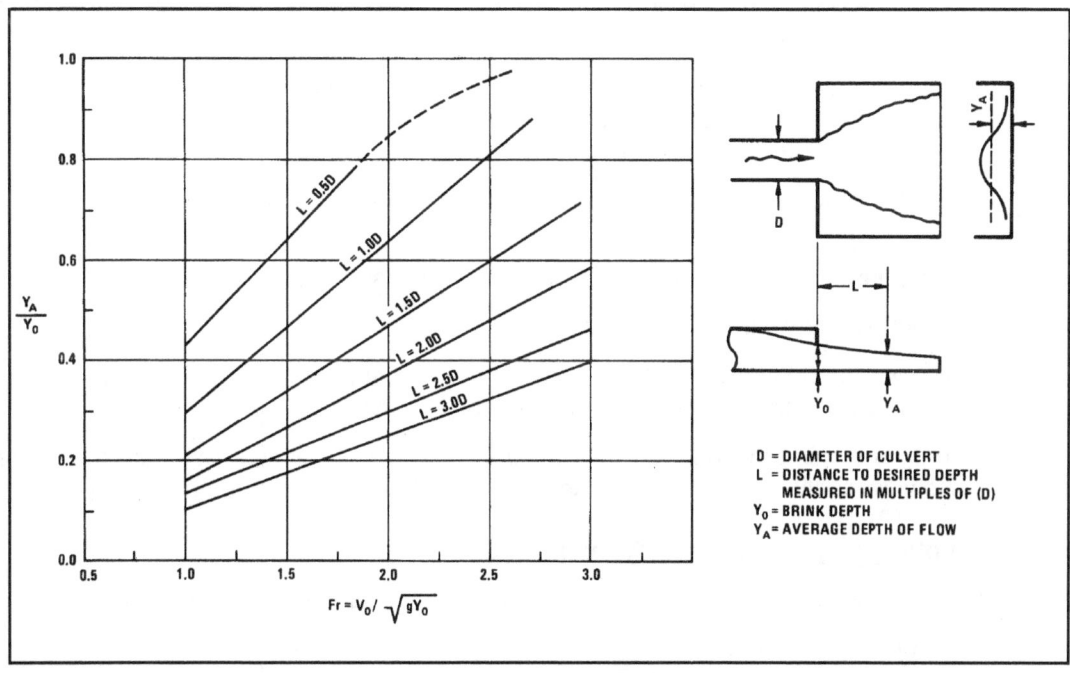

Figure 4.4. Average Depth for Abrupt Expansion Below Circular Culvert Outlet

4-3

Water cannot completely expand to fill the section between the wingwalls in an abrupt expansion. The majority of the flow will stay within an area whose boundaries are defined by:

$$\theta = \tan^{-1}(Fr/3) \tag{4.3}$$

where,

   θ =   optimum flare angle

The downstream width of the apron, W_2, is given by:

$$W_2 = W_o + 2L\tan\theta_w \tag{4.4}$$

where,

   W_2 =   width of apron at length, L, downstream from the culvert outlet, m (ft)
   L   =   distance downstream from culvert outlet, m (ft)
   θ_w =   wingwall flare angle

If $\theta_w > \theta$ then the designer should consider reducing θ_w to θ. As shown in Figure 4.2 flaring the wingwall more than 1/3Fr (for example 45°) provides unused space which is not completely filled with water.

The design procedure for an abrupt expansion may be summarized in the following steps:

   Step 1.   Determine the flow conditions at the culvert outlet: V_o and y_o (see Chapter 3).

   Step 2.   Calculate the Froude number: Fr = V_o /(g y_o)$^{0.5}$ at the culvert outlet.

   Step 3.   Find the optimum flare angle, θ, using Equation 4.3. If the chosen wingwall flare, θ_w, is greater than θ, consider reducing θ_w to θ.

   Step 4.   Find the average depth on the apron. For boxes, use Figure 4.3. For pipes, use Figure 4.4. The ratio y_A/y_o is obtained knowing the Froude number (Fr) and the desired distance downstream, L.

   Step 5.   Find average velocity on the apron, V_A, using Equation 4.1 or Equation 4.2. V_A = V_2.

   Step 6.   Calculate the downstream width, W_2, using Equation 4.4.

   Step 7.   Calculate downstream depth, y_2.

   If θ was used in Equation 4.4, calculate $y_2 = Q/(V_A W_2)$. This depth will be larger than y_A since the flow prism is now laterally confined.

   If θ_w was used in Equation 4.4, calculate $y_2 = y_A$. However, estimate the average flow width, W_A, = $Q/(V_A y_A)$. Check that $W_A < W_2$. If it is not, then $y_2 = Q/(V_A W_2)$.

Design Example: Abrupt Expansion Transition (SI)

Find the flow conditions (y_2 and V_2) at end of a 3.1 m apron. Assume negligible tailwater. Given:

   RCB = 1524 mm x 1524 mm
   Wingwall flare θ_w = 45°
   Culvert length = 61 m

S_o = 0.002 m/m

Q = 7.65 m^3/s

y_c = 1.37 m

Solution

Step 1. Find culvert outlet velocity from Figure 3.3 with TW/D ≈ 0.

Need quantity 1.811 Q/(BD$^{3/2}$) = 1.811(7.65)/(1.524(1.524)$^{3/2}$) = 4.83

y_o /D = 0.68

y_o = 0.68(1.524) = 1.036 m

V_o = Q/A = 7.65/(1.036 (1.524)) = 4.84 m/s

Step 2. Find culvert outlet Froude number.

Fr = V_o /(g y_o)$^{0.5}$ = 4.84/(9.81(1.036))$^{0.5}$ = 1.52

Step 3. Find θ

tanθ = 1/3 Fr = 1/3(1.52) = 0.22

θ = 12.37°

Step 4. Estimate average depth.

Apron Length/Diameter = 3.1/1.524 = 2 (Convert apron length to multiple of culvert diameter.)

Use Figure 4.3 for y_A/y_o = 0.26

y_A = 0.26(1.036) = 0.269 m

Step 5. Find average velocity, V_A, using Equation 4.1.

V_A/V_o = 1.65 - 0.3Fr = 1.65 – 0.3(1.52) = 1.2

V_A = 4.84(1.2) = 5.82 m/s

V_A = V_2 = 5.82 m/s

Step 6. Calculate downstream width using Equation 4.4.

$θ_w$ > θ use $θ_w$

W_2 = W_o + 2L tan ($θ_w$) = 1.524 + 2(3.1)(1.0) = 7.72 m

Step 7. Calculate downstream depth.

$θ_w$ was used, therefore,

y_2 = y_A = 0.269 m

Check W_A = Q/(V_A y_A) = 7.65/((5.82)(0.269))

W_A = 4.89 m < 7.72 m

Compare the above solution with two alternatives using the energy equation.

Alternative 1. Assume W_2 = full width between wingwalls at the end of the apron.

W_2 = W_o + 2L tan 45° = 7.72 m

$A_2 = W_2 \, y_2 = 7.62 \, y_2$

$V_2 = Q/A_2 = 7.65/(7.72y_2) = 0.99/y_2$

The energy balance between flow at the culvert outlet and the apron is given by:

$z_o + y_o + V_o^2/(2g) = z_2 + y_2 + V_2^2/(2g) + H_f$

Assuming $H_f = 0$ and $z_o = z_2$

$1.036 + (4.84)^2/(2(9.81)) = y_2 + (0.99/y_2)^2/(2(9.81))$

$1.036 + 1.194 = y_2 + 0.050/y_2^2$

$2.230 = y_2 + 0.050/y_2^2$

$y_2 = 0.157$ m, which is 41% lower than the original solution.

$V_2 = 0.99/0.157 = 6.31$ m/s which is 8% higher than the original solution.

Alternative 2. Assume W_2 is based on θ where $\tan\theta = 1/3$ Fr.

$W_2 = W_o + 2L \tan 12.41° = 1.524 + 6.2(0.22) = 2.89$ m

$A_2 = 2.89 \, y_2$ and $V_2 = 7.65/(2.89y_2) = 2.65/y_2$

$2.230 = y_2 + 0.360/y_2^2$

$y_2 = 0.45$ meters, which is 68% higher than the original solution

$V_2 = 2.65/0.45 = 5.89$ m/s, which is 2% higher than the original solution.

Design Example: Abrupt Expansion Transition (CU)

Find the flow condition (y_2 and V_2) at end of a 10 ft apron. Assume negligible tailwater. Given:

RCB = 5 ft x 5 ft
Wingwall flare $\theta_w = 45°$
Culvert length = 200 ft
$S_o = 0.002$ ft/ft
$Q = 270$ ft^3/s
$y_c = 4.5$ ft

Solution

Step 1. Find culvert outlet velocity from Figure 3.3 with TW/D $\approx$ 0

Need quantity $Q/(BD^{3/2}) = 270/(5(5)^{3/2}) = 4.83$

$y_o/D = 0.68$

$y_o = 0.68(5) = 3.4$ ft

$V_o = Q/A = 270/((5)\,3.4) = 15.9$ ft/s

Step 2. Find culvert outlet Froude number.

$Fr = V_o/(g \, y_o)^{0.5} = 15.9/(32.2(3.4))^{0.5} = 1.52$

Step 3. Find θ

$\tan\theta = 1/3 \, Fr = 1/3(1.52) = 0.22$

$\theta = 12.37°$

Step 4. Estimate average depth.

Apron Length/Diameter = 10/5 = 2 (Convert apron length to multiple of culvert diameter.)

Use Figure 4.3 for $y_A/y_o = 0.26$

$y_A = 0.26(3.4) = 0.88$ ft

Step 5. Find average velocity V_A using Equation 4.1.

$V_A/V_o = 1.65 - 0.3Fr = 1.65 - 0.3(1.52) = 1.2$

$V_A = 15.9(1.2) = 19.1$ ft/s

$V_A = V_2 = 19.1$ ft/s

Step 6. Calculate downstream width.

$\theta_w > \theta$ use θ_w

$W_2 = W_o + 2L \tan(\theta_w) = 5 + 2(10)(1.0) = 25$ ft

Step 7. Calculate downstream depth.

θ_w was used:

$y_2 = y_A = 0.88$ ft

Check $W_A = Q/(V_A \, y_A) = 270/((19.1)(0.88))$

$W_A = 16.1$ ft < 25 ft

Compare the above solution with two alternatives using the energy equation.

Alternative 1. Assume W_2 = full width between wingwalls at the end of the apron.

$W_2 = W_o + 2L \tan 45° = 25$ ft

$A_2 = W_2 \, y_2 = 25 \, y_2$

$V_2 = Q/A_2 = 270/(25y_2) = 10.8/y_2$

The energy balance between flow at the culvert outlet and the apron is given by:

$z_o + y_o + V_o^2/(2g) = z_2 + y_2 + V_2^2/(2g) + H_f$

Assuming $H_f = 0$ and $z_o = z_2$,

$3.4 + (15.9)^2/(2(32.2)) = y_2 + (10.8/y_2)^2/(2(32.2))$

$3.4 + 3.92 = y_2 + 1.81/y_2^2$

$7.32 = y_2 + 1.81/y_2^2$

$y_2 = 0.52$ ft, which is 41% lower than the original solution.

$V_2 = 10.8/0.52 = 20.8$ ft/s which is 10% higher than the original solution.

Alternative 2. Assume W_2 is based on θ where $\tan\theta = 1/3$ Fr.

$$W_2 = W_o + 2L \tan 12.41° = 5 + 20(0.22) = 9.4 \text{ ft}$$

$$A_2 = 9.4 \, y_2 \text{ and } V_2 = 270/(9.4y_2) = 28.7/y_2$$

$$7.32 = y_2 + 12.8/y_2{}^2$$

$y_2 = 1.48$ ft, which is 68% higher than the original solution

$V_2 = 28.7/1.48 = 19.4$ ft/s, which is 2% higher than the original solution.

4.2   SUBCRITICAL FLOW TRANSITION

Subcritical flow can be transitioned into and out of highway structures without causing adverse effect if subcritical flow is maintained throughout the structure. The flow cannot approach or pass through critical depth, y_c. The range of depths to avoid is $0.9y_c$ to $1.1y_c$. In this range, slight changes in specific energy are reflected in large changes in depth, i.e., wave problems develop. The straight line or wedge transition should be used if conservation of flow energy is required, for example, for an irrigation canal structure that traverses a highway. Warped and cylindrical transitions are more efficient, but the additional construction cost can only be justified for structures where backwater is critical.

Figure 4.5 illustrates the design problem. Starting upstream of section 1 where some backwater exists due to the culvert, the flow is transitioned from a canal into and then out of the highway culvert. The flare angle, θ_w, should be 12.5°, (1:4.5 (lateral:longitudinal) or smaller) according to Hinds (1928). This criterion provides a gradually varied transition that can be analyzed using the energy equation.

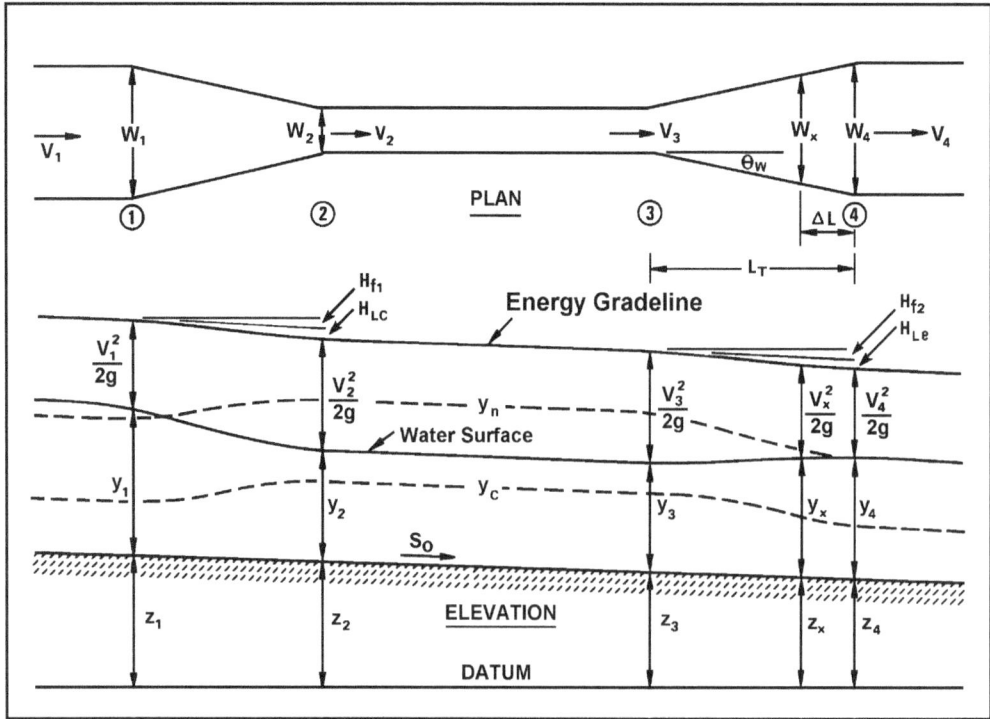

Figure 4.5. Subcritical Flow Transition

4-8

As the flow transitions into the culvert, the water surface approaches y_c. To minimize waves, y_2 should be equal to or greater than $1.1y_c$. In the culvert, the depth will increase and will reach y_n if the culvert is long enough. In the expansion, the depth increases to y_n of the downstream channel, Section 4. Associated with both transitions are energy losses that are proportional to the change in velocity head in the transitions. The energy loss in the contraction, H_{Lc}, and in the expansion, H_{Le}, are:

$$H_{Lc} = C_c \left(\frac{V_2^2}{2g} - \frac{V_1^2}{2g} \right) \tag{4.5}$$

$$H_{Le} = C_e \left(\frac{V_3^2}{2g} - \frac{V_4^2}{2g} \right) \tag{4.6}$$

where C_c and C_e are found from Table 4.1.

Table 4.1. Transition Loss Coefficients (USACE, 1994)

Transition Type	Contraction C_c	Expansion C_e
Warped	0.10	0.20
Cylindrical Quadrant	0.15	0.25
Wedge	0.30	0.50
Straight Line	0.30	0.50
Square End	0.30	0.75

The depth in the culvert, y_3, can be found by trial and error using the energy equation with $y_4 = y_n$ in the downstream channel and assuming $h_{f2} = 0$ (see Figure 4.5). The streambed elevation is equal to z. Writing the energy equation between sections 3 and 4 yields:

$$z_4 + y_4 + V_4^2/(2g) + H_{Le} + H_{f2} = z_3 + y_3 + V_3^2/(2g)$$

Assuming $H_{f2} \cong 0$, $V_3 = Q/(W_3 y_3)$, $V_4 = Q/(W_4 y_4)$, and substituting Equation 4.6 gives:

$$z_4 + y_4 + V_4^2/(2g) + C_e (V_3^2/(2g) - V_4^2/(2g)) = z_3 + y_3 + V_3^2/(2g)$$

$$z_4 + y_4 + (1 - C_e) V_4^2/(2g) = z_3 + y_3 + (1 - C_e) v_3^2/(2g)$$

$$z_4 - z_3 + y_4 + (1 - C_e) (Q/(W_4 y_4))^2/(2g) = y_3 + (1 - C_e) (Q/(W_3 y_3))^2/(2g) \tag{4.7}$$

After known values are substituted, Equation 4.7 reduces to

$$C_1 = y_3 + C_2/y_3^2$$

which can be solved by trial and error.

In a similar manner, y_1 can be determined by assuming $y_2 = y_3$ and $h_{f1} = 0$.

$$z_2 + y_2 + V_2^2 /(2g) + H_{LC} + H_{f1} = z_1 + y_1 + V_1^2 /(2g)$$

$$z_2 + y_2 + V_2^2 /(2g) + C_c (V_2^2 /(2g) - V_1^2 /(2g)) = z_1 + y_1 + V_1^2 /(2g)$$

$$z_2 + y_2 + (1 + C_c) V_2^2 /(2g) = z_1 + y_1 + (1 + C_c) V_1^2 /(2g)$$

$$z_2 - z_1 + y_2 + (1 + C_c) (Q/(W_2\, y_2))^2 /(2g) = y_1 + (1 + C_c) (Q/(W_1\, y_1))^2 /(2g) \qquad (4.8)$$

These depths are approximate because friction loss was neglected. They should be checked by computing the water surface profile using a standard step method computer program, like HEC-RAS (USACE, 2002).

4.3 SUPERCRITICAL FLOW CONTRACTION

The design of transitions for supercritical flow is difficult to manage without causing a hydraulic jump or other surface irregularity. Therefore, the full flow area should be maintained if at all possible. A smooth transition of supercritical flow requires a structure longer than typical wingwalls and should not be attempted unless the structure is of primary importance. A model study should be used to determine transition geometry where a hydraulic jump is not desired. If a hydraulic jump is acceptable, the inlet structure can be designed as shown in Figure 4.6. This design, which must be accomplished in a rectangular channel, yields a long transition. The design approach outlined below is from USACE (1994) and Ippen (1951).

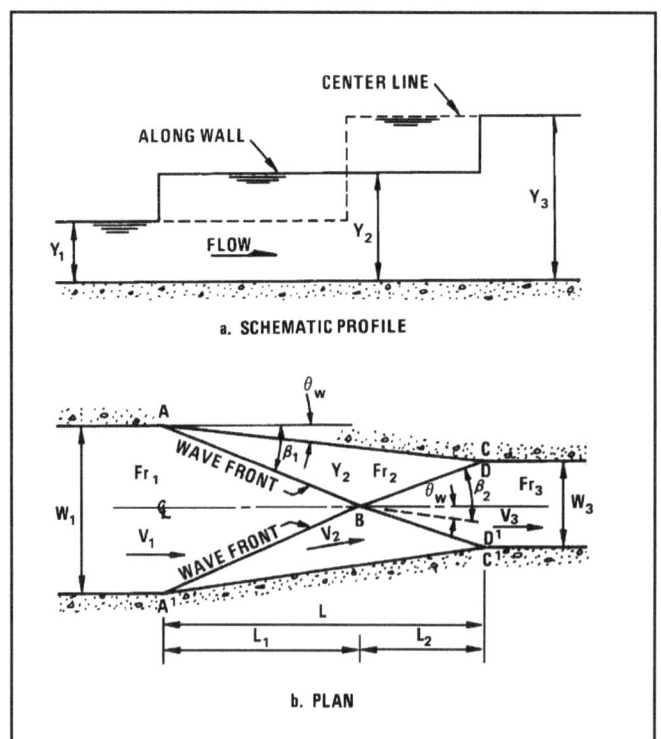

Figure 4.6. Supercritical Inlet Transition for Rectangular Channel (USACE, 1994)

4-10

The length, L, is defined by the channel contraction, $W_1 - W_3$, and the wall deflection angle, θ_w.

$$L = (W_1 - W_3)/(2 \operatorname{Tan}\theta_w) \qquad (4.9)$$

To minimize surface disturbances, L should also equal $L_1 + L_2$ where

$$L_1 = W_1 / (2\operatorname{Tan} \beta_1) \qquad (4.10)$$

$$L_2 = W_3 / (2\operatorname{Tan} (\beta_2 - \theta_w)) \qquad (4.11)$$

$$\tan \theta_w = \frac{\tan \beta_1 \left(\sqrt{1 + 8Fr_1^2 \sin^2 \beta_1} - 3 \right)}{2 \tan^2 \beta_1 + \sqrt{1 + 8Fr_1^2 \sin^2 \beta_1} - 1} \qquad (4.12)$$

The transition design requires a trial θ_w that fixes L as defined by Equation 4.9. This length is then checked by finding $L_1 + L_2$. To determine L_1, β_1 is found from Equation 4.12 by trial and error and then substituted into Equation 4.10. L_2 is calculated from Equation 4.11 with β_2 determined from Equation 4.12 by substituting β_2 for β_1 and Fr_1 for Fr_2. To find Fr_2 first calculate:

$$\frac{y_2}{y_1} = \frac{1}{2} \left(\sqrt{1 + 8Fr_1^2 \sin^2 \beta_1} - 1 \right) \qquad (4.13)$$

Then:

$$Fr_2^2 = \frac{y_1}{y_2} \left[Fr_1^2 - \left(\frac{y_1}{2y_2} \right) \left(\frac{y_2}{y_1} - 1 \right) \left(\frac{y_2}{y_1} + 1 \right)^2 \right] \qquad (4.14)$$

If the trial θ_w was chosen correctly the total length, L, will equal the sum of L_1 and L_2. If not, choose another trial θ_w and repeat the process until the lengths match. The depth, y_3, and Fr_3 in the culvert can now be calculated using Equation 4.13 and Equation 4.14 if the subscripts are increased by 1; i.e., y_2/y_1 is now y_3/y_2. The above design approach assumes that the width of the channel (W_1) and the width of the culvert (W_3) are known and L is found by trial and error. If W_3 has to be determined, the design problem is complicated by another trial and error process.

4.4   SUPERCRITICAL FLOW EXPANSION

Supercritical flow expansion design has, in part, been discussed in Section 4.1. The procedure outlined in Section 4.1 should be used to determine apron or expansion flow conditions if the culvert exit Froude number, Fr, is less than 3, if the location where the flow conditions are desired is within 3 culvert diameters of the outlet and S_o is less than 10%. For expansions outside these limits, the energy equation can be used to determine flow conditions leaving the transition. Normally, these parameters would then be used as the input values for a basin design.

This page intentionally left blank.

CHAPTER 5: ESTIMATING SCOUR AT CULVERT OUTLETS

This chapter presents a method for predicting local scour at the outlet of culverts based on discharge, culvert shape, soil type, duration of flow, culvert slope, culvert height above the bed, and tailwater depth. In addition to this local scour, channel degradation (discussed in Chapter 2) should be evaluated. The procedures in this chapter provide a good method for estimating the extent of the local scour hole. The designer should also review maintenance history, site reconnaissance and data on soils, flows, and flow duration to determine the best estimate of the potential scour hazard at a culvert outlet.

The prediction equations presented in this chapter are intended to serve along with field reconnaissance as guidance for determining the need for energy dissipators at culvert outlets. The designer should remember that the equations do not include long-term channel degradation of the downstream channel. The equations are based on tests that were conducted to determine maximum scour for the given condition and therefore represent what might be termed worst-case scour geometries. The equations were derived from tests conducted by the Corps of Engineers (Bohan, 1970) and Colorado State University (Abt, et al., 1985; Abt, et al., 1987; Abt, 1996; Doehring, 1994; Donnell and Abt, 1983; Ruff, et al., 1982).

5.1 COHESIONLESS SOILS

The general expression for determining scour geometry in a cohesionless soil at the culvert outlet is:

$$\left[\frac{h_s}{R_c}, \frac{W_s}{R_c}, \frac{L_s}{R_c}, \frac{V_s}{R_c^3}\right] = C_s C_h \left(\frac{\alpha}{\sigma^{1/3}}\right) \left(\frac{Q}{\sqrt{g}(R_c^{2.5})}\right)^\beta \left(\frac{t}{316}\right)^\theta \tag{5.1}$$

where,

h_s = depth of scour, m (ft)

W_s = width of scour, m (ft)

L_s = length of scour, m (ft)

V_s = volume of scour, m³ (ft³)

R_c = hydraulic radius at the end of the culvert (assuming full flow)

Q = discharge, m³/s (ft³/s)

g = acceleration of gravity, 9.81 m/s² (32.2 ft /s²)

t = time in minutes

σ = $(D_{84}/D_{16})^{0.5}$, material standard deviation

α, β, θ are coefficients, see Table 5.1

C_s = slope correction coefficient, see Table 5.2

C_h = drop height adjustment coefficient, see Table 5.3

The bed-material grain-size distribution is determined by performing a sieve analysis (ASTM DA22-63). The values of D_{84} and D_{16} are extracted from the grain size distribution. If σ <1.5, the material is considered to be uniform. If σ >1.5, the material is classified as graded. Typical values for σ are 2.10 for gravel and 1.87 for sand.

5.1.1 Scour Hole Geometry

Investigators Bohan (1970) and Fletcher and Grace (1972) indicate that the scour hole geometry varies with tailwater conditions with the maximum scour geometry occurring at tailwater depths less than half the culvert height (Bohan, 1970); and that the maximum depth of scour, h_s, occurs at a location approximately 0.4 L_s downstream of the culvert outlet (Fletcher and Grace, 1972) where L_s is the length of scour. The α, β, θ coefficients to determine scour geometry are shown in Table 5.1.

Table 5.1. Coefficients for Culvert Outlet Scour in Cohesionless Soils

	α	β	θ
Depth, h_S	2.27	0.39	0.06
Width, W_S	6.94	0.53	0.08
Length, L_S	17.10	0.47	0.10
Volume, V_S	127.08	1.24	0.18

5.1.2 Time of Scour

The time of scour is estimated based upon knowledge of peak flow duration. Lacking this knowledge, it is recommended that a time of 30 minutes be used in Equation 5.1. The tests indicate that approximately 2/3 to 3/4 of the maximum scour depth occurs in the first 30 minutes of the flow duration. The exponents for the time parameter in Table 5.1 reflect the relatively flat part of the scour-time relationship (t > 30 minutes) and are not applicable for the first 30 minutes of the scour process.

5.1.3 Headwalls

Installation of a perpendicular headwall at the culvert outlet moves the scour hole downstream (Ruff, et al., 1982). However, the magnitude of the scour geometry remains essentially the same as for the case without the headwall. If the culvert is installed with a headwall, the headwall should extend to a depth equal to the maximum depth of scour.

5.1.4 Drop Height

The scour hole dimensions will vary with the height of the culvert invert above the bed. The scour hole shape becomes deeper, wider, and shorter, as the culvert invert height is increased (Doehring, 1994). The coefficients, C_h, are derived from tests where the pipe invert is adjacent to the bed. In order to compensate for an elevated culvert invert, Equation 5.1 can be modified to where C_h, expressed in pipe diameters, is a coefficient for adjusting the compound scour hole geometry. The values of C_h are presented in Table 5.2.

Table 5.2. Coefficient C_h for Outlets above the Bed

$H_d{}^1$	Depth	Width	Length	Volume
0	1.00	1.00	1.00	1.00
1	1.22	1.51	0.73	1.28
2	1.26	1.54	0.73	1.47
4	1.34	1.66	0.73	1.55

^1H_d is the height above bed in pipe diameters.

5.1.5 Slope

The scour hole dimensions will vary with culvert slope. The scour hole becomes deeper, wider, and longer as the slope is increased (Abt, 1985). The coefficients presented are derived from tests where the pipe invert is adjacent to the bed. In order to compensate for a sloped culvert, Equation 5.1 can be adjusted with a coefficient, C_s, adjusting for scour hole geometry. The values of C_s are shown in Table 5.3.

Table 5.3. Coefficient C_s for Culvert Slope

Slope %	Depth	Width	Length	Volume
0	1.00	1.00	1.00	1.00
2	1.03	1.28	1.17	1.30
5	1.08	1.28	1.17	1.30
>7	1.12	1.28	1.17	1.30

5.1.6 Design Procedure

Step 1. Determine the magnitude and duration of the peak discharge. Express the discharge in m^3/s (ft^3/s) and the duration in minutes.

Step 2. Compute the full flow hydraulic radius, R_c

Step 3. Compute the culvert invert height above the bed ratio, H_d, for slopes > 0%.

$$H_d = \frac{\text{Drop Height}}{\text{Diameter}}$$

Step 4. Determine scour coefficients from Table 5.1 and coefficients for culvert drop height, C_h, from Table 5.2 and slope, C_s, from Table 5.3.

Step 5. Determine the material standard deviation, $\sigma = (D_{84}/D_{16})^{0.5}$ from a sieve analysis of a soil sample at the proposed culvert location.

Step 6. Compute the scour hole dimensions using Equation 5.1.

Step 7. Compute the location of maximum scour, $L_m = 0.4\ L_s$.

Design Example: Estimating Scour Hole Geometry in a Cohesionless Soil (SI)

Determine the scour geometry-maximum depth, width, length and volume of scour. Given:

D = 457 mm CMP Culvert

S = 2%

Drop height = 0.914 m from channel degradation

Q = 0.764 m³/s

σ = 1.87 for downstream channel which is graded sand

Solution

Step 1. Determine the magnitude and duration of the peak discharge: Q = 0.764 m³/s and the peak flow duration is estimated to be 30 minutes.

Step 2. Compute the full flow hydraulic radius, R_c:

$$R_c = \frac{D}{4} = \frac{0.457}{4} = 0.114 \text{ m}$$

Step 3. Compute the height above bed ratio, H_d, for slopes > 0%:

$$H_d = \frac{\text{Drop Height}}{\text{Diameter}} = \frac{0.914}{0.457} = 2$$

Step 4. The Coefficients of scour obtained from Table 5.1, Table 5.2, and Table 5.3 are:

	α	β	θ	C_s	C_h
Depth of scour	2.27	0.39	0.06	1.03	1.26
Width of scour	6.94	0.53	0.08	1.28	1.54
Length of scour	17.10	0.47	0.10	1.17	0.73
Volume of scour	127.08	1.24	0.18	1.30	1.47

Step 5. Determine the material standard deviation. σ = 1.87

Step 6. Compute the scour hole dimensions using Equation 5.1:

$$\left[\frac{h_s}{R_c}, \frac{W_s}{R_c}, \frac{L_s}{R_c}, \frac{V_s}{R_c^3}\right] = C_s C_h \left(\frac{\alpha}{\sigma^{1/3}}\right)\left(\frac{Q}{\sqrt{g(R_c^{2.5})}}\right)^\beta \left(\frac{t}{316}\right)^\theta$$

$$h_s = C_s C_h \left(\frac{\alpha}{1.87^{1/3}}\right)\left(\frac{Q}{\sqrt{g(R_c^{2.5})}}\right)^\beta \left(\frac{30}{316}\right)^\theta R_c$$

$$h_s = 1.03(1.26)\left(\frac{2.27}{1.23}\right)\left(\frac{0.764}{\sqrt{9.81(0.114)^{2.5}}}\right)^{0.39}(0.095)^{0.06}(0.114) = 1.14 \text{ m}$$

5-4

Similarly,

$$W_s = 1.28(1.54)\left(\frac{6.94}{1.23}\right)\left(\frac{0.764}{\sqrt{9.81}(0.114)^{2.5}}\right)^{0.53}(0.095)^{0.08}(0.114) = 8.82 \text{ m}$$

$$L_s = 1.17(0.73)\left(\frac{17.10}{1.23}\right)\left(\frac{0.764}{\sqrt{9.81}(0.114)^{2.5}}\right)^{0.47}(0.095)^{0.10}(0.114) = 7.06 \text{ m}$$

$$V_s = 1.30(1.47)\left(\frac{127.08}{1.23}\right)\left(\frac{0.764}{\sqrt{9.81}(0.114)^{2.5}}\right)^{1.24}(0.095)^{0.18}(0.114)^3 = 27.8 \text{ m}^3$$

Step 7. Compute the location of maximum scour. L_m = 0.4 L_s = 0.4 (7.06) = 2.82 m downstream of the culvert outlet.

Design Example: Estimating Scour Hole Geometry in a Cohesionless Soil (CU)

Determine the scour geometry-maximum depth, width, length and volume of scour. Given:

D = 18 in CMP Culvert
S = 2%
Drop height = 3 ft from channel degradation
Q = 27 ft³/s
σ = 1.87 for downstream channel which is graded sand

Solution

Step 1. Determine the magnitude and duration of the peak discharge: Q = 27 ft³/s and the peak flow duration is estimated to be 30 minutes.

Step 2. Compute the full flow hydraulic radius, R_c:

$$R_c = \frac{D}{4} = \frac{1.5}{4} = 0.375 \text{ ft}$$

Step 3. Compute the height above bed ratio, H_d, for slopes > 0%

$$H_d = \frac{\text{Drop Height}}{\text{Diameter}} = \frac{3}{1.5} = 2$$

Step 4. The coefficients of scour obtained from Table 5.1, Table 5.2, and Table 5.3 are:

	α	β	θ	C_s	C_h
Depth of scour	2.27	0.39	0.06	1.03	1.26
Width of scour	6.94	0.53	0.08	1.28	1.54
Length of scour	17.10	0.47	0.10	1.17	0.73
Volume of scour	127.08	1.24	0.18	1.30	1.47

Step 5. Determine the material standard deviation. σ = 1.87

Step 6. Compute the scour hole dimensions using Equation 5.1:

$$\left[\frac{h_s}{R_c}, \frac{W_s}{R_c}, \frac{L_s}{R_c}, \frac{V_s}{R_c^3}\right] = C_s C_h \left(\frac{\alpha}{\sigma^{1/3}}\right)\left(\frac{Q}{\sqrt{g(R_c^{2.5})}}\right)^\beta \left(\frac{t}{316}\right)^\theta$$

$$h_s = C_s C_h \left(\frac{\alpha}{1.87^{1/3}}\right)\left(\frac{Q}{\sqrt{g(R_c^{2.5})}}\right)^\beta \left(\frac{30}{316}\right)^\theta R_c$$

$$h_s = 1.03(1.26)\left(\frac{2.27}{1.23}\right)\left(\frac{27}{\sqrt{32.2(0.375)^{2.5}}}\right)^{0.39}(0.095)^{0.06}(0.375) = 3.7 \ \text{ft}$$

Similarly,

$$W_s = 1.28(1.54)\left(\frac{6.94}{1.23}\right)\left(\frac{27}{\sqrt{32.2(0.375)^{2.5}}}\right)^{0.53}(0.095)^{0.08}(0.375) = 29.0 \ \text{ft}$$

$$L_s = 1.17(0.73)\left(\frac{17.10}{1.23}\right)\left(\frac{27}{\sqrt{32.2(0.375)^{2.5}}}\right)^{0.47}(0.095)^{0.10}(0.375) = 23.2 \ \text{ft}$$

$$V_s = 1.30(1.47)\left(\frac{127.08}{1.23}\right)\left(\frac{27}{\sqrt{32.2(0.375)^{2.5}}}\right)^{1.24}(0.095)^{0.18}(0.375)^3 = 987 \ \text{ft}^3$$

Step 7. Compute the location of maximum scour. $L_m = 0.4 \ L_s = 0.4 \ (23.2) = 9.2$ ft downstream of the culvert outlet.

5.2 COHESIVE SOILS

If the soil is cohesive in nature, Equation 5.2 should be used to determine the scour hole dimensions. Shear number expressions, which relate scour to the critical shear stress of the soil, were derived to have a wider range of applicability for cohesive soils besides the one specific sandy clay that was tested. The sandy clay tested had 58 percent sand, 27 percent clay, 15 percent silt, and 1 percent organic matter; had a mean grain size of 0.15 mm (0.0059 in); and had a plasticity index, PI, of 15. The shear number expressions for circular culverts are:

$$\left[\frac{h_s}{D}, \frac{W_s}{D}, \frac{L_s}{D}, \frac{V_s}{D^3}\right] = C_s C_h (\alpha)\left(\frac{\rho V^2}{\tau_c}\right)^\beta \left(\frac{t}{316}\right)^\theta \tag{5.2}$$

and for other shaped culverts:

$$\left[\frac{h_s}{y_e}, \frac{W_s}{y_e}, \frac{L_s}{y_e}, \frac{V_s}{y_e^3}\right] = C_s C_h (\alpha_e)\left(\frac{\rho V^2}{\tau_c}\right)^\beta \left(\frac{t}{316}\right)^\theta \tag{5.3}$$

where,

D = culvert diameter, m (ft)

y_e = equivalent depth $(A/2)^{1/2}$, m (ft)

A = cross-sectional area of flow, m^2 (ft^2)

V = mean outlet velocity, m/s (ft/s)

τ_c = critical tractive shear stress, N/m^2 (lb/ft^2)

ρ = fluid density of water, 1000 kg/m^3 (1.94 $slugs/ft^3$)

$(\rho V^2)/\tau_c$ is the modified shear number

α_e = $\alpha_e = \alpha/0.63$ for h_s, W_s, and L_s and $\alpha_e = \alpha/(0.63)^3$ for V_s

α, β, θ, and α_e are coefficients found in Table 5.4

Use 30 minutes for t in Equation 5.2 and Equation 5.3 if it is not known.

The critical tractive shear stress is defined in Equation 5.4 (Dunn, 1959; Abt et al., 1996). Equations 5.2 and 5.3 should be limited to sandy clay soils with a plasticity index of 5 to 16.

$$\tau_c = 0.001 (S_v + \alpha_u) \tan (30 + 1.73\ PI) \tag{5.4}$$

where,

τ_c = critical tractive shear stress, N/m^2 (lb/ft^2)

S_v = the saturated shear strength, N/m^2 (lb/ft^2)

α_U = unit conversion constant, 8630 N/m^2 (SI), 180 lb/ft^2 (CU)

PI = Plasticity Index from the Atterberg limits

Table 5.4. Coefficients for Culvert Outlet Scour in Cohesive Soils

	α	β	θ	α_e
Depth, h_S	0.86	0.18	0.10	1.37
Width, W_S	3.55	0.17	0.07	5.63
Length, L_S	2.82	0.33	0.09	4.48
Volume, V_S	0.62	0.93	0.23	2.48

The design procedure for estimating scour in cohesive materials with PI from 5 to 16 may be summarized as follows.

Step 1. Determine the magnitude and duration of the peak discharge, Q. Express the discharge in m^3/s (ft^3/s) and the duration in minutes.

Step 2. Compute the culvert average outlet velocity, V.

Step 3. Obtain a soil sample at the proposed culvert location.

    a. Perform Atterberg limits tests and determine the plasticity index (PI) using ASTM D423-36.

    b. Saturate a sample and perform an unconfined compressive test (ASTM D211-66-76) to determine the saturated shear stress, S_v.

Step 4. Compute the critical tractive shear strength, τ_c, from Equation 5.4.

Step 5. Compute the modified shear number, S_{nm}, at the peak discharge and height above bed ratio, H_d, for slopes > 0%.

$$S_{nm} = \frac{\rho V^2}{\tau_c} \text{ and } H_d = \frac{\text{Drop Height}}{\text{Diameter}}$$

Step 6. Determine scour coefficients from Table 5.4 and, if appropriate, coefficients for culvert drop height, C_h, from Table 5.2 and slope, C_s, from Table 5.3.

Step 7. Compute the scour hole dimensions using Equation 5.2 for circular culverts and Equation 5.3 for other shapes.

Step 8. Compute the location of maximum scour. $L_m = 0.4 L_s$.

Design Example: Estimating Scour Hole Geometry in a Cohesive Soil (SI)

Determine the scour geometry: maximum depth, width, length and volume of scour. Given:

D   = 610 mm CMP Culvert
S   = 0%
Drop height = 0 m
Q   = 1.133 m^3/s
PI   = 12 and S_v = 23,970 N/m^2 for downstream channel

Solution

Step 1. Determine the magnitude and duration of the peak discharge: Q = 1.133 m^3/s and the peak flow duration is estimated to be 30 minutes.

Step 2. Compute the culvert average outlet velocity, V:

$$V = \frac{Q}{A} = \frac{1.133}{3.14(0.61)^2 / 4} = 3.88 \text{ m/s}$$

Step 3. Obtain a soil sample at the proposed culvert location: The sandy-clay soil was tested and found to have:

a.   Plasticity index (PI) = 12

b.   Saturated shear stress, S_v = 23,970 N/m^2.

Step 4. Compute the critical tractive shear strength, τ_c, from Equation 5.4.

τ_c = 0.001 ($S_v + \alpha_u$) tan (30 + 1.73 PI)

τ_c = 0.001 (23970 + 8630) tan [30 + 1.73(12)]

τ_c = 0.001 (32600) tan (50.76) = 39.9 N/m^2

Step 5. Compute the modified shear number, S_{nm} at the peak discharge and height above bed, H_d, ratio for slopes > 0%.

$$S_{nm} = \frac{\rho V^2}{\tau_c} = \frac{1000(3.88)^2}{39.9} = 377.3$$

Step 6. Determine scour coefficients from Table 5.4 and coefficients for culvert drop height, C_h, from Table 5.2 and slope from Table 5.3: C_h = 1 and C_s = 1

5-8

	α	β	θ
Depth	0.86	0.18	0.10
Width	3.55	0.17	0.07
Length	2.82	0.33	0.09
Volume	0.62	0.93	0.23

Step 7. Compute the scour hole dimensions using Equation 5.2 for circular culverts:

$$\left[\frac{h_s}{D}, \frac{W_s}{D}, \frac{L_s}{D}, \frac{V_s}{D^3}\right] = C_s C_h(\alpha)\left(\frac{\rho V^2}{\tau_c}\right)^\beta \left(\frac{t}{316}\right)^\theta$$

$$h_s = (1.0)(1.0)(\alpha)(377.3)^\beta \left(\frac{30}{316}\right)^\theta D$$

$h_s = (1.0)(1.0)(0.86)(377.3)^{0.18}(0.09)^{0.10}(0.61) = 1.2$ m

Similarly,

$W_s = (1.0)(1.0)(3.55)(377.3)^{0.17}(0.09)^{0.07}(0.61) = 5.02$ m

$L_s = (1.0)(1.0)(2.82)(377.3)^{0.33}(0.09)^{0.09}(0.61) = 9.81$ m

$V_s = (1.0)(1.0)(0.62)(377.3)^{0.93}(0.09)^{0.23}(0.61)^3 = 20.15$ m^3

Step 8. Compute the location of maximum scour. L_m = 0.4 L_s = 0.4(9.81) = 3.92 m downstream of culvert outlet.

Design Example: Estimating Scour Hole Geometry in a Cohesive Soil (CU)

Determine the scour geometry: maximum depth, width, length and volume of scour. Given:

D   = 24 in CMP Culvert
S   = 0%
Drop height = 0 ft
Q   = 40 ft^3/s
PI   = 12 and S_v = 500 lb/ft^2 for downstream channel

Solution

Step 1. Determine the magnitude and duration of the peak discharge: Q = 40 ft^3/s and the peak flow duration is estimated to be 30 minutes.

Step 2. Compute the culvert average outlet velocity, V:

$$V = \frac{Q}{A} = \frac{40}{3.14(2)^2 / 4} = 12.74 \text{ ft}/s$$

Step 3. Obtain a soil sample at the proposed culvert location: The sandy-clay soil was tested and found to have:

a.   Plasticity index, PI = 12

5-9

      b.   Saturated shear stress, $S_v = 500$ lb/ft^2.

Step 4. Compute the critical tractive shear strength, τ_c, from Equation 5.4.

$$\tau_c = 0.001\ (S_v + \alpha_u)\ \tan\ (30 + 1.73\ PI)$$

$$\tau_c = 0.001\ (500 + 180)\ \tan\ [30 + 1.73(12)]$$

$$\tau_c = 0.001\ (680)\ \tan\ (50.76) = 0.83\ \text{lb/ft}^2$$

Step 5. Compute the modified shear number, S_{nm}, at the peak discharge and height above bed ratio, H_d, for slopes > 0%.

$$S_{nm} = \frac{\rho V^2}{\tau_c} = \frac{1.94(12.74)^2}{0.83} = 379.4$$

Step 6. Determine scour coefficients from Table 5.4 and coefficients for culvert drop height, C_h, from Table 5.2 and slope from Table 5.3: $C_h = 1$ and $C_s = 1$

	α	β	θ
Depth	0.86	0.18	0.10
Width	3.55	0.17	0.07
Length	2.82	0.33	0.09
Volume	0.62	0.93	0.23

Step 7. Compute the scour hole dimensions using Equation 5.2 for circular culverts:

$$\left[\frac{h_s}{D}, \frac{W_s}{D}, \frac{L_s}{D}, \frac{V_s}{D^3}\right] = C_s C_h (\alpha)\left(\frac{\rho V^2}{\tau_c}\right)^\beta \left(\frac{t}{316}\right)^\theta$$

$$h_s = (1.0)(1.0)(\alpha)(379.4)^\beta \left(\frac{30}{316}\right)^\theta D$$

$$h_s = (1.0)(1.0)(0.86)(379.4)^{0.18}(0.09)^{0.10}(2) = 3.9\ \text{ft}$$

Similarly,

$$W_s = (1.0)(1.0)(3.55)(379.4)^{0.17}(0.09)^{0.07}(2) = 16.5\ \text{ft}$$

$$L_s = (1.0)(1.0)(2.82)(379.4)^{0.33}(0.09)^{0.09}(2) = 32.2\ \text{ft}$$

$$V_s = (1.0)(1.0)(0.62)(379.4)^{0.93}(0.09)^{0.23}(2)^3 = 713.7\ \text{ft}^3$$

Step 8. Compute the location of maximum scour. $L_m = 0.4\ L_s = 0.4(32.2) = 12.9$ ft downstream of culvert outlet.

CHAPTER 6: HYDRAULIC JUMP

The hydraulic jump is a natural phenomenon that occurs when supercritical flow is forced to change to subcritical flow by an obstruction to the flow. This abrupt change in flow condition is accompanied by considerable turbulence and loss of energy. The hydraulic jump can be illustrated by use of a specific energy diagram as shown in Figure 6.1. The flow enters the jump at supercritical velocity, V_1, and depth, y_1, that has a specific energy of $E = y_1 + V_1^2/(2g)$. The kinetic energy term, $V^2/(2g)$, is predominant. As the depth of flow increases through the jump, the specific energy decreases. Flow leaves the jump area at subcritical velocity with the potential energy, y, predominant.

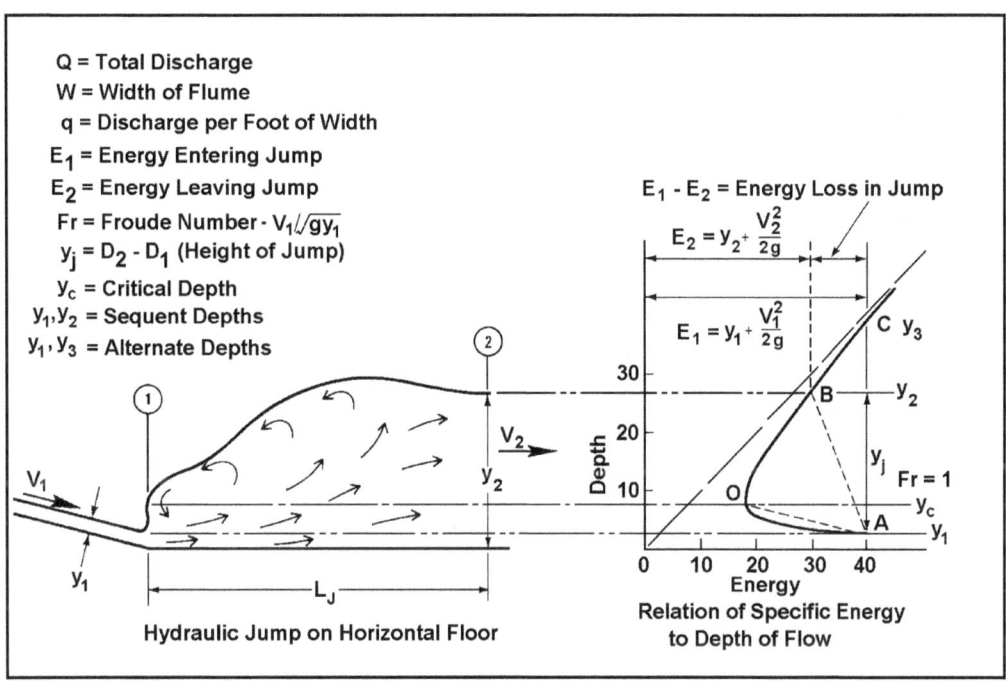

Figure 6.1. Hydraulic Jump

6.1 TYPES OF HYDRAULIC JUMP

When the upstream Froude number, Fr, is 1.0, the flow is at critical and a jump cannot form. For Froude numbers greater than 1.0, but less than 1.7, the upstream flow is only slightly below critical depth and the change from supercritical to subcritical flow will result in only a slight disturbance of the water surface. On the high end of this range, Fr approaching 1.7, the downstream depth will be about twice the incoming depth and the exit velocity about half the upstream velocity.

The Bureau of Reclamation (USBR, 1987) has related the jump form and flow characteristics to the Froude number for Froude numbers greater than 1.7, as shown in Figure 6.2. When the upstream Froude number is between 1.7 and 2.5, a roller begins to appear, becoming more intense as the Froude number increases. This is the prejump range with very low energy loss. The water surface is quite smooth, the velocity throughout the cross section uniform, and the energy loss in the range of 20 percent.

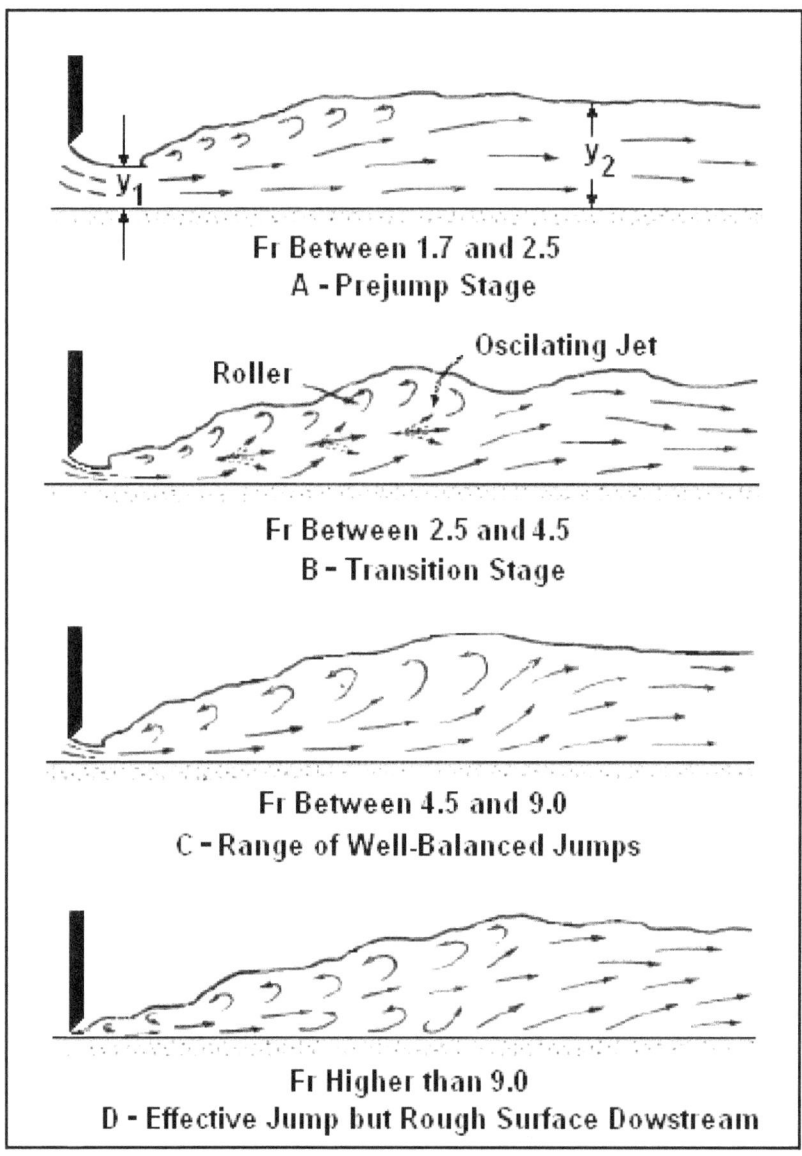

Figure 6.2. Jump Forms Related to Froude Number (USBR, 1987)

An oscillating form of jump occurs for Froude numbers between 2.5 and 4.5. The incoming jet alternately flows near the bottom and then along the surface. This results in objectionable surface waves that can cause erosion problems downstream from the jump.

A well balanced and stable jump occurs where the incoming flow Froude number is greater than 4.5. Fluid turbulence is mostly confined to the jump, and for Froude numbers up to 9.0 the downstream water surface is comparatively smooth. Jump energy loss of 45 to 70 percent can be expected.

With Froude numbers greater than 9.0, a highly efficient jump results but the rough water surface may cause downstream erosion problems.

The hydraulic jump commonly occurs with natural flow conditions and with proper design can be an effective means of dissipating energy at hydraulic structures. Expressions for computing the before and after jump depth ratio (conjugate depths) and the length of jump are needed to design energy dissipators that induce a hydraulic jump. These expressions are related to culvert outlet Froude number, which for many culverts falls within the range 1.5 to 4.5.

6.2 HYDRAULIC JUMP IN HORIZONTAL CHANNELS

The hydraulic jump in any shape of horizontal channel is relatively simple to analyze (Sylvester, 1964). Figure 6.3 indicates the control volume used and the forces involved. Control section 1 is before the jump where the flow is undisturbed, and control section 2 is after the jump, far enough downstream for the flow to be again taken as parallel. Distribution of pressure in both sections is assumed hydrostatic. The change in momentum of the entering and exiting stream is balanced by the resultant of the forces acting on the control volume, i.e., pressure and boundary frictional forces. Since the length of the jump is relatively short, the external energy losses (boundary frictional forces) may be ignored without introducing serious error. Also, a channel may be considered horizontal up to a slope of 18 percent (10 degree angle with the horizontal) without introducing serious error. The momentum equation provides for solution of the sequent depth, y_2, and downstream velocity, V_2. Once these are known, the internal energy losses and jump efficiency can be determined by application of the energy equation.

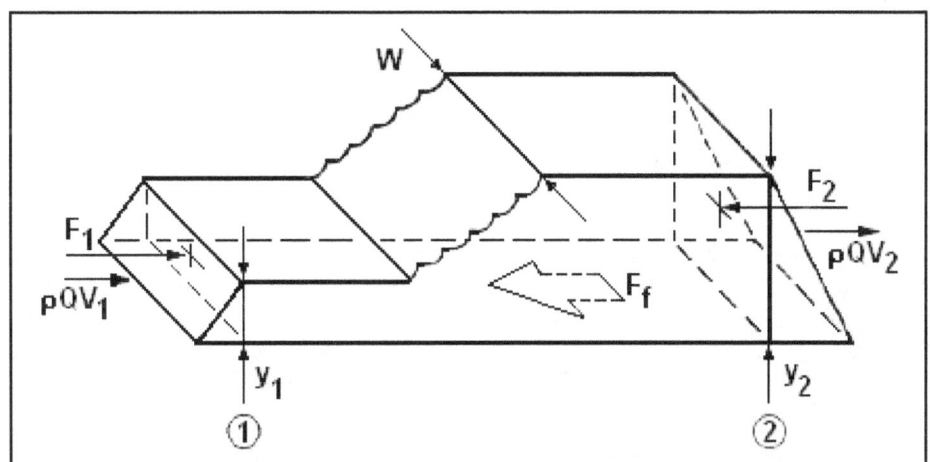

Figure 6.3. Hydraulic Jump in a Horizontal Channel

The general form of the momentum equation can be used for the solution of the hydraulic jump sequent-depth relationship in any shape of channel with a horizontal floor. Defining a momentum quantity as, $M = Q^2/(gA) + AY$ and recognizing that momentum is conserved through a hydraulic jump, the following can be written:

$$Q^2/(gA_1) + A_1Y_1 = Q^2/(gA_2) + A_2Y_2 \qquad (6.1)$$

where,

Q = channel discharge, m^3/s (ft^3/s)

A_1, A_2 = cross-sectional flow areas in sections 1 and 2, respectively, m^2 (ft^2)

Y_1, Y_2 = depth from water surface to centroid of cross-section area, m (ft)

The depth from the water surface to the centroid of the cross-section area can be defined as a function of the channel shape and the maximum depth: $Y = Ky$. In this relationship, K is a parameter representing the channel shape while y is the maximum depth in the channel. Substituting this quantity into Equation 6.1 and rearranging terms yields:

$$A_1 K_1 y_1 - A_2 K_2 y_2 = (1/A_2 - 1/A_1)Q^2/g$$

Rearranging and using $Fr_1^2 = V_1^2/(gy_1) = Q^2/(A_1^2 gy_1)$, gives:

$$A_1 K_1 y_1 - A_2 K_2 y_2 = Fr_1^2 A_1 y_1 (A_1/A_2 - 1).$$

Dividing this by $A_1 y_1$ provides:

$$K_2 A_2 y_2/(A_1 y_1) - K_1 = Fr_1^2 (1 - A_1/A_2) \qquad (6.2)$$

This is a general expression for the hydraulic jump in a horizontal channel. The constants K_1 and K_2 and the ratio A_1/A_2 have been determined for rectangular, triangular, parabolic, circular, and trapezoidal shaped channels by Sylvester (1964). The relationships for rectangular and circular shapes are summarized in the following sections.

6.2.1   Rectangular Channels

For a rectangular channel, substituting $K_1 = K_2 = 1/2$ and $A_1/A_2 = y_1/y_2$ into Equation 6.2, the expression becomes:

$$y_2^2/y_1^2 - 1 = 2Fr_1^2 (1 - y_1/y_2)$$

If $y_2/y_1 = J$, the expression for a hydraulic jump in a horizontal, rectangular channel becomes Equation 6.3, which is plotted as Figure 6.4.

$$J^2 - 1 = 2Fr_1^2 \left(1 - \frac{1}{J}\right) \qquad (6.3)$$

The length of the hydraulic jump can be determined from Figure 6.5. The jump length is measured to the downstream section at which the mean water surface attains the maximum depth and becomes reasonably level. Errors may be introduced in determining length since the water surface is rather flat near the end of the jump. This is undoubtedly one of the reasons so many empirical formulas for determining jump length are found in the literature. The jump length for rectangular basins has been extensively studied.

Stilling basin design is a common application for hydraulic jumps in rectangular channels (see Chapter 8). Free jump basins can be designed for any flow conditions; but because of economic and performance characteristics they are, in general, only employed in the lower range of Froude numbers. Flows with Froude numbers below 1.7 may not require stilling basins but may require protection such as riprap and wingwalls and apron. For Froude numbers between 1.7 and 2.5, the free jump basin may be all that is required. In this range, loss of energy is less than 20 percent; the conjugate depth is about three times the incoming flow depth; and, the length of basin required is less than about 5 times the conjugate depth. Many highway culverts operate in this flow range. At higher Froude numbers, the use of baffles and sills make it possible to reduce the basin length and stabilize the jump over a wider range of flow situations.

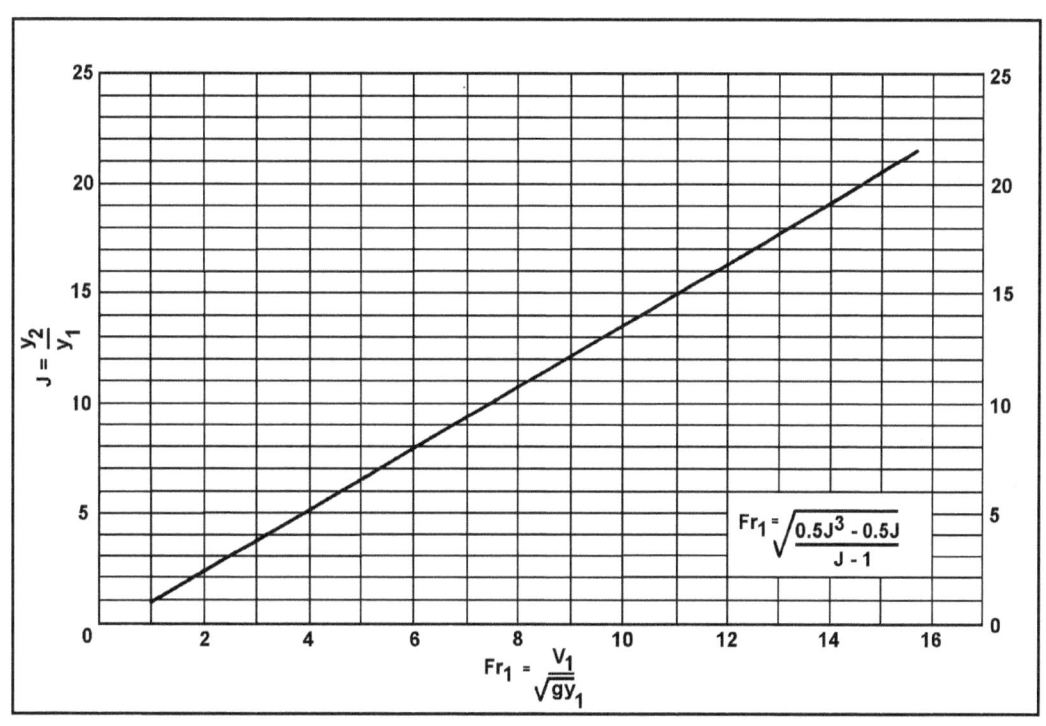

Figure 6.4. Hydraulic Jump - Horizontal, Rectangular Channel

$$Fr_1 = \sqrt{\frac{0.5J^3 - 0.5J}{J - 1}}$$

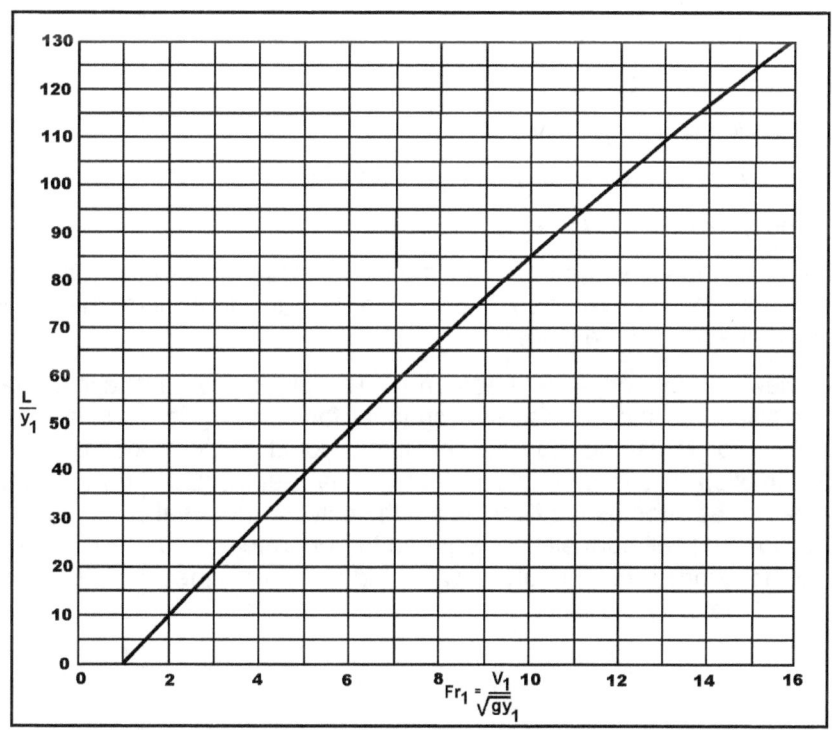

Figure 6.5. Length of Jump for a Rectangular Channel

Design Example: Hydraulic Jump in a Horizontal, Rectangular Channel (SI)

Determine the height and length of a hydraulic jump in a box culvert with a 2.134 m span. Also, estimate the range of flows for which a jump would be triggered as discharged to a trapezoidal channel. Given:

S = 0.2%

Q = 11.33 m³/s

V_1 = 5.79 m/s

y_1 = 0.914 m

Fr = 1.9

For the trapezoidal channel:

B = 3.04 m

Side slopes = 1V:2H

n = 0.03

S = 0.04%

Solution

Step 1. Find the conjugate depth from Figure 6.4.

$J = y_2 / y_1 = 2.2$

$y_2 = 2.2(0.914) = 2.011$ m

Step 2. Find the Length of jump from Figure 6.5

$L / y_1 = 9.0$

$L = 0.914(9.0) = 8.226$ m

Step 3. Calculate the after jump velocity

$V_2 = Q/A_2 = 11.33/ [2.134(2.011)] = 2.64$ m/s

Velocity reduction is $(5.79 - 2.64)/5.79 = 54.4\%$.

Step 4. Develop a Q vs. stage curve for the downstream trapezoidal channel using either HDS No. 3 (FHWA, 1961) or Table B.1 to determine the relationship with conjugate depth (see below).

Step 5. Review sequent depth requirements. The plot shows that excess tailwater depth is available in the downstream channel for discharges up to approximately 13.6 m³/s. For larger discharges, the jump would begin to move downstream. The assumption in this example is that normal depth in the downstream channel is obtained as soon as the flow leaves the culvert. In a real case, a stilling basin (see Section 8.1) will normally be required to generate enough tailwater depth to cause a jump to occur or the culvert will need to be designed as a broken-back culvert (see Chapter 7).

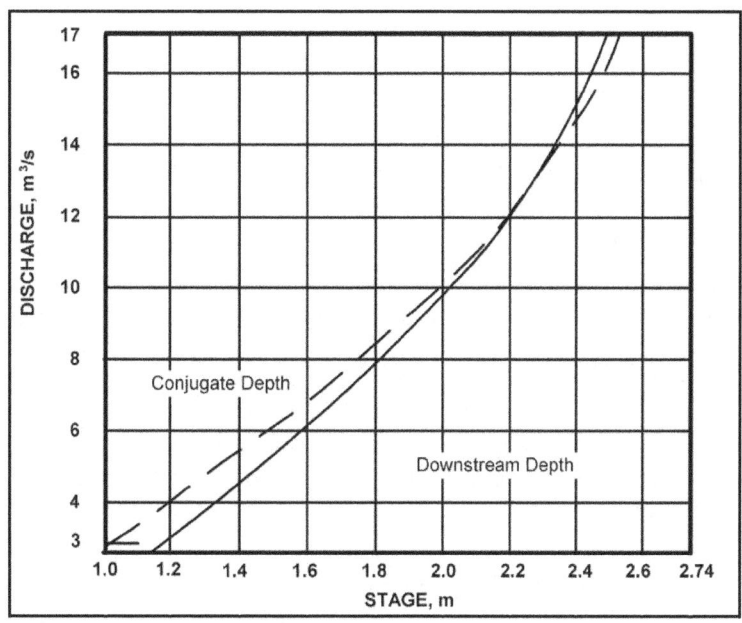

Normal Channel Depth - Conjugate Depth Relationship

Design Example: Hydraulic Jump in a Horizontal, Rectangular Channel (CU)

Determine the height and length of a hydraulic jump in a box culvert with a 7 ft span. Also, estimate the range of flows for which a jump would be triggered as discharged to a trapezoidal channel. Given:

S   = 0.2%

Q   = 400 ft^3/s

V_1 = 19 ft/s

y_1 = 3.0 ft

Fr   = 1.9

For the trapezoidal channel:

B   = 10 ft

Side slopes = 1V:2H

n   = 0.03

S   = 0.04%

Solution

Step 1. Find the conjugate depth in a rectangular basin from Figure 6.4.

$J = y_2 / y_1 = 2.2$

$y_2 = 2.2(3.0) = 6.6$ ft

Step 2. Find the Length of jump from Figure 6.5

$L / y_1 = 9.0$

$L = 3.0(9.0) = 27$ ft

Step 3. Calculate the after jump velocity

$V_2 = Q/A_2 = 400/[7(6.6)] = 8.7$ ft/s

Velocity reduction is $(19 - 8.7)/19 = 54.2\%$.

Step 4. Develop a Q vs. stage curve for the downstream trapezoidal channel using either HDS 3 (FHWA, 1961) or Table B.1 to determine the relationship with conjugate depth (see below).

Step 5. Review sequent depth requirements. The plot shows that excess tailwater depth is available in the downstream channel for discharges up to approximately 480 ft³/s. For larger discharges, the jump would begin to move downstream. The assumption in this example is that normal depth in the downstream channel is obtained as soon as the flow leaves the culvert. In a real case, a stilling basin (see Section 8.1) will normally be required to generate enough tailwater depth to cause a jump to occur or the culvert will need to be designed as a broken-back culvert (see Chapter 7).

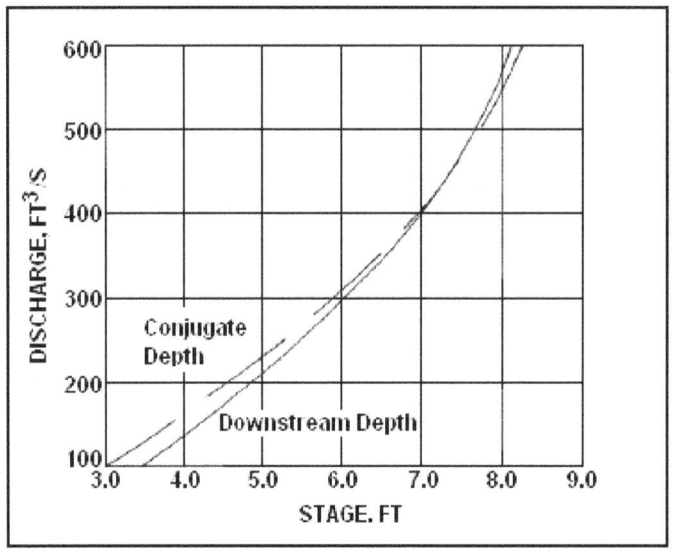

Normal Channel Depth - Conjugate Depth Relationship

6.2.2 Circular Channels

Circular channels are divided into two cases: where y_2 is greater than the diameter, D, and where y_2 is less than D. For y_2 less than D:

$$(K_2 y_2 C_2 /(y_1 C_1)) - K_1 = Fr_1^2 (1 - C_1 /C_2) \tag{6.4}$$

For y_2 greater than or equal to D:

$$(y_2 C_2/(y_1 C_1)) - 0.5 (C_2 D/(C_1 y_1)) - K_1 = Fr_1^2 (1 - C_1/C_2) \tag{6.5}$$

C and K are functions of y/D and may be evaluated from the Table 6.1.

Table 6.1. Coefficients for Horizontal, Circular Channels

Y/D	0.1	0.2	0.3	0.4	0.5	0.6	0.7	0.8	0.9	1.0
K	0.410	0.413	0.416	0.419	0.424	0.432	0.445	0.462	0.473	0.500
C	0.041	0.112	0.198	0.293	0.393	0.494	0.587	0.674	0.745	0.748
C'	0.600	0.800	0.917	0.980	1.000	0.980	0.917	0.800	0.600	

In Equations 6.4 and 6.5, Fr_1 is computed using the maximum depth in the channel. Figure 6.6 may be used as an alternative to these equations.

Alternatively, the designer may calculate a Froude number based on hydraulic depth, $Fr_m = V/(gy_m)^{1/2}$. Where $y_m = (C/C')D$ or $y_m = A/T$. For the first expression, C' is taken from Table 6.1. For the second expression, A is the cross-sectional area of flow and T is the water surface width. Figure 6.7 is the design chart for horizontal, circular channels using the hydraulic depth in computing the Froude number.

The length of the hydraulic jump is generally measured to the downstream section at which the mean water surface attains the maximum depth and becomes reasonably level. The jump length in circular channels is determined using Figure 6.8. This curve is for the case where y_2 is less than D. For the case where y_2 is greater than D, the length should be taken as seven times the difference in depths, i.e., $L_J = 7(y_2 - y_1)$.

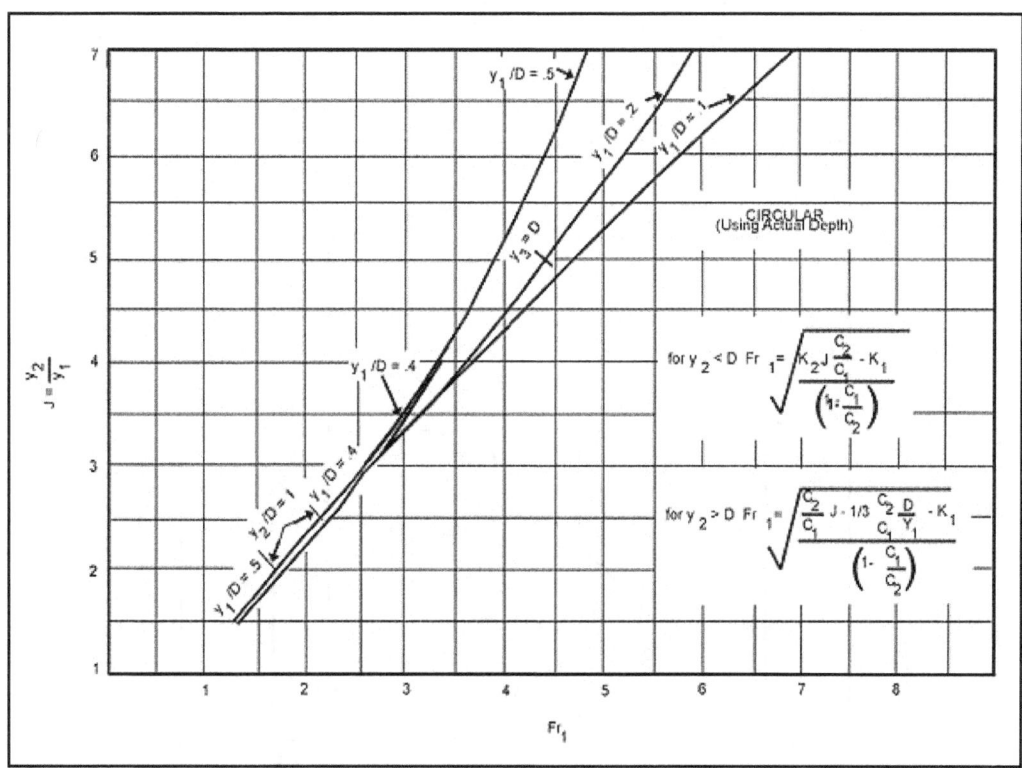

Figure 6.6. Hydraulic Jump - Horizontal, Circular Channel (actual depth)

6-9

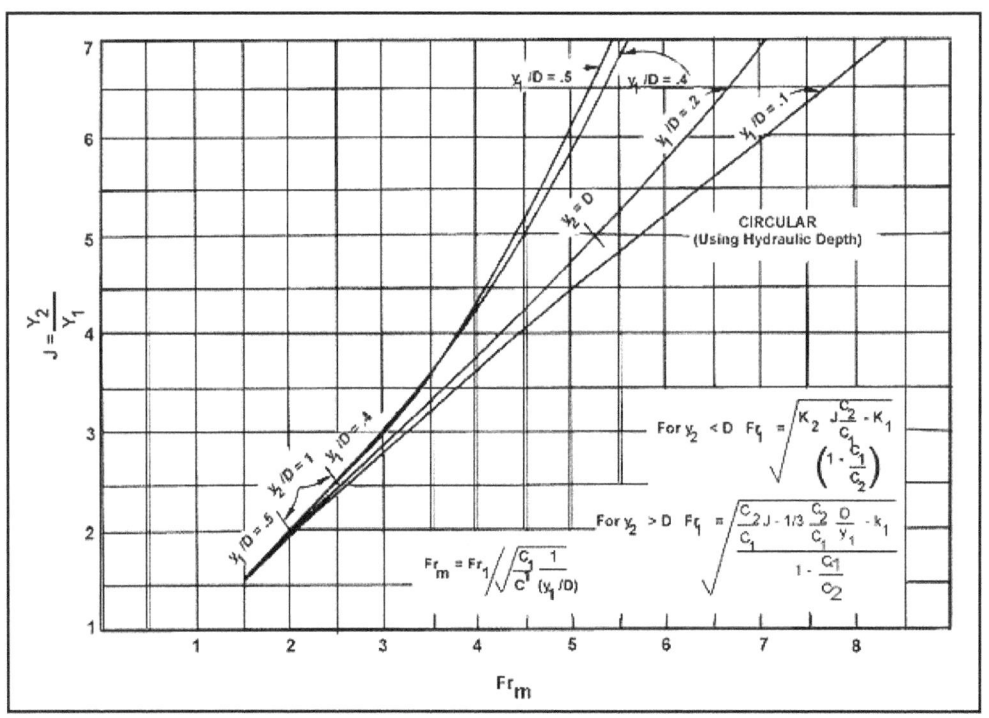

Figure 6.7. Hydraulic Jump - Horizontal, Circular Channel (hydraulic depth)

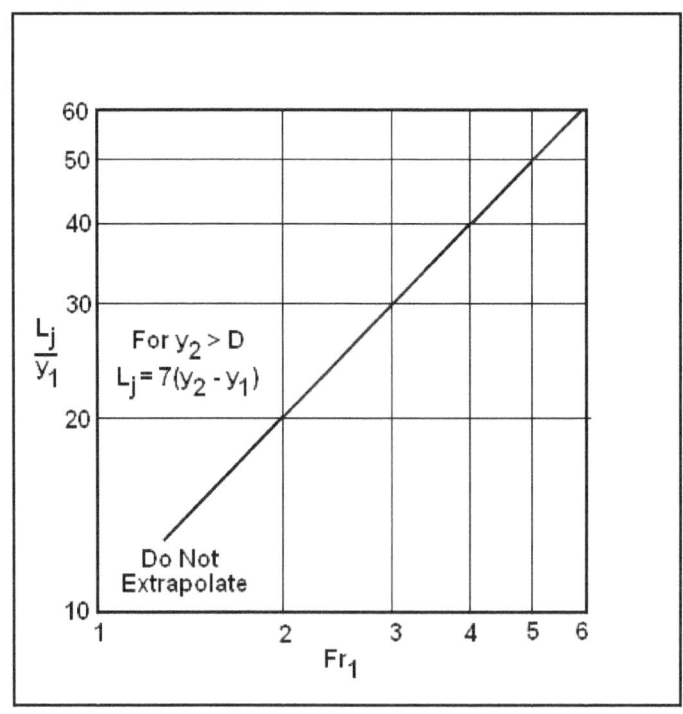

Figure 6.8. Jump Length Circular Channel with $y_2 < D$

Design Example: Hydraulic Jump in a Horizontal, Circular Channel (SI)

Determine the height and length of a hydraulic jump in an RCP culvert with a 2.134 m diameter. Given:

S = 2%
Q = 5.664 m³/s
V_1 = 5.182 m/s
y_1 = 0.732 m
Fr_1 = 1.9

Solution

Step 1. Find the conjugate depth in a circular channel.

y_1/D = 0.732/2.134 = 0.34 (use 0.4)

$J = y_2/y_1$ = 2.3 from Figure 6.6

y_2 = 2.3(0.732) = 1.684 m and y_2/D = 0.78 (use 0.8)

(Using Equation 6.4 with C_1 = 0.293, K_1 = 0.419, C_2 = 0.674, K_2 = 0.462 yields the same result.)

Step 2. Find the Length of jump from Figure 6.8

L_j/y_1 = 19

L_j = 0.732 (19) = 13.9 m

Step 3. Calculate the after jump velocity

For y_2/D = 0.78, A/D^2 = 0.6573 from Table 3.3 and A = 2.99 m²

$V_2 = Q/A_2$ = 5.664/2.99 = 1.89 m/s

Velocity reduction is (5.182 - 1.89)/ 5.182 = 63.5%.

Design Example: Hydraulic Jump in a Horizontal, Circular Channel (CU)

Determine the height and length of a hydraulic jump in an RCP culvert with a 7 ft diameter. Given:

S = 2%
Q = 200 ft³/s
V_1 = 17 ft/s
y_1 = 2.4 ft
Fr_1 = 1.9

Solution

Step 1. Find the conjugate depth in a circular channel.

y_1/D = 2.4/7 = 0.34 (use 0.4)

$J = y_2/y_1$ = 2.3 from Figure 6.6

$y_2 = 2.3(2.4) = 5.5$ ft and $y_2/D = 0.78$ (use 0.8)

(Using Equation 6.4 with $C_1 = 0.293$, $K_1 = 0.419$, $C_2 = 0.674$, $K_2 = 0.462$ yields the same result.)

Step 2. Find the Length of jump from Figure 6.8

$L_j /y_1 = 19$

$L_j = 2.4(19) = 46$ ft

Step 3. Calculate the after jump velocity

For $y_2/D = 0.78$, $A/D^2 = 0.6573$ from Table 3.3 and $A = 32.2$ ft^2

$V_2 = Q/A_2 = 200/ 32.2 = 6.2$ ft/s

Velocity reduction is $(17 - 6.2)/17 = 63.5\%$.

6.2.3 Jump Efficiency

A general expression for the energy loss (H_L/H_1) in any shape channel is:

$$H_L/H_1 = 2 - 2(y_2) + Fr_m^2 [1 - A_1^2 /A_2^2] / (2 + Fr^2) \tag{6.6}$$

where,

Fr_m = upstream Froude number at section 1, $Fr_m^2 = V^2/(gy_m)$

y_m = hydraulic depth, m (ft)

This equation is plotted for the various channel shapes as Figure 6.9. Even though Figure 6.9 indicates that the non-rectangular sections are more efficient for the higher Froude numbers, it should be remembered that these sections also involve longer jumps, stability problems, and a rough downstream water surface.

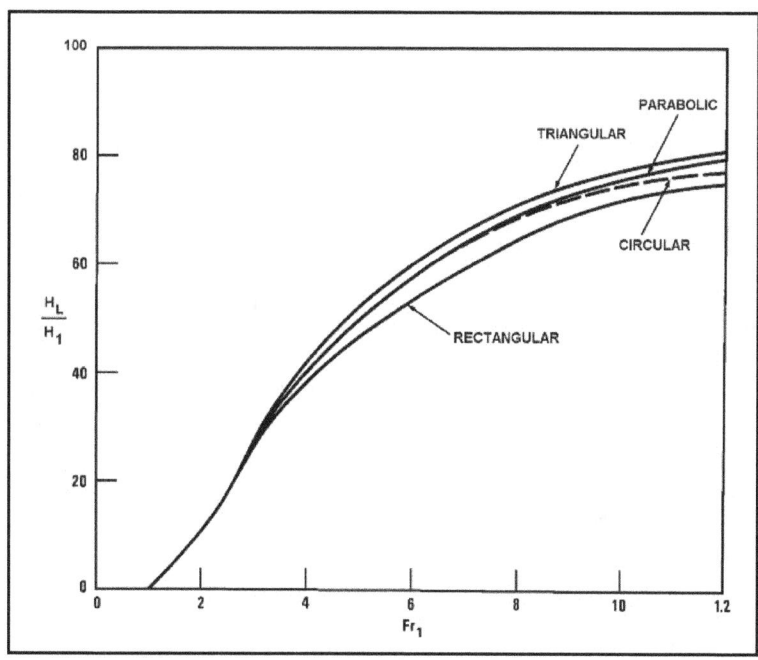

Figure 6.9. Relative Energy Loss for Various Channel Shapes

6.3 HYDRAULIC JUMP IN SLOPING CHANNELS

Figure 6.10 (Bradley, 1961) indicates a method of delineating hydraulic jumps in horizontal and sloping channels. Horizontal channels (case A) were discussed in the previous section. Sloping channels are discussed in this section. If the channel bottom is selected as a datum, the momentum equation becomes:

$$\frac{Q(V_2 - V_1)}{g} = 0.5B(y_1^2 - y_2^2)\cos\phi + \frac{w(\sin\phi)}{\gamma} \qquad (6.7)$$

where,

γ = unit weight of water, N/m^3 (lb/ft^3)

ϕ = angle of channel with the horizontal

B = channel bottom width (rectangular channel), m (ft)

w = weight of water in jump control volume, N (lb)

The momentum equation used for the horizontal channels cannot be applied directly to hydraulic jumps in sloping channels since the weight of water within the jump must be considered. The difficulty encountered is in defining the water surface profile to determine the volume of water within the jumps for various channel slopes. This volume may be neglected for slopes less than 10 percent and the jump analyzed as a horizontal channel.

The Bureau of Reclamation (Bradley, 1961) conducted extensive model tests on case B and C type jumps to define the length and depth relationships. This reference should be consulted if a hydraulic jump in a sloping rectangular channel is being considered. Model tests should be considered if other channel shapes are being considered.

6-13

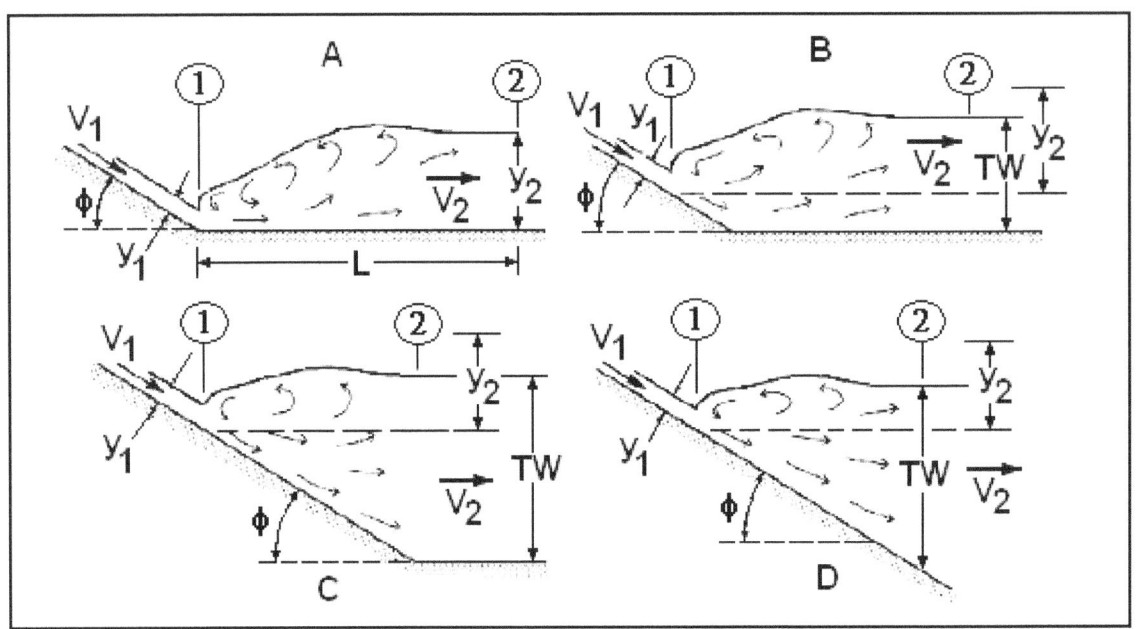

Figure 6.10. Hydraulic Jump Types Sloping Channels (Bradley, 1961)

CHAPTER 7: INTERNAL (INTEGRATED) DISSIPATORS

In situations where there is limited right-of-way for an energy dissipator beyond the discharge point of a culvert or chute there are several options for internal or integrated dissipators, including adding internal roughness elements throughout the culvert or chute or just prior to the outlet. These approaches may be applicable within closed culvert barrels (conventional or broken-back) as well as in open, usually rectangular, chutes.

Roughness elements are sometimes a convenient way of controlling outlet velocities for culvert installations where the culvert barrel is not used to capacity because it is operating in inlet control. These roughness elements may be designed to slow the velocity in the culvert including, at the limit, creation of a condition of tumbling flow, where the outlet velocity is reduced to critical velocity. Such internal roughness elements may be placed throughout the entire length of the culvert or chute, or simply near the end prior to the outlet, depending on the hydraulic conditions and desired outlet conditions.

This chapter describes a series of strategies for increasing roughness including tumbling flow, increased roughness, the USBR Type IX baffled apron, broken-back culvert runout sections, outlet weirs, and outlet drop/weirs. Their applicability and limitations are discussed in the following sections.

7.1 TUMBLING FLOW

Roughness elements placed in the culvert barrel or open chute may be used to decrease velocities by creating a series of hydraulic jumps in a phenomenon known as tumbling flow (Peterson and Mohanty, 1960). Tumbling flow is an optimum dissipator on steep slopes. It is essentially a series of hydraulic jumps and overfalls that maintain the predominant flow paths at approximately critical velocity even on slopes that would otherwise be characterized by high supercritical velocities.

A major concern with tumbling flow is that silt may accumulate in front of the roughness elements and render them ineffective. This is perhaps unwarranted as the element enhances sediment transport capacity and tends to be self-cleansing. In their original list of possible applications, Peterson and Mohanty (1960) noted that by "using roughness elements to induce greater turbulence, the sediment-carrying capacity of a channel may be increased."

Tumbling flow is uniform flow in a cyclical sense, with the same patterns of depth and velocity repeated at each roughness element. It is not necessary to line the entire length of the culvert with roughness elements to get outlet velocity control. Five rows of roughness elements are sufficient to establish the cyclical uniform flow pattern.

The basic premise of the tumbling flow regime is that it will maintain essentially critical flow even on very steep slopes. The last element is located a distance L/2 upstream of the outlet so the flow reattaches to the channel bed right at the outlet. The first element for an enlarged section, as shown in Figure 7.1, should also be located a distance L/2 downstream of the start of the enlarged section. The distance L/2 for both the first and last elements should be considered a minimum. Sizing and spacing for the roughness elements are described in subsequent sections. Outlet velocity will approach critical velocity, unless backwater exists. It is not unreasonable to expect to provide additional culvert height in the roughened region of culverts as shown in Figure 7.1.

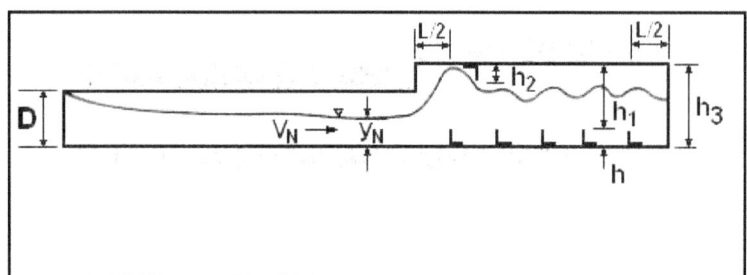

Figure 7.1. Definition Sketch for Tumbling Flow in a Culvert

The design procedure for tumbling flow in box and circular culverts may be summarized in the following steps:

Step 1. Verify the culvert is in inlet control.

Step 2. Compute normal flow conditions in the culvert to determine if the discharge conditions at the outlet require mitigation.

Step 3. Compute critical depth and velocity. If the critical velocity is less than or equal to the desired outlet velocity, tumbling flow may be an appropriate energy dissipation approach.

Step 4. Size the element heights, element spacing, and other design features. Design details differ for box and circular culverts and are described in the following sections.

7.1.1   Tumbling Flow in Box Culverts/Chutes

The tumbling flow phenomenon was investigated as a means of dissipating energy in box culverts and embankment chutes at Virginia Polytechnic Institute (VPI), (Morris, 1968; Morris, 1969; Mohanty, 1959). Slopes up to 20 percent were tested at VPI and up to 35 percent in subsequent tests by the Federal Highway Administration (Jones, 1975).

Drainage chutes on highway cut and fill slopes are candidate sites for roughness element energy dissipators. Use of roughness elements is reasonable for slopes up to 10 or 15 percent. Beyond this, flow separation and the trajectory of the flow that is out of contact with the channel bed are so exaggerated that provisions must be incorporated to counter splashing.

One of the major limitations of tumbling flow as an energy dissipator is that the required height of the roughness elements is closely related to the discharge for a given size culvert. Conversely, the required element height is less sensitive to the culvert slope. For example, given a slope and culvert size, doubling the discharge increases the required height of roughness elements by approximately 50 percent in box culverts; whereas, for a given discharge, increasing the slope from four percent to eight percent increases the required element height by less than six percent. There will be many situations where the element height may have to be half the culvert height to maintain tumbling flow. Practical applications of tumbling flow are likely to be limited to low-discharge per unit width, high-velocity culverts.

Tumbling flow is established in box culverts and rectangular chutes with roughness elements placed on the bottom of the culvert as shown in Figure 7.1.   Critical depth in a box culvert is calculated from the unit discharge (discharge divided by culvert/chute width, B).

$$y_c = \left(\frac{q^2}{g}\right)^{\frac{1}{3}}$$

(7.1)

where,

y_c = critical depth, m (ft)

q = unit discharge (Q/B), m (ft)

Critical velocity, which will be the outlet velocity, may be determined using the critical depth and the continuity equation.

Tumbling flow can be established rather quickly by using one of two configurations. The first configuration is to use five rows of uniformly sized roughness elements as shown in Figure 7.2a. This configuration is recommended for use in box culverts. The second, alternative, configuration uses a larger initial element with four additional rows of uniformly sized roughness elements as shown in Figure 7.2b. (Note that only one of the uniform elements is shown.) The alternative configuration is not considered to be a practical solution in box culverts since the element size is likely to be excessive. However, it may be useful for open chutes.

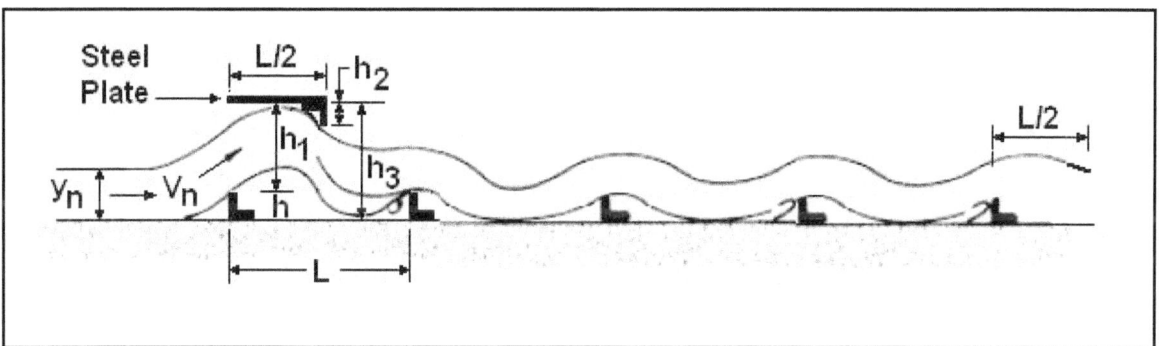

Figure 7.2a. Tumbling Flow in a Box Culvert or Open Chute: Recommended Configuration

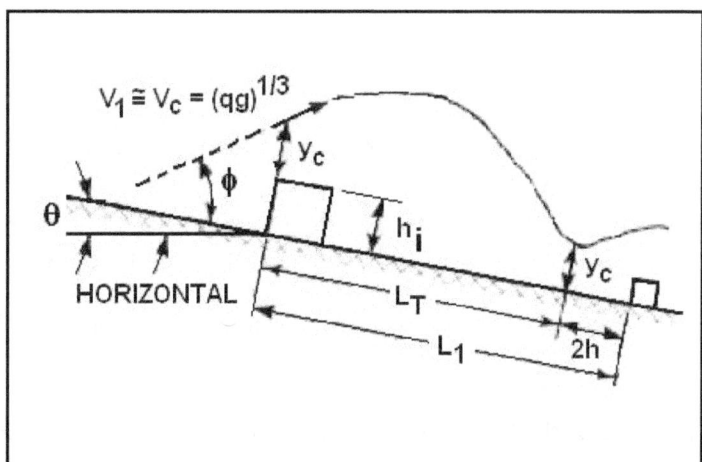

Figure 7.2b. Tumbling Flow in a Box Culvert or Open Chute: Alternative Configuration

7-3

The recommended configuration requires a splash shield to reverse the flow jet between the first and second rows of elements. The splash shield has merit since it deflects the so-called "rooster tail" jet against the channel bed and brings the flow under control very quickly without using a large leading roughness element. For box culverts, the top of the culvert can serve as the shield. However there should be a top baffle to help redirect the flow. The top baffle need not be the same size as the bed elements.

The recommended configuration is to use 5 rows of elements all the same height, where height is determined from the following equation:

$$h = \frac{y_c}{(3 - 3.7S_o)^{\frac{2}{3}}} \tag{7.2}$$

where,

h	=	element height, m (ft)
y_c	=	critical depth, m (ft)
S_o	=	culvert slope, m/m (ft/ft)

Spacing between the roughness element rows, L, is set by choosing the ratio of L/h to be between 8.5 and 10, inclusive.

The alternative configuration is to use a large initial roughness element followed by four smaller elements as shown in Figure 7.2b. The large initial roughness element must meet the following requirement:

$$h_i > y_2 - y_c \tag{7.3}$$

where,

h_i	=	large initial element height, m (ft)
y_2	=	sequent depth required for a hydraulic jump, m (ft)

The sequent depth, y_2, required for the hydraulic jump is computed as follows:

$$y_2 = 7.5(Fr)y_n(S_o + 0.153) \tag{7.4}$$

where,

Fr	=	Froude number at the approach condition at the toe of the jump, dimensionless

The large initial element is followed by four smaller elements with a height computed by Equation 7.2, as before.

Spacing between the small elements is determined by selecting an L/h ratio between 8.5 and 10 as before. Spacing, L_1, between the large initial element and the first small element is:

$$L_1 = 2h + 2y_c\left(\frac{\cos(\phi - \theta)}{\cos\theta}\right)(\tan\theta\cos(\phi - \theta) + \sin(\phi - \theta)) \tag{7.5}$$

where,

θ	=	slope of the culvert bottom expressed in degrees (see Figure 7.2)
ϕ	=	jet angle, taken as 45 degrees
y_c	=	critical depth (see Equation 7.1), m (ft)

For either configuration, continuous elements across the bottom of the culvert will trap water and tend to collect sediment and debris. Slots may be provided in the roughness elements as shown in Figure 7.3. The slot width, W_2, should be:

$$W_2 = \frac{h}{2} \tag{7.6}$$

where,

h = height of the small elements

The width of the elements is then calculated based on the width of the culvert and the number of slots.

$$W_1 = \frac{B - N_s W_2}{3} \tag{7.7}$$

where,

N_s = number of slots

B = culvert width, m (ft)

For rows 1, 3, and 5, the element width is calculated with 2 slots and for rows 2 and 4 the element width is calculated with three slots. An alternating pattern is recommended to disrupt streamlines between roughness elements as shown in Figure 7.3.

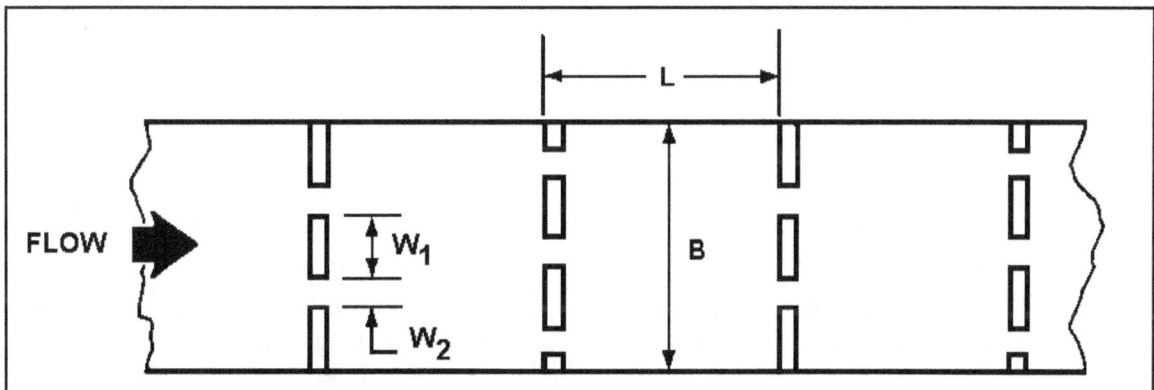

Figure 7.3. Definition Sketch for Slotted Roughness Elements

The culvert rise must be checked to insure sufficient space for the jet tumbling over the roughness elements. For culverts, the jet should just clear the culvert top. The jet height, h_1, is:

$$h_1 = 1.25 y_c \tag{7.8}$$

Referring to Figure 7.1, if D < (h_1+h) an enlarged culvert height, h_3, equal to h_1+h is required. If D > (h_1+h), a splash shield with height, h_2, is required to redirect the flow and should be located downstream of the first roughness element.

$$h_2 = 1.5\left(D - \left(h_1 + h\right)\right) \qquad (7.9)$$

If the value for h_2 is less than 50 mm (2 in), then 50 mm (2 in) should be used for h_2. The splash shield length is taken as the element spacing, L, divided by 2 as shown in Figure 7.2a. The splash shield should span the entire flow width. No splash shield is necessary when the large initial roughness element design is used.

Design Example: Tumbling Flow in a Box Culvert (SI)

Design a concrete box culvert for tumbling flow (See Figure 7.2). Determine if the outlet velocity is less than 3 m/s. Given:

Q	=	2.8 m³/s
B	=	1.2 m
D	=	1.2 m
n	=	0.013
S_o	=	0.06 m/m

Solution

Step 1. Verify the culvert is in inlet control. In this example the culvert is in inlet control.

Step 2. Compute normal flow conditions. Using trial and error with Manning's Equation:

y_n = 0.342 m

V_n = 6.8 m/s (Since this is greater than 3 m/s, energy dissipation is required.)

Step 3. Compute critical depth and velocity. First compute the unit discharge,

q = Q/B = 2.8/1.2 = 2.333 m²/s

Using Equation 7.1,

$$y_c = \left(\frac{q^2}{g}\right)^{\frac{1}{3}} = \left(\frac{2.333^2}{9.81}\right)^{\frac{1}{3}} = 0.822 \text{ m}$$

V_c = Q/(y_cB) = 2.8/(0.822 (1.2)) = 2.8 m/s (Since the outlet velocity will be critical velocity, a proper design will meet the design criterion of 3 m/s)

Step 4. Size the element height, element spacing, and splash shield. We will use the recommended procedure of 5 rows of equal height roughness elements. Roughness element height, h, longitudinal spacing, L, and transverse spacing, W_1 and W_2, are as follows.

From Equation 7.2:

$$h = \frac{y_c}{(3 - 3.7S_o)^{\frac{2}{3}}} = \frac{0.822}{(3 - 3.7(0.06))^{\frac{2}{3}}} = 0.42 \text{ m}$$

L = 8.5h = 8.5(0.42) = 3.57 m

From Equations 7.6 and 7.7:

$$W_2 = \frac{h}{2} = \frac{0.42}{2} = 0.21 \, m$$

$$W_1 = \frac{B - N_s W_2}{3} = \frac{1.2 - 2(0.21)}{3} = 0.26 \, m \quad \text{for rows 1, 3, and 5}$$

$$W_1 = \frac{B - N_s W_2}{3} = \frac{1.2 - 3(0.21)}{3} = 0.19 \, m \quad \text{for rows 2 and 4}$$

The splash shield height and length are calculated by first calculating the jet height:

$h_1 = 1.25 y_c = 1.25(0.822) = 1.03 \, m$

$h_1 + h = 1.03 + 0.42 = 1.45 \, m$

Since culvert rise is only 1.2 m, an enlarged section of culvert with a 1.45 m rise is required. Use 1.5 m based on constructibility and materials availability.

With culvert rise set to 1.5, use Equation 7.9 to calculate splash shield height:

$h_2 = 1.5(D-(h_1+h)) = 1.5(1.5-1.45) = 0.075 \, m$

Take $h_2 = 0.075$ m since this is greater than the minimum (0.05 m).

Length of splash shield = L/2 = 3.57/2 = 1.78 m.

Design summary:

- 5 rows of roughness elements, h = 0.42 m

- length of roughened and enlarged section =17.85 m

- outlet velocity = 2.8 m/s. (Velocity reduction = 58%)

Design Example: Tumbling Flow in a Box Culvert (CU)

Design a concrete box culvert for tumbling flow (see Figure 7.2). Determine if the outlet velocity is less than 10 ft/s. Given:

Q	=	100 ft³/s
B	=	4.0 ft
D	=	4.0 ft
n	=	0.013
S_o	=	0.06 ft/ft

Solution

Step 1. Verify the culvert is in inlet control. In this example the culvert is in inlet control.

Step 2. Compute normal flow conditions. Using trial and error with Manning's Equation.

$y_n = 1.12$ ft

$V_n = 22.4$ ft/s (Since this is greater than 10 ft/s, energy dissipation is required.)

Step 3. Compute critical depth and velocity. First compute the unit discharge,

$q = Q/B = 100/4 = 25 \ \text{ft}^2/\text{s}$

Using Equation 7.1,

$$y_c = \left(\frac{q^2}{g}\right)^{\frac{1}{3}} = \left(\frac{25^2}{32.2}\right)^{\frac{1}{3}} = 2.69 \ \text{ft}$$

$V_c = Q/(y_c B) = 100/(2.69 \ (4)) = 9.3 \ \text{ft/s}$ (Since the outlet velocity will be critical velocity, a proper design will meet the design criterion of 10 ft/s.)

Step 4. Size the element height, element spacing, and splash shield. We will use the recommended procedure of 5 rows of equal height roughness elements. Roughness element height, h, longitudinal spacing, L, and transverse spacing, W_1 and W_2, are as follows:

From Equation 7.2:

$$h = \frac{y_c}{(3 - 3.7S_o)^{\frac{2}{3}}} = \frac{2.69}{(3 - 3.7(0.06))^{\frac{2}{3}}} = 1.36 \ \text{ft}$$

$L = 8.5h = 8.5(1.36) = 11.56 \ \text{ft}$

From Equations 7.6 and 7.7:

$$W_2 = \frac{h}{2} = \frac{1.36}{2} = 0.68 \ \text{ft}$$

$$W_1 = \frac{B - N_s W_2}{3} = \frac{4 - 2(0.68)}{3} = 0.88 \ \text{ft} \quad \text{for rows 1, 3, and 5}$$

$$W_1 = \frac{B - N_s W_2}{3} = \frac{4 - 3(0.68)}{3} = 0.65 \ \text{ft} \quad \text{for rows 2 and 4}$$

The splash shield height and length are calculated by first calculating the jet height:

$h_1 = 1.25y_c = 1.25(2.69) = 3.36 \ \text{ft}$

$h_1 + h = 3.36 + 1.36 = 4.72 \ \text{ft}$

Since culvert rise is only 4 ft, an enlarged section of culvert with a 4.72 ft rise is required. Use 5.0 ft based on constructibility and materials availability.

With culvert rise set to 5.0, use Equation 7.9 to calculate splash shield height:

$h_2 = 1.5(D-(h_1+h)) = 1.5(5.0-4.72) = 0.42 \ \text{ft}$

Take $h_2 = 0.42$ ft since this is greater than the minimum (0.2 ft).

Length of splash shield = L/2 = 11.56/2 = 5.78 ft.

Design summary:

- 5 rows of roughness elements, h = 1.36 ft

- length of roughened and enlarged section = 57.8 ft

- outlet velocity = 9.3 ft/s. (Velocity reduction = 58%)

7.1.2 Tumbling Flow in Circular Culverts

Tumbling flow in circular culverts can be attained by inserting circular rings inside the barrel as shown in Figure 7.4. Geometrical considerations are more complex, but the phenomenon of tumbling flow is the same as for box culverts. For box culverts, only bottom roughness elements were considered, whereas in circular culverts the elements are complete rings. The culvert is treated as an open channel, which greatly simplifies the discussion, and the diameter is varied to obtain vertical clearance for free surface flow.

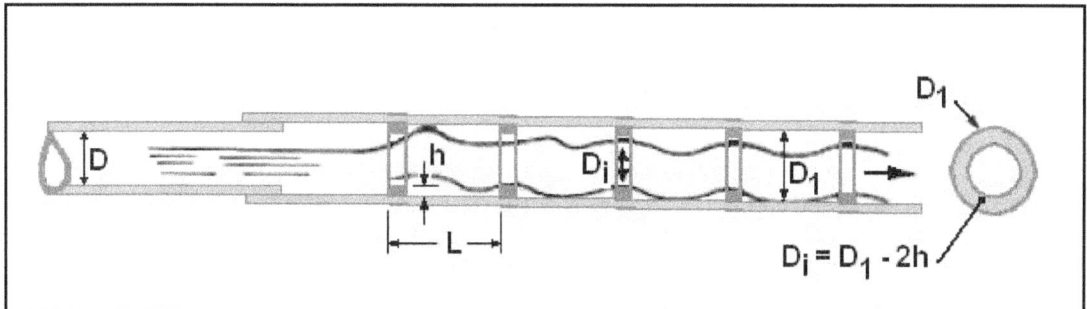

Figure 7.4. Definition Sketch for Tumbling Flow in Circular Culverts

Design procedures have been described by Wiggert and Erfle (1971). Their experiments for tumbling flow in circular culverts were run with a 152 mm (6 in) plexiglass model and a 457 mm (18 in) concrete prototype culvert. Slopes ranged from 0 to 25 percent, h/D_1 ranged from 0.06 to 0.15 and L/D_1 ranged from 0.3 to 3.0 (L/h from 5 to 20). The experimental variables are illustrated in Figure 7.4. The variables that determine whether or not tumbling flow will occur are: roughness height, h, spacing, L, slope, S_o, discharge, Q, and diameter, D_1.

A functional relationship for the roughness height can be described as:

$$h = f (L, S_o, Q, D_1, g)$$

Establishing dimensionless groupings yields:

$$h/D_1 = f(L/D_1, S_o, Q/(gD_1^5)^{1/2})$$

Practical design limits can be assigned to h/D_1 and L/D_1 to simplify the functional relationship. Based on qualitative laboratory observations, tumbling flow is easiest to maintain when L/D_1 is between 1.5 and 2.5 and when h/D_1 is between 0.10 and 0.15. Assigning these limits for circular culverts is analogous to assigning values for L/h in the design procedure for box culverts. The previous functional relationship can be rewritten:

$$Constant = f(S_o, Q/(g D_1^5)^{1/2})$$

or

$$Q/(gD_1^5)^{1/2} = f(S_o)$$

Theoretically $f(S_o)$ could be any function involving the slope term. Empirically $f(S_o)$ was found to be approximately a constant. The slight observed dependence of $f(S_o)$ on slope is considered to be much less significant than the inaccuracies associated with measuring flow characteristics over the large roughness elements. Based on model and prototype data, $f(S_o)$ ranges from 0.21 to 0.32 if the slope is between 4 percent and 25 percent. For slopes less than four percent, the culvert should be designed for full flow rather than tumbling flow. (See section 7.2.)

With the observed limits on $f(S_o)$, the following expression is developed:

$$0.21 < Q/(g\,D_1^5)^{1/2} < 0.32$$

Rewriting for use in design:

$$1.6\left(\frac{Q^2}{g}\right)^{1/5} < D_1 < 1.9\left(\frac{Q^2}{g}\right)^{1/5} \tag{7.10}$$

where,

D_1 = Diameter of the enlarged culvert section, m (ft)

Equation 7.10 is the basic design equation for tumbling flow in steep circular culverts. If the diameter of the roughened section of the culvert is sized according to this equation, tumbling flow will occur and the outlet velocity will be approximately critical velocity. This design is limited to the following conditions:

1. $L/D_1 \approx 2.0$ (tolerance plus/minus 25%)

2. $h/D_1 \approx 0.125$ (tolerance plus/minus 20%)

3. S_o greater than 4% and less than 25%

Since tumbling flow is an open channel phenomenon, gravity forces prevail and the Froude number, $V/(gy)^{1/2}$, should be used as the basis for design (or interpretation of model results). Watts (1968) established, by reference to several publications, that h/y is an important scaling parameter for roughness elements in open channel flow. In both of these dimensionless terms, y is a characteristic flow depth. The validity of using D_1 in lieu of a characteristic flow depth in $Q/(gD_1^5)^{1/2}$ must be carefully examined for culverts flowing less than full. The characteristic depth for tumbling flow, however, is critical depth, which is uniquely defined by Q and D_1; so D_1 can be substituted for y in this _special_ case of partially full culverts.

Furthermore, the higher coefficient in Equation 7.10 resulted from the 152 mm (6 in) model data rather than from the 457 mm (18 in) prototype. Differences in model and prototype data were attributed to experimental difficulties with the prototype; nevertheless, if there are scaling errors, they appear to be on the conservative side.

As with box culverts, a major concern is that silt may accumulate in front of the roughness elements and render them ineffective. This is perhaps unwarranted as the element enhances sediment transport capacity and tends to be self-cleansing. In their original list of possible applications, Peterson and Mohanty (1960) noted that by "using roughness elements to induce greater turbulence, the sediment-carrying capacity of a channel may be increased."

Water trapped between elements may cause difficulties during dry periods due to freezing and thawing and insect breeding. Narrow slots in the roughness rings (less than 0.5h) can be used to allow complete drainage without changing the design criteria. Sarikelle and Simon (1980) performed field studies of internal rings on circular pipes and found that modifications to ease installation (effectively adding slots) did not impair energy dissipation performance.

Five roughness rings at the outlet end of the culvert are sufficient to establish tumbling flow. The diameter computed from Equation 7.10 is for the roughened section only, and will not necessarily be the same as the rest of the culvert. The American Concrete Pipe Association (ACPA, 1972) introduced the telescoping concept in which the main section of the culvert is governed by the usual design parameters (presumably inlet control) and the roughened section is designed by Equation 7.10. They suggest telescoping the larger diameter pipe over the smaller "for at least the length of a normal joint and using normal sealing materials in the annular space." This concept is shown in Figure 7.4.

The design procedure requires computation of both the normal depth in the culvert based on the culvert diameter D and the critical depth based on the internal diameter of the roughened section, D_i. Using the definition sketch of Figure 7.5, the following geometric relationships are determined given a depth.

$$\theta = a\cos\left(1 - \frac{2y}{D}\right) \tag{7.11}$$

where,

θ = internal angle, degrees

$$T = D\sin\theta \tag{7.12}$$

$$A = \frac{\theta}{180}\left(\frac{\pi D^2}{4}\right) - \left(\frac{D}{2}\sin\theta\right)\left(\frac{D}{2} - y\right) \tag{7.13}$$

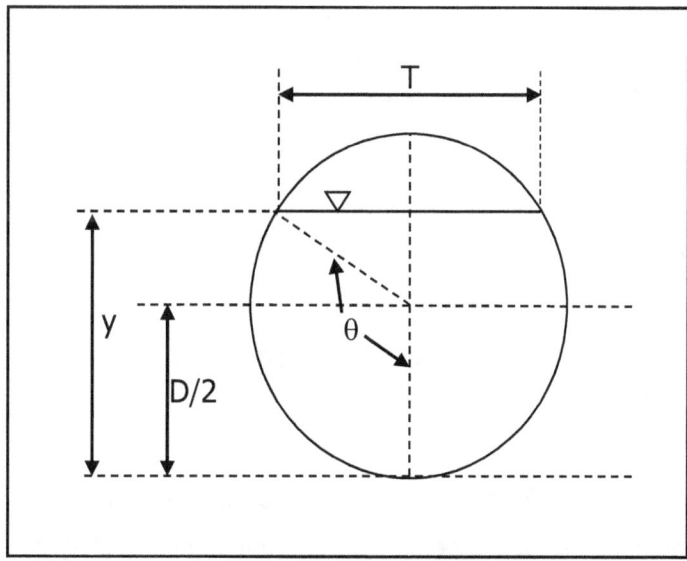

Figure 7.5. Definition Sketch for Flow in Circular Pipes

The outlet velocity for tumbling flow is approximately critical velocity. It can be computed by determining the critical depth, y_c, for the inside diameter of the roughness rings. Critical flow for an open channel of any shape will occur when:

$$\frac{Q^2 T_c}{g A_c^3} = 1 \tag{7.14}$$

where,

T_c = water surface width at critical flow condition, m (ft)

A_c = flow area at critical flow condition, m (ft)

Once the critical depth is found, critical velocity is determined using the continuity equation.

Design Example: Tumbling Flow in a Circular Culvert (SI)

Design concrete pipe culvert for tumbling flow. Determine if the outlet velocity is less than 3 m/s. Given:

Q = 2.8 m³/s

D = 1.2 m

n = 0.013

S_o = 0.06 m/m

Solution

Step 1. Verify the culvert is in inlet control. In this example the culvert is in inlet control.

Step 2. Compute normal flow conditions. Using trial and error and the geometric relations of Equations 7.11, 7.12, and 7.13:

y_n = 0.445 m

V_n = 7.3 m/s (Since this is greater than 3 m/s, energy dissipation is required.)

Step 3. Compute critical depth and velocity. First we need to compute the diameter of the roughened section using Equation 7.10 and taking the range midpoint:

$$D_1 = 1.75 \left(\frac{Q^2}{g} \right)^{\frac{1}{5}} = 1.75 \left(\frac{2.8^2}{9.81} \right)^{\frac{1}{5}} = 1.67 \text{ m}$$

h = 0.125D_1 = 0.125(1.67) = 0.21 m

D_i = D_1-2h = 1.67-2(0.21) = 1.25 m

By trial and error, using Equation 7.14 and D = D_i,

y_c = 0.913 m

A_c = 0.961 m²

V_c = Q/A_c = 2.8/0.961 = 2.9 m/s (Meets design criteria of 3 m/s)

Step 4. Determine the remaining design component: element spacing.

L = 2D_1 = 2(1.67) = 3.34 m

Design summary (see figure below; all dimensions not otherwise indicated are in meters):

- 5 rows of roughness elements, h = 0.21 m

- length of roughened and enlarged section, L_m = 16.7 m

- outlet velocity = critical velocity = 2.9 m/s. (Velocity reduction = 60%)

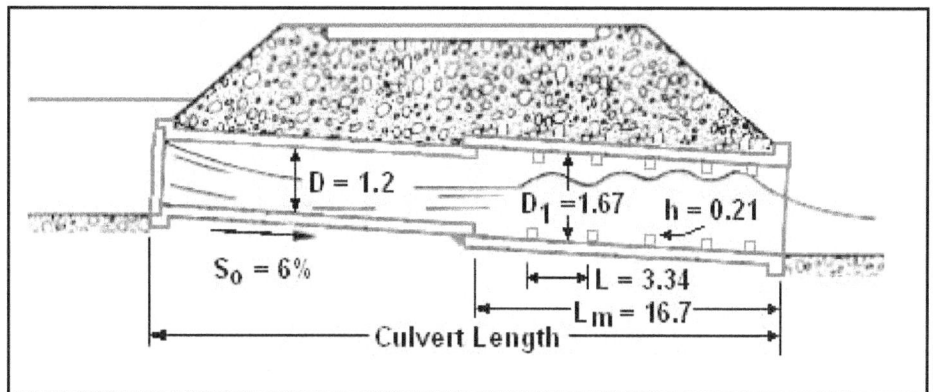

Solution for Tumbling Flow Example for a Circular Culvert (SI)

Design Example: Tumbling Flow in a Circular Culvert (CU)

Design a concrete pipe culvert for tumbling flow. Determine if the outlet velocity is less than 10 ft/s. Given:

Q = 100 ft³/s
D = 4.0 ft
n = 0.013
S_o = 0.06 ft/ft

Solution

Step 1. Verify the culvert is in inlet control. In this example the culvert is in inlet control.

Step 2. Compute normal flow conditions. Using trial and error and the geometric relations of Equations 7.11, 7.12, and 7.13:

y_n = 1.46 ft

V_n = 24.2 ft/s (Since this is greater than 10 ft/s, energy dissipation is required.)

Step 3. Compute critical depth and velocity. First we need to compute the diameter of the roughened section using Equation 7.10 and taking the range midpoint:

$$D_1 = 1.75\left(\frac{Q^2}{g}\right)^{\frac{1}{5}} = 1.75\left(\frac{100^2}{32.2}\right)^{\frac{1}{5}} = 5.51\,\text{ft}$$

h = 0.125D₁ = 0.125(5.51) = 0.69 ft

$D_i = D_1 - 2h = 5.51 - 2(0.69) = 4.13$ ft

By trial and error, using Equation 7.14 and $D = D_i$,

$y_c = 3.01$ ft

$A_c = 10.45$ ft^2

$V_c = Q/A_c = 100/10.45 = 9.6$ ft/s (Meets design criteria of 10 ft/s)

Step 4. Determine the remaining design component: element spacing.

$L = 2D_1 = 2(5.51) = 11.0$ ft

Design summary (see figure below; all dimensions not otherwise indicated are in feet):

- 5 rows of roughness elements, $h = 0.69$ ft

- length of roughened and enlarged section, $L_m = 55$ ft

- outlet velocity = critical velocity = 9.6 ft/s. (Velocity reduction = 60%)

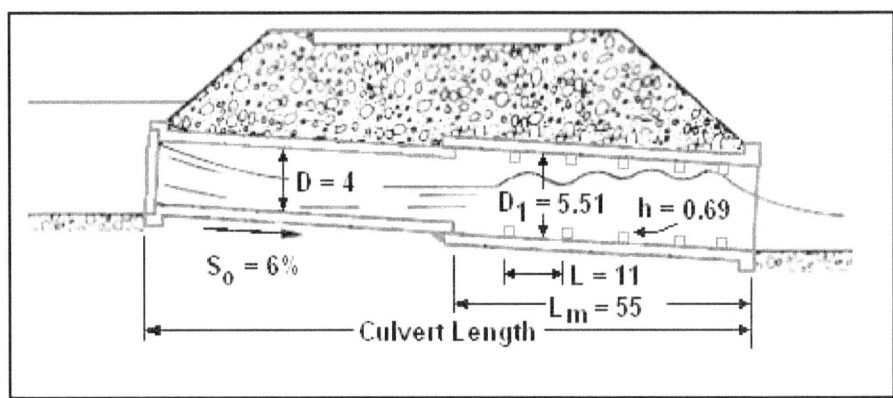

Solution for Tumbling Flow Example for a Circular Culvert (CU)

7.2 INCREASED RESISTANCE

The methodology described in this section involves using roughness elements to increase resistance and induce velocity reductions. Increasing resistance may cause a culvert to change from partial flow to full flow in the roughened zone. Velocity reduction is accomplished by increasing the wetted surfaces as well as by increasing drag and turbulence by the use of roughness elements.

Tumbling flow, as described in the previous section, is the limiting design condition for roughness elements on steep slopes. Tumbling flow essentially delivers the outlet flow at critical velocity. If the requirement is for outlet velocities between critical and the normal culvert velocity, designing increased resistance into the barrel is a viable alternative.

The most obvious situation for application of increased barrel resistance is a culvert flowing partially full with inlet control. The objective is to force full flow near the culvert outlet without creating additional headwater. Although based on the same principles, the design approaches for circular and box culverts differ.

Morris (1963) studied all pertinent rough pipe flow data available and concluded that there are three flow regimes and each has a different resistance relationship. Conceptually, the description of these regimes also applies to box culverts. The three regimes illustrated in Figure 7.6 are:

1. Quasi-smooth flow: Occurs only when there are depressions or when roughness elements are spaced very close (L/h approximately equal to 2). Quasi-smooth flow is not important for this discussion.

2. Hyper-turbulent flow: Occurs when roughness elements are sufficiently close so each element is in the wake of the previous element and rough surface vortices are the primary source of the overall friction drag.

3. Isolated roughness flow: Occurs when roughness spacing is large and overall resistance is due to drag on the culvert surface plus form drag on the roughness elements.

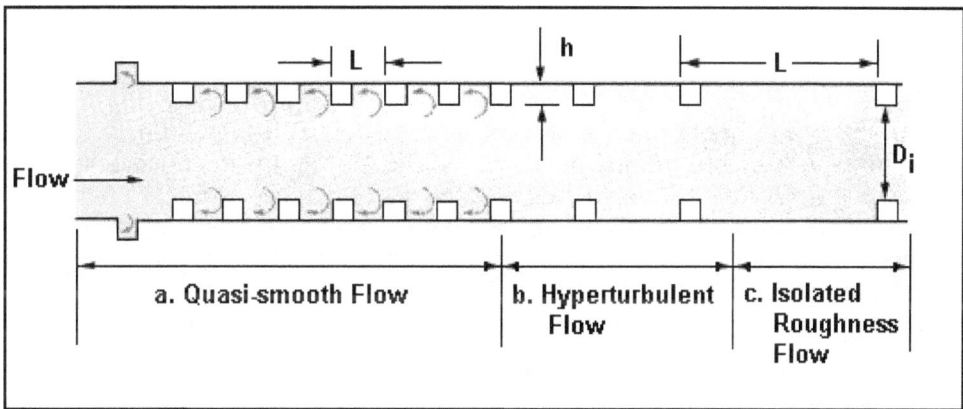

Figure 7.6. Flow Regimes in Rough Pipes

The design procedure for increased resistance in box and circular culverts may be summarized in the following steps:

Step 1. Verify the culvert is in inlet control.

Step 2. Compute normal flow conditions in the culvert to determine if the discharge conditions at the outlet require mitigation.

Step 3. Select initial design scale ratios. Determine Manning's n value for the roughened section of culvert.

Step 4. Compute mitigated depth and velocity. Check mitigated velocity against design criteria. One of three conditions will be observed:

1. The computed depth will exceed the culvert rise meaning the culvert will flow full and, potentially, increase headwater. In such cases, a larger roughened section is required and step 3 is repeated.

2. The computed depth is less than the culvert rise and the velocity is lower than the design criteria. This is an acceptable design. Verify that full flow capacity is greater than design discharge. If not, repeat step 3.

3. The computed depth is less than the culvert rise, but the velocity is higher than the design criteria, then one of three alternatives may be pursued:

    a. Increase the roughness element height to approach full flow (and therefore lower the velocity). Repeat step 3.

    b. Use the tumbling flow design discussed in Section 7.1.1.

    c. Use another type of dissipator in lieu of or in addition to increased roughness.

Step 5. Complete sizing the element heights, element spacing, and other design features. Design details differ for box and circular culverts and are described in the following sections.

7.2.1 Increased Resistance in Circular Culverts

Wiggert and Erfle (1971) studied the effectiveness of roughness rings as energy dissipators in circular culverts. Although their study was primarily a tumbling flow study, they observed in many tests that they could get velocity reductions greater than 50 percent without reaching the roughness level necessary for tumbling flow. They did not derive resistance equations, but they did establish approximate design limits.

From these studies, good performance was observed when h/D was 0.06 to 0.09 using five rings. (See Figure 7.7.) Doubling the height, h_1, of the first ring was effective in triggering full flow in the roughened zone. Adequate performance was obtained with four identical rings, but with double spacing between the first two. However, the same pipe length is involved if a constant spacing is maintained and five rings used, with the first double the height of the other four. The additional ring should help establish the assumed full flow condition. In addition, the last (downstream) ring must be located no closer than one-half the ring spacing from the end of the culvert.

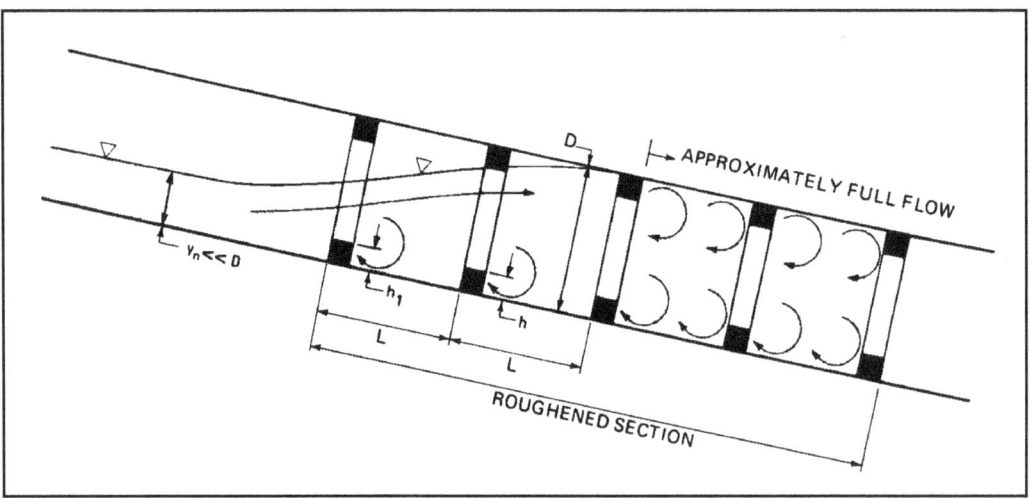

Figure 7.7. Conceptual Sketch of Roughness Elements to Increase Resistance

Subsequent experience reported by the American Concrete Pipe Association (ACPA, 1972) indicated a need to consider lower values of h/D, and to establish approximate resistance curves for evaluating a design in order to avoid installations that will propagate full flow upstream to the culvert inlet.

Based on experience with large elements used to force tumbling flow (see Section 7.1) and the work of Wiggert and Erfle (1971), five rows of roughness elements with heights ranging from 5 to 10 percent of the culvert diameter are sufficient.

A key element in the design of increased roughness elements is determination of the roughness regime and, subsequently, the appropriate Manning's n value. Although much of the literature relative to large roughness elements in circular pipes expresses resistance in terms of the friction factor, "f", all resistance equations are converted to Manning's "n" expressions for this manual.

7.2.1.1 Isolated-Roughness Flow

Isolated roughness flow was introduced in Section 7.2. The overall friction or resistance, f_{IR}, is made up of two parts:

$$f_{IR} = f_s + f_d \tag{7.15}$$

where,

f_s = friction on the culvert surface.

f_d = friction due to form drag on the roughness elements.

The friction due to form drag is a function of the drag coefficient for the particular shape, the percentage of the wetted perimeter that is roughened, the roughness dimensions and spacing and the velocity impinging on the roughness elements. Morris (1963) related the velocity to surface drag and derived the following equation:

$$f_{IR} = f_s \left(1 + 67.2 C_D \left(\frac{L_r}{P} \right) \left(\frac{h}{r_i} \right) \left(\frac{r_i}{L} \right) \right) \tag{7.16}$$

where,

C_D = drag coefficient for the roughness shape

L_r/P = ratio of total peripheral length of roughness elements to total wetted perimeter

r_i = pipe radius based on the inside diameter of roughness rings measured from crest to crest

L_r may be less than P to facilitate constructibility of the rings or to permit a low flow opening at the bottom of the ring.

Throughout Morris' work, he used measurements from crest to crest of a roughness element ring as the effective diameter, D_i. To convert the expression for roughness to Manning's n, the following expressions are needed:

$$f_s = \alpha \, (n/D^{1/6})^2$$

$$f_{IR} = \alpha \, (n_{IR}/D_i^{1/6})^2$$

α represents a unit conversion constant equal to 124 in SI and 184 in CU. Equation 7.16 can then be converted to Manning's n:

7-17

$$n_{IR} = n\left(\frac{D_i}{D}\right)^{1/6}\left(1 + 67.2C_D\left(\frac{L_r}{P}\right)\left(\frac{h}{L}\right)\right)^{1/2} \tag{7.17}$$

where,

n_{IR} = overall Manning's "n" for isolated roughness flow

n = Manning's "n" for the culvert surface without roughness rings

D = nominal diameter of the culvert, m (ft)

D_i = inside diameter of roughness rings, m (ft) (D_i = D-2h)

For sharp edge rectangular roughness shapes, a constant value of 1.9 can be used for C_D. It is noteworthy that the overall resistance, n_{IR}, decreases as the relative spacing, L/D_i, increases for this regime.

7.2.1.2 Hyperturbulent Flow

The friction in this regime is independent of friction on the culvert surface:

$$f_{HT} = \left(\frac{1}{2\log\left(\frac{r_i}{L}\right) + 1.75 + \phi}\right)^2 \tag{7.18}$$

where,

f_{HT} = overall friction for hyper-turbulent flow

ϕ = function of Reynolds number, element shape, and relative spacing

By restricting application of Equation 7.18 to sharp edged roughness rings and to a spacing greater than the pipe radius, ϕ can be neglected.

Substituting the following expression:

$$f_{HT} = \alpha \ (n_{HT}/D_i^{1/6})^2$$

Equation 7.18 can then be converted to Manning's n:

$$n_{HT} = \frac{\alpha D_i^{1/6}}{2\log\left(\frac{r_i}{L}\right) + 1.75} \tag{7.19}$$

where,

n_{HT} = Manning's n for hyper-turbulent flow

α = unit conversion constant, 0.0898 (SI) and 0.0737 (CU)

The effect of the roughness height, h, is included inherently in D_i. From Equation 7.19 it can be seen that n_{HT} increases as the spacing increases for this regime.

7.2.1.3 Regime Boundaries

Since resistance increases when the spacing increases for the hyper-turbulent regime and when the spacing decreases for the isolated roughness regime, the boundary between the regimes occurs when the resistance equations are the same. The boundary is determined by equating n_{IR} in Equation 7.17 to n_{HT} in Equation 7.19. For design, both are calculated; the lowest n value is used and indicates which regime is applicable.

The recommended design is limited to the following conditions:

1. $0.5 < L/D_i < 1.5$ (1.0 to 1.1 is a suggested starting point)

2. $0.05 < h/D_i < 0.10$ (0.06 is a suggested starting point using sharp-edged roughness rings)

Once these ranges are selected, the roughness element height is computed as follows:

$$h = \frac{D}{2 + \dfrac{1}{c}} \qquad (7.20)$$

where,

h = roughness element height, m (ft)

c = ratio of h/D_i

Once h is calculated, values of D_i and L follow directly and the roughness values are calculated.

Design Example: Increased Resistance in a Circular Culvert (SI)

Design a concrete pipe culvert for increased roughness. Determine if the outlet velocity is less than 3 m/s. Given:

Q = 2.8 m³/s
D = 1.2 m
n = 0.013
S_o = 0.06 m/m

Solution

Step 1. Verify the culvert is in inlet control. In this case, the culvert is in inlet control.

Step 2. Compute normal flow conditions. Using trial and error and the geometric relations of Equations 7.11, 7.12, and 7.13:

y_n = 0.445 m

V_n = 7.3 m/s (Since this is greater than 3 m/s, energy dissipation is required.)

Step 3. Select initial design scale ratios and determine Manning's n value for the roughened section of culvert.

Try $L/D_i = 1.1$ and $h/D_i = 0.06$

Calculate h from Equation 7.20:

$$h = \frac{D}{2 + \dfrac{1}{c}} = \frac{1.2}{2 + \dfrac{1}{0.06}} = 0.064 \text{ m} \quad \text{(round to 0.06 m)}$$

$D_i = D - 2h = 1.2 - 2(0.06) = 1.08$ m

$L = 1.1D_i = 1.1(1.08) = 1.19$ m

For this design, we will not have gaps in the rings, therefore, $L_r/P = 1$.

Now, we calculate Manning's n for the isolated roughness (Equation 7.17) and hyperturbulent flow (Equation 7.19) to determine flow regime and Manning's n.

$$n_{IR} = n\left(\frac{D_i}{D}\right)^{\frac{1}{6}}\left(1 + 67.2C_D\left(\frac{L_r}{P}\right)\left(\frac{h}{L}\right)\right)^{\frac{1}{2}} = 0.013\left(\frac{1.08}{1.2}\right)^{\frac{1}{6}}\left(1 + 67.2(1.9)(1)\left(\frac{0.060}{1.19}\right)\right)^{\frac{1}{2}} = 0.035$$

$$n_{HT} = \frac{\alpha D_i^{\frac{1}{6}}}{2\log\left(\frac{r_i}{L}\right) + 1.75} = \frac{0.0898(1.08)^{\frac{1}{6}}}{2\log\left(\frac{0.54}{1.19}\right) + 1.75} = 0.085$$

Since $n_{IR} < n_{HT}$ the roughness is characterized as isolated roughness and n = 0.035

Step 4. Compute mitigated depth and velocity. Check mitigated velocity against design criteria.

With the internal diameter, D_i, and the Manning's n values calculated in step 3, the normal depth for the roughened condition is calculated to be (by trial and error);

$y_n = 0.932$ m

$V_n = 3.3$ m/s

Compared with the design goal of 3.0 m/s, this velocity is unacceptable even though it has been reduced significantly from the unmitigated velocity. Since the depth is less than D_i, we can increase h to further slow the velocity. (We also need to increase the culvert size because the full flow velocity of 3.1 m/s still exceeds our design criteria.) Steps 3 and 4 must be repeated.

Step 3 (2nd iteration). Select trial design scale ratios and determine Manning's n value for the roughened section of culvert.

Maintain $L/D_i = 1.1$, increase D = 1.50 m (next available size) and try h = 0.1 m

$D_i = D - 2h = 1.50 - 2(0.10) = 1.30$ m ($h/D_i = 0.077$)

$L = 1.1D_i = 1.1(1.3) = 1.43$ m

For this design, we will not have gaps in the rings, therefore, $L_r/P = 1$.

Now, we calculate Manning's n for the isolated roughness (Equation 7.17) and hyperturbulent flow (Equation 7.19) to determine flow regime and Manning's n.

$$n_{IR} = n\left(\frac{D_i}{D}\right)^{\frac{1}{6}}\left(1 + 67.2C_D\left(\frac{L_r}{P}\right)\left(\frac{h}{L}\right)\right)^{\frac{1}{2}} = 0.013\left(\frac{1.30}{1.50}\right)^{\frac{1}{6}}\left(1 + 67.2(1.9)(1)\left(\frac{0.10}{1.43}\right)\right)^{\frac{1}{2}} = 0.040$$

$$n_{HT} = \frac{\alpha D_i^{\frac{1}{6}}}{2\log\left(\dfrac{r_i}{L}\right) + 1.75} = \frac{0.0898(1.30)^{\frac{1}{6}}}{2\log\left(\dfrac{0.65}{1.43}\right) + 1.75} = 0.088$$

Since $n_{IR} < n_{HT}$ the roughness is characterized as isolated roughness and n = 0.040.

Step 4 (second iteration). Compute trial of the mitigated depth and velocity. Check mitigated velocity against design criteria.

With the internal diameter, D_i, and the Manning's n values calculated in step 3, the normal depth for the roughened condition is calculated to be (by trial and error);

$y_n = 1.025$ m

$V_n = 2.5$ m/s

Compared with the design goal of 3.0 m/s, this velocity is acceptable. However, we must check the culvert capacity using Manning's Equation.

We assume that the culvert is flowing full and estimate the wetted perimeter.

$A = \pi D_i^2/4 = \pi(1.30)^2/4 = 1.33$ m²

$P = \pi D_i = \pi(1.30) = 4.08$ m

$R = A/P = 1.33/4.08 = 0.326$ m

Using Manning's Equation,

$$Q = \frac{1}{n} AR^{\frac{2}{3}}S^{\frac{1}{2}} = \frac{1}{0.040} 1.33(0.326)^{\frac{2}{3}}(0.06)^{\frac{1}{2}} = 3.9 \text{ m}^3/\text{s}$$

Since this flow is greater than the design flow of 2.8 m³/s we know that the design is acceptable. If this had not been the case, a larger culvert barrel could be evaluated going back to step 3 or tumbling flow or another type of dissipator could be considered.

Step 5. Complete sizing the element heights, element spacing, and other design features.

Roughness height and spacing have been established as well as an oversized culvert section. For 5 rows of roughness elements, the length of the oversized section with increased roughness is 5.06 m.

Design summary (see figure below):

- 5 rows of roughness elements, h = 0.1 m

- length of roughened section = 7.15 m. (Roughened length considered from L/2 before the first element and with the last roughness element no less than L/2 from the culvert outlet.)

- outlet velocity = 2.9 m/s. (Velocity reduction = 60%)

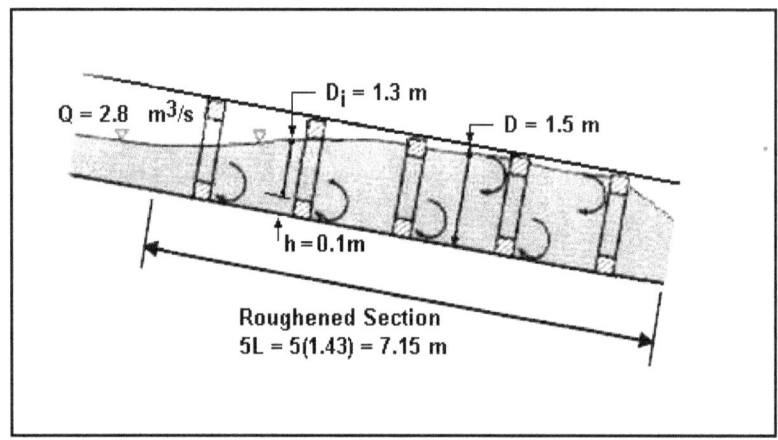

Sketch for Increased Resistance In a Circular Culvert Design Example (SI)

Design Example: Increased Resistance in a Circular Culvert (CU)

Design a concrete pipe culvert for increased roughness. Determine if the outlet velocity is less than 10 ft/s. Given:

Q = 100 ft³/s

D = 4.0 ft

n = 0.013

S_o = 0.06 ft/ft

Solution

Step 1. Verify the culvert is in inlet control. In this case the culvert is in inlet control.

Step 2. Compute normal flow conditions. Using trial and error and the geometric relations of Equations 7.11, 7.12, and 7.13:

y_n = 1.45 ft

V_n = 24.1 ft/s (Since this is greater than 10 ft/s, energy dissipation is required.)

Step 3. Select initial design scale ratios and determine Manning's n value for the roughened section of culvert.

Try L/D_i = 1.1 and h/D_i = 0.06

Calculate h from Equation 7.20:

$$h = \frac{D}{2 + \dfrac{1}{c}} = \frac{4.0}{2 + \dfrac{1}{0.06}} = 0.214 \text{ ft} \quad \text{(round to 0.21 ft)}$$

$D_i = D - 2h = 4.0 - 2(0.21) = 3.58$ ft

$L = 1.1D_i = 1.1(3.58) = 3.94$ ft

For this design, we will not have gaps in the rings, therefore, $L_r/P = 1$.

Now, we calculate Manning's n for the isolated roughness (Equation 7.17) and hyperturbulent flow (Equation 7.19) to determine flow regime and Manning's n.

$$n_{IR} = n\left(\frac{D_i}{D}\right)^{\frac{1}{6}}\left(1 + 67.2C_D\left(\frac{L_r}{P}\right)\left(\frac{h}{L}\right)\right)^{\frac{1}{2}} = 0.013\left(\frac{3.58}{4.0}\right)^{\frac{1}{6}}\left(1 + 67.2(1.9)(1)\left(\frac{0.21}{3.94}\right)\right)^{\frac{1}{2}} = 0.036$$

$$n_{HT} = \frac{\alpha D_i^{\frac{1}{6}}}{2\log\left(\frac{r_i}{L}\right) + 1.75} = \frac{0.0737(3.58)^{\frac{1}{6}}}{2\log\left(\frac{1.79}{3.94}\right) + 1.75} = 0.086$$

Since $n_{IR} < n_{HT}$ the roughness is characterized as isolated roughness and n = 0.036

Step 4. Compute mitigated depth and velocity. Check mitigated velocity against design criteria.

With the internal diameter, D_i, and the Manning's n values calculated in step 3, the normal depth for the roughened condition is calculated using Manning's Equation to be (by trial and error);

y_n = 3.10 ft

V_n = 10.7 ft/s

Compared with the design goal of 10 ft/s, this velocity is unacceptable even though it has been reduced significantly from the unmitigated velocity. Since the depth is less than D_i, we can increase h to further slow the velocity. (We also need to increase the culvert size because the full flow velocity exceeds our design criteria.) Steps 3 and 4 must be repeated.

Step 3 (2nd iteration). Select trial design scale ratios and determine Manning's n value for the roughened section of culvert.

Maintain L/D_i = 1.1, increase D = 4.5 ft and try h = 0.32 ft

D_i = D – 2h = 4.5 – 2(0.32) = 3.86 ft (h/D_i = 0.083)

L = 1.1D_i = 1.1(3.86) = 4.25 ft

For this design, we will not have gaps in the rings, therefore, L_r/P = 1.

Now, we calculate Manning's n for the isolated roughness (Equation 7.17) and hyperturbulent flow (Equation 7.19) to determine flow regime and Manning's n.

$$n_{IR} = n\left(\frac{D_i}{D}\right)^{\frac{1}{6}}\left(1 + 67.2C_D\left(\frac{L_r}{P}\right)\left(\frac{h}{L}\right)\right)^{\frac{1}{2}} = 0.013\left(\frac{3.86}{4.5}\right)^{\frac{1}{6}}\left(1 + 67.2(1.9)(1)\left(\frac{0.32}{4.25}\right)\right)^{\frac{1}{2}} = 0.041$$

$$n_{HT} = \frac{\alpha D_i^{\frac{1}{6}}}{2\log\left(\frac{r_i}{L}\right) + 1.75} = \frac{0.0737(3.86)^{\frac{1}{6}}}{2\log\left(\frac{1.93}{4.25}\right) + 1.75} = 0.087$$

Since $n_{IR} < n_{HT}$ the roughness is characterized as isolated roughness and n = 0.041

Step 4 (2nd iteration). Compute trial of the mitigated depth and velocity. Check mitigated velocity against design criteria.

With the internal diameter, D_i, and the Manning's n values calculated in step 3, the normal depth for the roughened condition is calculated to be (by trial and error);

y_n = 3.07 ft

V_n = 9.9 ft/s

Compared with the design goal of 10 ft/s, this velocity is acceptable. However, we must check the culvert capacity using Manning's Equation.

We assume that the culvert is flowing full and estimate the wetted perimeter.

$A = \pi D_i^2/4 = \pi(3.86)^2/4 = 11.70\ ft^2$

$P = \pi D_i = \pi(3.86) = 12.13\ ft$

$R = A/P = 11.70/12.13 = 0.964\ ft$

Using Manning's Equation,

$$Q = \frac{1.49}{n}AR^{\frac{2}{3}}S^{\frac{1}{2}} = \frac{1.49}{0.041}11.70(0.964)^{\frac{2}{3}}(0.06)^{\frac{1}{2}} = 102\ ft^3/s$$

Since this flow is greater than the design flow of 100 ft^3/s we know that the design is acceptable. If this had not been the case, a larger culvert barrel could be evaluated going back to step 3 or tumbling flow or another type of dissipator could be considered.

Step 5. Complete sizing the element heights, element spacing, and other design features.

Roughness height and spacing have been established as well as an oversized culvert section. For 5 rows of roughness elements, the length of the oversized section with increased roughness is 17.0 ft.

Design summary (see figure below):

- 5 rows of roughness elements, h = 0.32 ft

- length of roughened and enlarged section = 21.25 ft. (Roughened length considered from L/2 before the first element and with the last roughness element no less than L/2 from the culvert outlet.)

- outlet velocity = 9.9 ft/s. (Velocity reduction = 59%)

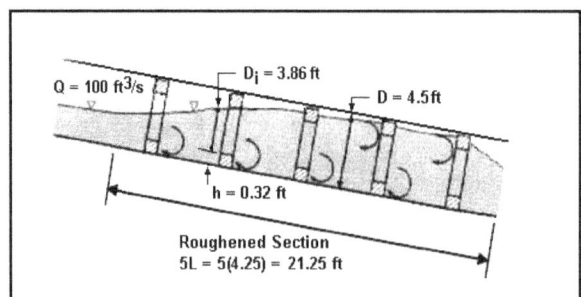

Sketch for Increased Resistance In a Circular Culvert Design Example (CU)

7.2.2 Increased Resistance in Box Culverts

Material for this section was drawn primarily from a preliminary FHWA report on fish baffles in box culverts (Normann, 1974). This report used Morris' categorization of flow regimes and basic friction equations (Morris, 1963), but a more representative approach velocity, V_A, in one of the regimes. Experimental data by Shoemaker (1956) were also utilized to define the transition curves. For several reasons, modifications to the fish baffle development were necessary to adapt to energy dissipator design. In fish baffle design, the interest is in a conservative estimate of resistance in order to size a culvert; whereas, in this manual, a conservative estimate of the outlet velocity is also important. Also, fish baffle design curves involve bottom roughness only.

As before, both the hyperturbulent and isolated roughness flow regimes are considered. For box culverts in hyperturbulent flow, Manning's roughness may be estimated by:

$$n_{HT} = n\left[\left(1-\frac{L_r}{P}\right) + \frac{70.6\left(\frac{L_r}{P}\right)}{\left(2\log\left(\left(\frac{R_i}{h}\right)\left(\frac{h}{L}\right)\right) + 1.75\right)^2}\right]^{\frac{1}{2}} \qquad (7.21)$$

where,

n_{HT} = Manning's n for the hyperturbulent flow regime

n = Manning's roughness coefficient for the culvert without the roughness elements (maximum n value is 0.015 for this equation)

L_r/P = ratio of total peripheral length of roughness elements to total wetted perimeter

For the isolated roughness regime, a high and low Manning's range are considered as shown by the following equations:

$$n_{IR,LOW} = n\left[1 + 200\left(\frac{h}{L}\right)\left(\frac{L_r}{P}\right)\right]^{\frac{1}{2}} \qquad (7.22a)$$

$$n_{IR,HIGH} = n\left[1 + 390\left(\frac{h}{L}\right)\left(\frac{L_r}{P}\right)\right]^{\frac{1}{2}} \qquad (7.22b)$$

where,

$n_{IR,LOW}$ = Manning's n for the isolated roughness flow regime, low design range for estimating velocity.

$n_{IR,HIGH}$ = Manning's n for the isolated roughness flow regime, high design range for estimating depth.

n = Manning's roughness coefficient for the culvert without the roughness elements (maximum n value is 0.015 for these equations)

L_r will equal the bottom width, B, for installations with bottom roughness only and will equal the wetted perimeter, P, when roughness elements are attached to all sides or when the roughness elements extend through the flow. The presence of drainage slots in the roughness elements is ignored in estimating L_r.

The equations are based on $C_D = 1.9$, $f = 0.14$ (where f is the Darcy friction factor for the culvert surface without roughness elements), and $V_A/V = 0.60$ or 0.85. The lower value of V_A/V is implicitly included in Equation 7.22a and the higher value in Equation 7.22b. The use of a representative approach velocity, V_A, allows an opportunity to input culvert parameters that will lean towards either an overprediction or an underprediction of resistance. It is assumed that $(R/R_i)^{1/3}$ is approximately one to simplify the analysis. R is the hydraulic radius of the culvert proper and R_i is the hydraulic radius taken inside the crests of the roughness elements. For designs considered in this section, the approximation is reasonable.

For this manual, it is appropriate to develop high as well as low resistance curves. Rather than attempt to define the transition between these curves, an abrupt transition is used as the worst condition for the high curves, and a straight-line transition is assumed as the mildest condition for the low curves. This is illustrated in Figure 7.8. Observations by Powell (1946) are the basis for assuming the 6 to 12 range of L/h for the transition curve between isolated roughness and hyperturbulent flow. An L/h=10 is chosen for design because it yields the largest n value.

For estimating velocities, it is appropriate to estimate resistance based on the straight-line (low) relationship shown in Figure 7.8. Therefore, Equation 7.21 (hyperturbulent) is evaluated at L/h=6 and Equation 7.22a (isolated roughness) is evaluated at L/h=12. A linear interpolation between the two for L/h=10 results in the relationship provided as follows:

$$n_{LOW} = n \left\{ \frac{2}{3}\left(1 + 16.7\left(\frac{L_r}{P}\right)\right)^{1/2} + \frac{1}{3}\left(\left(1 - \frac{L_r}{P}\right) + \frac{70.6\left(\frac{L_r}{P}\right)}{\left(2\log\left(\frac{R_i}{h}\right) + 0.194\right)^2} \right)^{1/2} \right\}$$

(7.23)

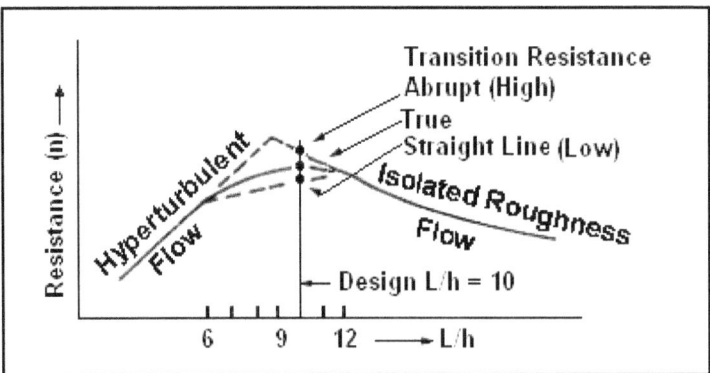

Figure 7.8. Transition Curves between Flow and Regimes

For determining the lower value of Manning's n, n_{LOW}, for the purpose of estimating outlet velocity, Equation 7.23 applies to values of h/R_i less than or equal to 0.3. For ratios above 0.3, n_{LOW} is calculated directly from Equation 7.22a evaluated for L/h = 10.

For determining the upper value of Manning's n, n_{HIGH}, for the purpose of estimating flow depths, the abrupt (high) value indicated in Figure 7.8 is desired. For h/R_i greater than 0.2, Equation 7.22b is evaluated at L/h = 10 and used for n_{HIGH}. For ratios less than or equal to 0.2, Equations 7.22b and 7.21 are both evaluated at L/h = 10 and the lower value is taken for n_{HIGH}. Both are compared to avoid unrealistic values from Equation 7.22b.

Since the above equations are normal flow equations and roughness elements may be relatively small using this method, it is necessary to compute the length of the culvert to be roughened. The momentum equation, written for the roughened section of culvert, is used to compute the number of rows of roughness element needed.

$$N = \frac{gB\left(y_n^2 - y_i^2\right) + 2Q\left(V_n - V_i\right)}{C_D A_f V_w^2}$$

(7.24)

where,

N = number of roughness element rows

B = culvert bottom width, m (ft)

y_n = normal depth in the culvert approaching the roughened section, m (ft)

y_i = normal depth in the roughened section of the culvert, m (ft)

V_n = normal velocity in the culvert approaching the roughened section, m/s (ft/s)

V_i = normal velocity in the roughened section of the culvert, m/s (ft/s)

C_D = coefficient of drag (taken as 1.9)

A_f = wetted frontal area of a roughness row, m^2 (ft^2), equal to B(h) for bottom roughness

V_w = average wall velocity action on the roughness elements, m/s (ft/s), equal to $(V_n + V_i)/6$

Regardless of the result of Equation 7.24, the number of rows should never be less than five. Furthermore, it is recommended that one large element be used at the beginning of the roughened zone to accelerate the asymptotic approach to normal flow. The recommended height of the larger element is twice the height of the regular elements. The spacing is the same for all rows of elements.

Slots in the roughness elements are provided for low flow drainage. The slot opening should not exceed h/2.

The procedure is limited to solid strip roughness elements with sharp upstream edges. Rectangular cross section roughness elements will best fit the assumptions made.

Due to the assumed velocity distribution, application of the procedure must be limited to small roughness heights and to relatively flat slopes. The roughness height should not exceed ten percent of the flow depth. This restriction is inherently included in the suggested range of h/R_i in the design procedure.

The recommended design is limited to the following conditions:

1. $S_o \leq 6\,\%$
2. $0.1 < h/R_i < 0.4$
3. $L/h = 10$

7-27

Design Example: Increased Resistance for a Box Culvert (SI)

Design a concrete box culvert for increased roughness using bottom roughness elements. Determine if the outlet velocity is less than 3 m/s. Given:

$$Q = 2.8 \text{ m}^3\text{/s}$$
$$D = 1.2 \text{ m}$$
$$B = 1.2 \text{ m}$$
$$n = 0.013$$
$$S_o = 0.06 \text{ m/m}$$

Solution

Step 1. Verify the culvert is in inlet control. In this case the culvert is in inlet control.

Step 2. Compute normal flow conditions. Using trial and error:

$y_n = 0.34$ m

$V_n = 6.8$ m/s (Since this is greater than 3 m/s, energy dissipation is required.)

Step 3. Select initial design scale ratios. Determine Manning's n value for the roughened section of culvert for estimating the mitigated depth and velocity.

Try $L/h = 10$ and $h/R_i = 0.3$

We will assume that the culvert is not flowing full for computation of the wetted perimeter. (If the culvert did flow full for this computation, it will be surcharged when we compute the high Manning's n for the capacity check. Using the sides and bottom of the culvert will provide the lowest (short of full flow), and therefore, conservative value of n for estimating velocity in the roughened section. This assumption may be revised in subsequent iterations.

$P = 2D+B = 2(1.2)+1.2 = 3.6$ m

$L_r = B = 1.2$ m (bottom roughness only)

Using Equation 7.23 and n = 0.013 (maximum value is 0.015):

$$n_{LOW} = n\left\{\frac{2}{3}\left(1+16.7\left(\frac{L_r}{P}\right)\right)^{\frac{1}{2}} + \frac{1}{3}\left(1-\frac{L_r}{P}\right) + \frac{70.6\left(\frac{L_r}{P}\right)}{\left(2\log\left(\frac{R_i}{h}\right)+0.194\right)^2}\right\}^{\frac{1}{2}} = 0.040$$

Step 4. Compute mitigated depth and velocity and roughness height, h. Check mitigated velocity against design criteria. Using trial and error:

$y_i = 0.783$ m (note that culvert is not flowing full)

$A_i = 0.783(1.2) = 0.940$ m^2

$P_i = 2(0.783)+1.2 = 2.77$ m

$R_i = 0.940/2.77 = 0.340$ m

$h = (h/R_i)(R_i) = 0.3(0.340) = 0.102$ m (round to 0.10 m)

V_i = 2.98 m/s

Since this velocity is less than or equal to 3 m/s, the dissipation design is satisfactory. However, we must check the culvert capacity using the high estimate of Manning's n from Equation 7.22b. Two possible flow limiting scenarios exist.

First, assume the culvert is flowing nearly full. Using Equation 7.22b and n = 0.013 (maximum value is 0.015):

$$n_{IR,HIGH} = n\left[1 + 390\left(\frac{h}{L}\right)\left(\frac{L_r}{P}\right)\right]^{\frac{1}{2}} = 0.013\left[1 + 390(0.1)\left(\frac{1.2}{3.6}\right)\right]^{\frac{1}{2}} = 0.049$$

Area and hydraulic radius are calculated as:

y = D – h = 1.2 - 0.10 = 1.10 m

A = yB = 1.10(1.2) = 1.32 m²

P = 2y+B = 2(1.10)+1.2 = 3.40 m

R = A/P = 1.32/3.40 = 0.388 m

Using Manning's Equation,

$$Q = \frac{1}{n}AR^{\frac{2}{3}}S^{\frac{1}{2}} = \frac{1}{0.049}1.32(0.388)^{\frac{2}{3}}(0.06)^{\frac{1}{2}} = 3.5 \text{ m}^3/\text{s}$$

Q is greater than the design flow using this scenario.

Second, assume the culvert is flowing full. Using Equation 7.22b and n = 0.013 (maximum value is 0.015):

$$n_{IR,HIGH} = n\left[1 + 390\left(\frac{h}{L}\right)\left(\frac{L_r}{P}\right)\right]^{\frac{1}{2}} = 0.013\left[1 + 390(0.1)\left(\frac{1.2}{4.8}\right)\right]^{\frac{1}{2}} = 0.043$$

Area and hydraulic radius are calculated as:

A = yB = 1.10(1.2) = 1.32 m²

P = 2(y+B) = 2(1.10+1.2) = 4.6 m

R = A/P = 1.32/4.6 = 0.287 m

Using Manning's Equation,

$$Q = \frac{1}{n}AR^{\frac{2}{3}}S^{\frac{1}{2}} = \frac{1}{0.043}1.32(0.287)^{\frac{2}{3}}(0.06)^{\frac{1}{2}} = 3.3 \text{ m}^3/\text{s}$$

Since both flows are greater than the design flow of 2.8 m³/s we know that the design is acceptable. If this had not been the case, a larger culvert barrel could be evaluated going back to step 3 or tumbling flow or another type of dissipator could be considered.

Step 5. Complete sizing the element heights, element spacing, and other design features.

Spacing between roughness elements is calculated to be:

L = (L/h)h = (10)0.10 = 1.0 m

Number of rows of roughness elements is estimated using Equation 7.24. First calculate:

$A_f = B(h) = 1.2(0.10) = 0.12 \text{ m}^2$

$V_w = (V_n+V_i)/6 = (6.8+2.98)/6 = 1.63 \text{ m/s}$

$$N = \frac{gB\left(y_n^2 - y_i^2\right) + 2Q\left(V_n - V_i\right)}{C_D A_f V_w^2} = \frac{9.81(1.2)\left(0.34^2 - 0.783^2\right) + 2(2.80)(6.8 - 2.98)}{1.9(0.12)1.63^2} = 25.7$$

Round up to the nearest whole number, N = 26

Design summary:

- 26 rows of roughness elements, h = 0.10 m

- length of roughened section = 25.0 m. The last roughness element should be no less than L/2 from the culvert outlet.

- outlet velocity = 2.98 m/s. (Velocity reduction = 57%)

Design Example: Increased Resistance for a Box Culvert (CU)

Design a concrete box culvert for increased roughness using bottom roughness elements. Determine if the outlet velocity is less than 10 ft/s. Given:

Q = 100 ft³/s
D = 4.0 ft
B = 4.0 ft
n = 0.013
S_o = 0.06 ft/ft

Solution

Step 1. Verify the culvert is in inlet control. In this case the culvert is in inlet control.

Step 2. Compute normal flow conditions. Using trial and error:

y_n = 1.11 ft

V_n = 22.5 ft/s (Since this is greater than 10 ft/s, energy dissipation is required.)

Step 3. Select initial design scale ratios. Determine Manning's n value for the roughened section of culvert for estimating the mitigated depth and velocity.

Try L/h = 10 and h/R_i = 0.3

We will assume that the culvert is not flowing full for computation of the wetted perimeter. (If the culvert did flow full for this computation, it will be surcharged when we compute the high Manning's n for the capacity check. Using the sides and bottom of the culvert will provide the lowest (short of full flow), and therefore, conservative value of n for estimating velocity in the roughened section. This assumption may be revised in subsequent iterations.

P = 2D+B = 2(4.0)+4.0 = 12.0 ft

L_r = B = 4.0 ft (bottom roughness only)

Using Equation 7.23 and n = 0.013 (maximum value is 0.015):

$$n_{LOW} = n \left\{ \frac{2}{3}\left(1+16.7\left(\frac{L_r}{P}\right)\right)^{\frac{1}{2}} + \frac{1}{3}\left(\left(1-\frac{L_r}{P}\right) + \frac{70.6\left(\frac{L_r}{P}\right)}{\left(2\log\left(\frac{R_i}{h}\right)+0.194\right)^2}\right)^{\frac{1}{2}} \right\} = 0.040$$

Step 4. Compute mitigated depth and velocity and roughness height, h. Check mitigated velocity against design criteria. Using trial and error:

y_i = 2.54 ft (note that culvert is not flowing full)

A_i = 2.54(4.0) = 10.17 ft^2

P_i = 2(2.54)+4.0 = 9.083 ft

R_i = 10.17/9.083 = 1.119 ft

h = (h/R_i)(R_i) = 0.3(1.119) = 0.336 ft (round to 0.34 ft)

V_i = 9.84 ft/s

Since this velocity is less than or equal to 10 ft/s, the dissipation design is satisfactory. However, we must check the culvert capacity using the high estimate of Manning's n from Equation 7.22b. Two flow limiting scenarios exist.

First, assume the culvert is flowing nearly full. Using Equation 7.22b and n = 0.013 (maximum value is 0.015):

$$n_{IR,HIGH} = n\left[1+390\left(\frac{h}{L}\right)\left(\frac{L_r}{P}\right)\right]^{\frac{1}{2}} = 0.013\left[1+390(0.1)\frac{4.0}{12.0}\right]^{\frac{1}{2}} = 0.049$$

Area and hydraulic radius are calculated as:

y = D – h = 4.0 - 0.34 = 3.66 ft

A = yB = 3.66(4.0) = 14.64 ft^2

P = 2y+B = 2(3.66)+4.0 = 11.32 ft

R = A/P = 14.64/11.32 = 1.293 ft

Using Manning's Equation,

$$Q = \frac{1.49}{n} AR^{\frac{2}{3}}S^{\frac{1}{2}} = \frac{1.49}{0.049} 14.64(1.293)^{\frac{2}{3}}(0.06)^{\frac{1}{2}} = 129 \text{ ft}^3/s$$

Q is greater than the design flow using this scenario.

Second, assume the culvert is flowing full. Using Equation 7.22b and n = 0.013 (maximum value is 0.015):

$$n_{IR,HIGH} = n\left[1+390\left(\frac{h}{L}\right)\left(\frac{L_r}{P}\right)\right]^{\frac{1}{2}} = 0.013\left[1+390(0.1)\frac{4.0}{16.0}\right]^{\frac{1}{2}} = 0.043$$

Area and hydraulic radius are calculated as:

A = yB = 3.66(4.0) = 14.64 ft^2

P = 2(y+B) = 2(3.66+4.0) = 15.32 ft

R = A/P = 14.64/15.32 = 0.956 ft

Using Manning's Equation,

$$Q = \frac{1.49}{n} AR^{\frac{2}{3}}S^{\frac{1}{2}} = \frac{1.49}{0.043} 14.64(0.956)^{\frac{2}{3}}(0.06)^{\frac{1}{2}} = 121 \, ft^3/s$$

Since both flows are greater than the design flow of 100 ft^3/s we know that the design is acceptable. If this had not been the case, a larger culvert barrel could be evaluated going back to step 3 or tumbling flow or another type of dissipator could be considered.

Step 5. Complete sizing the element heights, element spacing, and other design features.

Spacing between roughness elements is calculated to be:

L = (L/h)h = (10)0.34 = 3.4 ft

Number of rows of roughness elements is estimated using Equation 7.24. First calculate:

A$_f$ = B(h) = 4.0(0.34) = 1.36 ft^2

V$_w$ = (V$_n$+V$_i$)/6 = (22.5+9.84)/6 = 5.39 ft/s

$$N = \frac{gB(y_n^2 - y_i^2) + 2Q(V_n - V_i)}{C_D A_f V_w^2} = \frac{32.2(4.0)(1.11^2 - 2.54^2) + 2(100)(22.5 - 9.84)}{1.9(1.36)5.39^2} = 24.8$$

Round up to the next whole number, N = 25

Design summary:

- 25 rows of roughness elements, h = 0.34 ft

- length of roughened section = 81.6 ft. The last roughness element should be no less than L/2 from the culvert outlet.

- outlet velocity = 9.84 ft/s. (Velocity reduction = 56%)

7.3   USBR TYPE IX BAFFLED APRON

Peterka (1978) has described the design process for a baffled apron that makes use of roughness elements on the floor of a box culvert or chute as shown in Figure 7.9. The roughness elements, referred to as baffles or blocks, perturb the flow pattern such that flow slows as it approaches each block and then accelerates as it passes each block and approaches the next row. By placing the baffles from the top of the culvert or chute to the bottom, the baffles prevent excessive acceleration of flows regardless of the total drop height. Based on model studies reported by Peterka, the baffled apron design produces velocities at the bottom of the apron equal to no more than one-third of the critical velocity if the design guidance is followed. This approach works satisfactorily with or without downstream tailwater and is generally not susceptible to trash or debris accumulation.

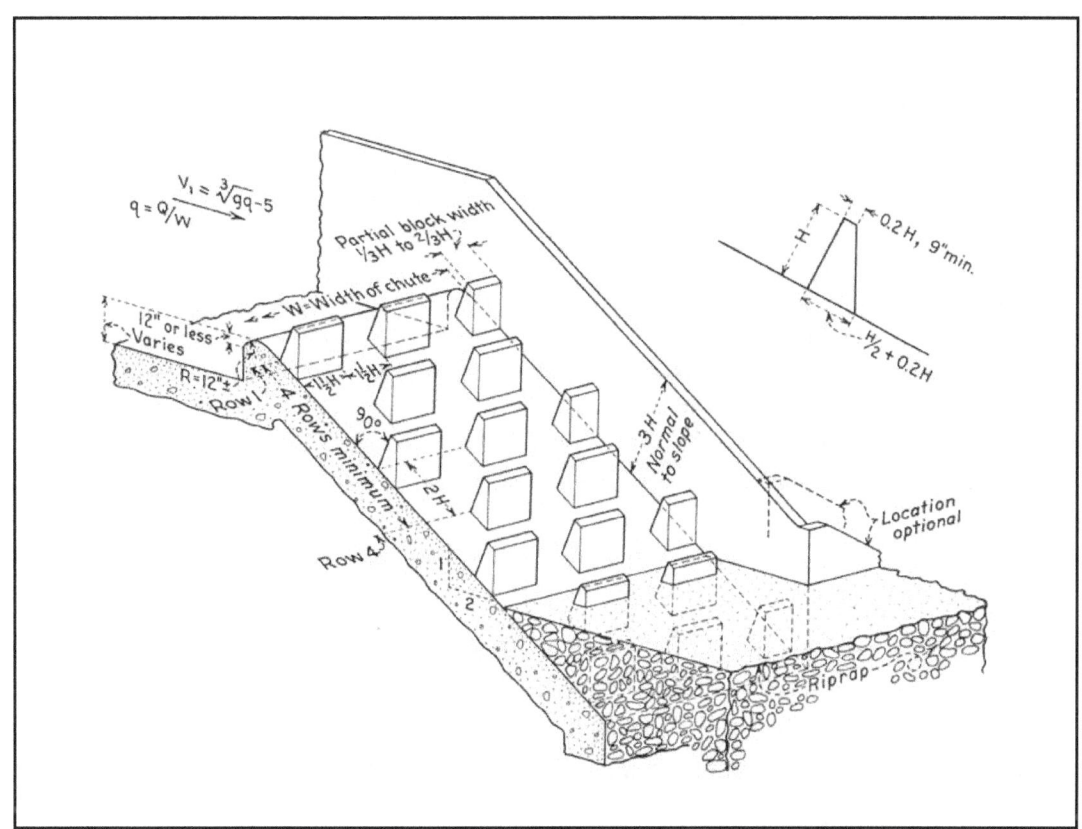

Figure 7.9. USBR Type IX Baffled Apron (Peterka, 1978)

The USBR Type IX baffled apron is limited to the following site conditions and design limits:

1. Culvert/chute slopes of no greater than 50 percent (1:2) and no less than 25 percent (1:4).

2. Unit discharge less than or equal to 5.6 $m^3/s/m$ (60 $ft^3/s/ft$).

3. Approach velocity less than critical velocity (Froude number prior to drop less than 1).

The baffled apron is not a device intended to slow excessive approach velocity, but to prevent excessive acceleration during the vertical drop. According to Peterka (1978), the recommended approach velocity is 1.5 m/s (5 ft/s) less than critical velocity. Velocities near or above critical velocity tend to cause the flow to be thrown into the air after striking the first row of baffles and jumping past the first two or three baffle rows. This is of particular concern for relatively short aprons. One strategy for reducing the approach velocity is providing a recessed approach prior to the entrance of the apron as shown in Figure 7.9.

Another key design element is the selection of the baffle dimensions (height, width, and spacing). Based on model testing and prototype observations by Peterka, baffle height, H, should be about 0.8 times the critical depth, y_c. The height may be increased as high as $0.9y_c$, but should not be less than $0.8y_c$.

7-33

As shown in Figure 7.9, the baffle widths and horizontal baffle spacing should be equal to 1.5H, but not less than H. Each row should be alternating and partial blocks will likely be necessary at the culvert/chute walls.

Longitudinal spacing between rows is based on an assumed apron slope at the maximum slope of 50%. Under these conditions, successive rows of baffles are placed 2H apart as measured along the slope. For baffles less than 0.9 m (3 ft) in height, the row spacing may be greater, than 2H, but not exceeding 1.8 m (6 ft). For apron slopes less than 50 percent, the spacing along the apron slope may be increased such that the vertical drop between baffle rows is 0.89H.

Four rows of baffles are required to establish full control of the flow, although fewer rows have been successful. As shown in Figure 7.9, the chute is generally extended below the bed level of the downstream channel with the lower row of baffles buried to control scour. Riprap consisting of 6 to 12-inch rock should be placed at the downstream ends of the sidewalls to prevent turbulence from undermining the walls, but should not extend appreciably into the channel.

The design procedure for the USBR Type IX baffled apron may be summarized in the following steps:

Step 1. Compute normal flow conditions in the culvert/chute to determine if the discharge conditions at the outlet require mitigation.

Step 2. Verify that the approach flow conditions are acceptable.

Step 3. Compute discharge velocity. If this velocity meets criteria, the USBR Type IX may be an appropriate energy dissipation approach.

Step 4. Size the baffle height, spacing, and other design features.

Design Example: USBR Type IX Baffled Apron (SI)

Design a USBR Type IX baffled apron for energy dissipation in a concrete box culvert with an overall vertical drop of 8 m. Determine if the outlet velocity is less than 3 m/s. Given:

Q = 2.8 m³/s

D = 1.2 m

B = 1.2 m

Approach channel:

n = 0.020

S_o = 0.01 m/m

Box/Chute:

n = 0.013

S_o = 0.333 m/m

Solution

Step 1. Compute normal flow conditions in the box/chute to determine if a baffled apron is needed. Using trial and error and the geometric relations of Equations 7.11, 7.12, and 7.13:

y_n = 0.19 m

V_n = 12.2 m/s (Since this is greater than 3 m/s, energy dissipation is required.)

Step 2. Verify that the approach flow conditions are acceptable. Estimate the approach flow conditions using trial and error and Manning's Equation:

$y_n = 0.92$ m

$V_n = 2.54$ m/s

This approach velocity must be compared with critical velocity computed from:

$q = Q/B = 2.8/1.2 = 2.33$ m^3/s/m

$V_c = (qg)^{1/3} = (2.33(9.8))^{1/3} = 2.84$ m/s

Since $V_c > V_n$ (approach velocity) and the unit discharge is less than 5.6 m^3/s/m, the baffled apron is applicable.

Step 3. Compute discharge velocity. Discharge velocity is no more than one-third of critical velocity; in this case discharge velocity = (2.84 m/s)/3 = 0.95 m/s. This is well below the design requirement of 3 m/s, therefore, the USBR Type IX may be an appropriate energy dissipation approach.

Step 4. Size the baffle height, spacing, and other design features. Critical depth must be computed first.

$y_c = q/V_c = 2.33/2.84 = 0.82$ m

Baffle height, $H = 0.8y_c = 0.8 (0.82) = 0.66$ m

Baffle width $= 1.5H = 1.5 (0.66) = 0.99$ m

Baffle spacing (horizontal) $= 1.5H = 0.99$ m

Vertical drop between baffle rows, $\Delta h = 0.89H = 0.89 (0.66) = 0.59$ m

Spacing (measured along apron) between baffle rows, $L = 1.87$ m

Minimum sidewall height $= 3H = 3(0.66) = 1.98$ m (Since this is a closed box with a rise = 1.2 m, splash is not a concern.)

Design summary:

- 13 rows of baffles, h = 0.66 m

- length of apron = 25.3 m

- outlet velocity = 0.95 m/s. (Velocity reduction = 92%)

Design Example: USBR Type IX Baffled Apron (CU)

Design a USBR Type IX baffled apron for energy dissipation in a concrete box culvert with an overall vertical drop of 26.2 ft. Determine if the outlet velocity is less than 10 ft/s. Given:

Q   = 100 ft^3/s

D   = 4.0 ft

B   = 4.0 ft

Approach channel:

n   = 0.020

S_o = 0.01 ft/ft

Box/Chute:

n = 0.013

S_o = 0.333 ft/ft

Solution

Step 1. Compute normal flow conditions in the box/chute to determine if a baffled apron is needed. Using trial and error and the geometric relations of Equations 7.11, 7.12, and 7.13:

y_n = 0.62 ft

V_n = 40.2 ft/s (Since this is greater than 10 ft/s, energy dissipation is required.)

Step 2. Verify that the approach flow conditions are acceptable. Estimate the approach flow conditions using trial and error and Manning's Equation:

y_n = 2.98 ft

V_n = 8.4 ft/s

This approach velocity must be compared with critical velocity computed from:

$q = Q/B = 100/4 = 25$ ft³/s/ft

$V_c = (qg)^{1/3} = (25.0(32.2))^{1/3} = 9.31$ ft/s

Since $V_c > V_n$ (approach velocity) and the unit discharge is less than 60 ft³/s/ft, the baffled apron is applicable.

Step 3. Compute discharge velocity. Discharge velocity is no more than one-third of critical velocity; in this case discharge velocity = (9.3 ft/s)/3 = 3.1 ft/s. This is well below the design requirement of 10 ft/s, therefore, the USBR Type IX may be an appropriate energy dissipation approach.

Step 4. Size the baffle height, spacing, and other design features. Critical depth must be computed first.

$y_c = q/V_c = 25/9.3 = 2.69$ ft

Baffle height, $H = 0.8y_c = 0.8(2.69) = 2.15$ ft

Baffle width = $1.5H = 1.5(2.15) = 3.23$ ft

Baffle spacing (horizontal) = $1.5H = 3.23$ ft

Vertical drop between baffle rows, $\Delta h = 0.89H = 0.89(2.15) = 1.91$ ft

Spacing (measured along apron) between baffle rows, $L = 6.08$ ft

Minimum sidewall height = $3H = 3(2.15) = 6.45$ ft (Since this is a closed box with a rise = 4.0 ft, splash is not a concern.)

Design summary:

- 13 rows of baffles, h = 2.15 ft

- length of apron = 82.9 ft

- outlet velocity = 3.1 ft/s. (Velocity reduction = 92%)

7.4 BROKEN-BACK CULVERTS/OUTLET MODIFICATION

An alternative to installing a steeply sloped culvert is to break the slope into a steeper portion near the inlet followed by a horizontal runout section. This configuration is referred to as a broken-back culvert and may be considered another internal (integrated) energy dissipator strategy if it is designed so that a hydraulic jump occurs in the runout section to dissipate energy. Figure 7.10 illustrates two cases: a double broken-back culvert, and a single broken-back culvert. In both cases, the exit or runout section is assumed to be horizontal. Under certain conditions of culvert properties and tailwater levels, a hydraulic jump will form in the runout section and reduce the outlet velocity from that associated with a supercritical depth to that associated with a subcritical depth. Modifications to the runout section may be used to induce a hydraulic jump within the culvert.

7.4.1 Broken-back Culvert Hydraulics

A hydraulic jump will form in a channel if either of the following two conditions occurs: (1) the momentum in the tailwater downstream from the culvert exceeds that in the barrel, or (2) the supercritical Froude number in the barrel is reduced to approximately 1.7 in a decelerating flow environment (Chow, 1959).

To solve for the hydraulics of a broken-back culvert, a gradually varied water surface profile is calculated within the culvert from the entrance down to the flat runout section. This supercritical profile is compared to the tailwater elevation and the sequent depth to determine whether or not a hydraulic jump will occur in the runout section.

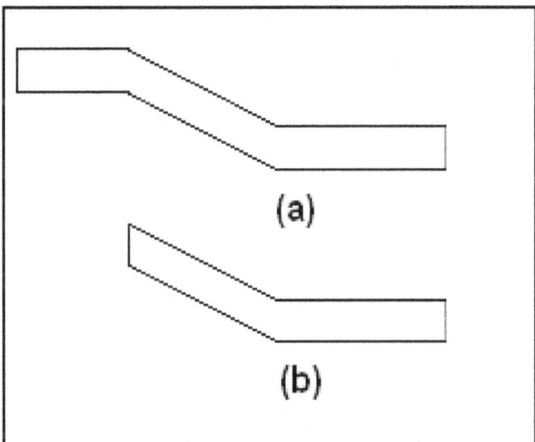

Figure 7.10. Elevation view of (a) Double and (b) Single Broken-back Culvert

If a jump does occur, the design should ensure that the jump is confined to the runout section. First, the location of the jump referenced from the beginning of the runout section must be determined. This is accomplished by computing the profiles upstream and downstream of the jump to find where the momenta are the same, that is, where the jump is located.

Second, the length of the jump is estimated. The sum of these two quantities must be less than the runout section length. The jump length for a rectangular culvert or channel is given by:

$$L = 220(y_1)\tanh\left(\frac{Fr-1}{22}\right) \tag{7.25}$$

where,

L = jump length, m (ft)
y_1 = supercritical flow depth, m (ft)
Fr = supercritical Froude number

For a circular barrel, the jump length is equal to six times the subcritical sequent depth, where the sequent depth is computed using an empirical formulation (French, 1985).

The hydraulic analysis of broken-back culverts has been simplified by the computer application entitled Broken-back Computer Analysis Program, or BCAP (Hotchkiss et al. 2004).

The recommended design is limited to the following conditions:

1. Slope of the steep section must be less than or equal to 1.4:1 (V:H)

2. Hydraulic jump must be completed within the culvert barrel

For situations where the runout section is too short and/or there is insufficient tailwater for a jump to be completed (or initiated) within the barrel, modifications may be made to the outlet that will induce a jump. Two modification alternatives are presented in the following sections.

7.4.2 Outlet Weir

Placing a weir near the outlet of a culvert will induce a hydraulic jump under certain flow conditions (see Figure 7.11). The weir spans the width of a box culvert and is located approximately 3 m (10 ft) upstream from the culvert outlet. This location will facilitate debris removal from the upstream side of the weir. Drain holes in the weir prevent water from standing upstream. The distance L_w is referenced to the break in slope from the more steeply sloped section of the culvert. The rise of the culvert must be greater than y_2.

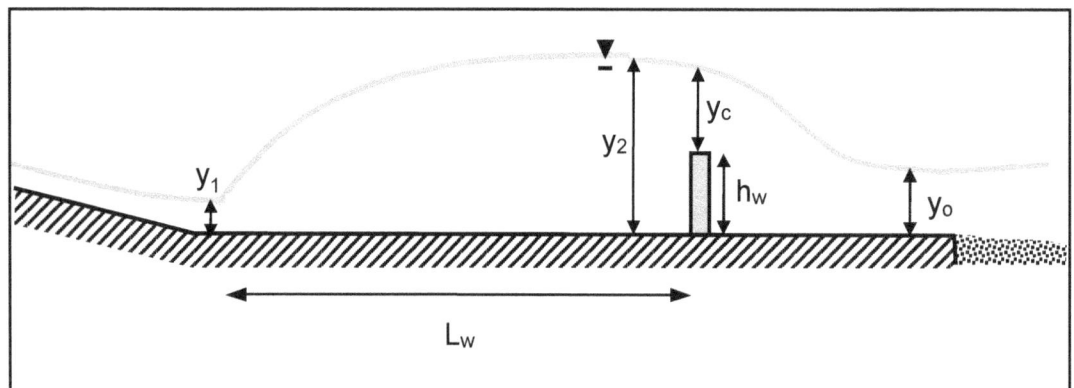

Figure 7.11. Weir Placed near Outlet of Box Culvert

Weirs of this nature are intended for use in conjunction with broken-back culverts, but may be used for chutes. They are placed in the horizontal runout section downstream from the change in slope exiting the steep section a distance to be determined during the design process. The weir is best used when there will be no standing water or design tailwater downstream from the culvert. Because flow will pass over the weir without the mitigating effect of tailwater, the flow will pass through critical depth and become supercritical as it approaches the culvert outlet. The need for downstream channel protection will be decreased due to the presence of the weir.

Hotchkiss, et al. (2005) tested conditions similar to those investigated by Forster and Skrinde (1950). Weirs near culvert outlets will induce hydraulic jumps for approach Froude numbers between 2 and 7. Designers interested in this dissipator may also wish to compare with the stilling basin designs found in Chapter 8.

The recommended design is limited to the following conditions:

1. Approach Froude number between 2 and 7
2. Weir heights between $0.7y_1$ and $4.2y_1$
3. Rectangular culverts

The approach hydraulic conditions may be determined for broken-back culverts (see Section 7.4) or for chutes or any other steep approach to a horizontal runout section. However, the design procedure that follows has only been developed for rectangular shapes. Future extensions of the methodology will need to be supported by additional experimental testing.

The procedure makes use of the critical depth and the sequent depth for the hydraulic jump. The critical depth for a rectangular culvert was given earlier by Equation 7.1. The sequent depth is as follows:

$$y_2 = \frac{y_1}{2}\left(\sqrt{1 + 8Fr_1^2} - 1\right) \tag{7.26}$$

where,

y_2 = sequent depth, m (ft)

Design of the weir primarily involves selecting its location and height. The relationship between weir height, approach depth, and Froude number is given by:

$$h = \left(0.0331Fr^2 + 0.4385Fr - 0.6534\right)y \tag{7.27}$$

where,

h_w = weir height, m (ft)
y_1 = depth at the beginning of the runout section, m (ft)

The distance from the break in slope to the weir, approximately equal to the length of the hydraulic jump, is calculated as follows:

$$L_W = 5y_2 \tag{7.28}$$

Equation 7.28 is empirically based on the experimental data. For this reason, and because of its simplicity, it is used in this design procedure rather than Equation 7.25.

To calculate conditions downstream of the weir, near the culvert outlet, it is necessary to solve the energy equation iteratively for the depth downstream from the weir assuming no losses:

$$h_w + y_c + \frac{\left(\dfrac{Q}{By_c}\right)^2}{2g} = y_3 + \frac{\left(\dfrac{Q}{By_3}\right)^2}{2g} \qquad (7.29)$$

where,

y_3 = theoretical depth leaving the culvert, m (ft)

B   = culvert width, m (ft)

Equation 7.29 has two solutions: subcritical and supercritical. The supercritical solution is taken because after passing through critical depth going over the weir the flow will be supercritical. The theoretical depth is adjusted for energy losses from the experimental data of Hotchkiss, et al. (2005):

$$y_o = 1.23y_3 + \alpha \qquad (7.30)$$

where,

y_o = outlet depth, m (ft)

α = constant equal to 0.015 m in SI and 0.05 ft in CU

The corresponding outlet velocity is computed from:

$$V_o = \frac{Q}{By_o} \qquad (7.31)$$

The following design procedure may be used:

Step 1.   Find the depth of flow, y_1, velocity, and Froude number at the beginning of the horizontal runout section. (This can be calculated using BCAP (Hotchkiss et al. 2004), HY8 (FHWA culvert analysis software), or other calculation tool to determine the depth entering the runout section.)

Step 2.   Find critical depth, y_c, using Equation 7.1.

Step 3.   Find the weir height (Equation 7.27) and location (Equation 7.28).

Step 4.   Solve the energy equation (Equation 7.29) iteratively for the depth downstream from the weir.

Step 5.   Compute the outlet depth (Equation 7.30) and velocity (Equation 7.31). Evaluate if energy dissipation is sufficient. Check culvert height for sufficient clearance.

Design Example: Outlet Weir in a Box Culvert (SI)

Design an outlet weir in the runout section of a RCB and determine the outlet conditions. The approach depth to the runout section is $y_1 = 0.375$ m. Given:

Q   = 14.2 m³/s

D   = 2.44 m

B   = 4.3 m

Solution

Step 1. Find the depth of flow, y_1, velocity, and Froude number at the beginning of the horizontal runout section. y_1 was given.

$$V_1 = Q/(By_1) = 14.2/((4.3)(0.375)) = 8.81 \text{ m/s}$$

$$Fr_1 = \frac{V_1}{\sqrt{gy_1}} = \frac{8.81}{\sqrt{9.81(0.375)}} = 4.6$$

Step 2. Find critical depth, y_c, using Equation 7.1. Unit discharge, $q = Q/B = 14.2/4.3 = 3.302 \text{ m}^2/\text{s}$.

$$y_c = \left(\frac{q^2}{g}\right)^{\frac{1}{3}} = \left(\frac{3.302^2}{9.81}\right)^{\frac{1}{3}} = 1.036 \text{ m}$$

Step 3. Find the weir height (Equation 7.27) and location (Equations 7.27 and 7.28).

$$h_w = \left(0.0331Fr_1^2 + 0.4385Fr_1 - 0.6534\right)y_1 = \left(0.0331(4.6)^2 + 0.4385(4.6) - 0.6534\right)0.375 = 0.774 \text{ m}$$

$$y_2 = \frac{y_1}{2}\left(\sqrt{1 + 8Fr_1^2} - 1\right) = \frac{0.375}{2}\left(\sqrt{1 + 8(4.6)^2} - 1\right) = 2.255 \text{ m}$$

$$L_W = 5y_2 = 5(2.255) = 11.27 \text{ m} \quad \text{(round to 11.3 m)}$$

Step 4. Solve the energy equation (Equation 7.29) iteratively for the depth downstream from the weir. From this trial and error process, $y_3 = 0.561$ m

Step 5. Compute the outlet depth (Equation 7.30) and velocity (Equation 7.31). Evaluate if energy dissipation is sufficient.

$$y_o = 1.23y_3 + \alpha = 1.23(0.561) + 0.015 = 0.705 \text{ m}$$

$$V_o = \frac{Q}{By_o} = \frac{14.2}{4.3(0.705)} = 4.7 \text{ m/s}$$

If this velocity is acceptable, then the weir design is appropriate. Also, verify that the depth inside the culvert does not touch the top of the culvert. In this case, the rise of the culvert (2.44 m) is higher than the jump height (2.25 m) and the design is acceptable.

Design Example: Outlet Weir in a Box Culvert (CU)

Design an outlet weir in the runout section of a RCB and determine the outlet conditions. The approach depth to the runout section is $y_1 = 1.23$ ft. Given:

Q = 500 ft³/s
D = 8.0 ft
B = 14.0 ft

Solution

Step 1. Find the depth of flow, y_1, velocity, and Froude number at the beginning of the horizontal runout section. y_1 was given.

$$V_1 = Q/(By_1) = 500/((14.0)(1.23)) = 29.0 \text{ ft/s}$$

$$Fr_1 = \frac{V_1}{\sqrt{gy_1}} = \frac{29.0}{\sqrt{32.2(1.23)}} = 4.6$$

Step 2. Find critical depth, y_c, using Equation 7.1. Unit discharge, $q = Q/B = 500/14.0 = 35.71 \text{ ft}^2/\text{s}$.

$$y_c = \left(\frac{q^2}{g}\right)^{\frac{1}{3}} = \left(\frac{35.71^2}{32.2}\right)^{\frac{1}{3}} = 3.41 \text{ ft}$$

Step 3. Find the weir height (Equation 7.27) and location (Equations 7.27 and 7.28).

$$h_w = \left(0.0331Fr_1^2 + 0.4385Fr_1 - 0.6534\right)y_1 = \left(0.0331(4.6)^2 + 0.4385(4.6) - 0.6534\right)1.23 = 2.54 \text{ ft}$$

$$y_2 = \frac{y_1}{2}\left(\sqrt{1 + 8Fr_1^2} - 1\right) = \frac{1.23}{2}\left(\sqrt{1 + 8(4.6)^2} - 1\right) = 7.43 \text{ ft}$$

$$L_w = 5y_2 = 5(7.43) = 37.15 \text{ ft} \quad \text{(round to 37.2 ft)}$$

Step 4. Solve the energy equation (Equation 7.29) iteratively for the depth downstream from the weir. From this trial and error process, $y_3 = 1.847$ ft

Step 5. Compute the outlet depth (Equation 7.30) and velocity (Equation 7.31). Evaluate if energy dissipation is sufficient.

$$y_o = 1.23y_3 + \alpha = 1.23(1.847) + 0.05 = 2.32 \text{ ft}$$

$$V_o = \frac{Q}{By_o} = \frac{500}{14.0(2.32)} = 15.4 \text{ ft/s}$$

If this velocity is acceptable, then the weir design is appropriate. Also, verify that the depth inside the culvert does not touch the top of the culvert. In this case, the rise of the culvert (8.0 ft) is higher than the jump height (7.43 ft) and the design is acceptable.

7.4.3 Outlet Drop Followed by a Weir

A drop in the culvert invert followed by a weir is shown in Figure 7.12. As with the weir near the culvert outlet (Section 7.5), this installation is intended to be used in conjunction with a broken-back culvert or chute where a steeply sloped section terminates with a horizontal runout section. The location of the drop beyond the break in slope from the steep barrel, L_d, is about 1.5 m (5 ft). The drop effectively decreases the slope of the steep culvert section, while the weir induces a hydraulic jump between the drop and weir. The drop may also be used if the height of the hydraulic jump for the design in Section 7.5 reaches the top of the culvert.

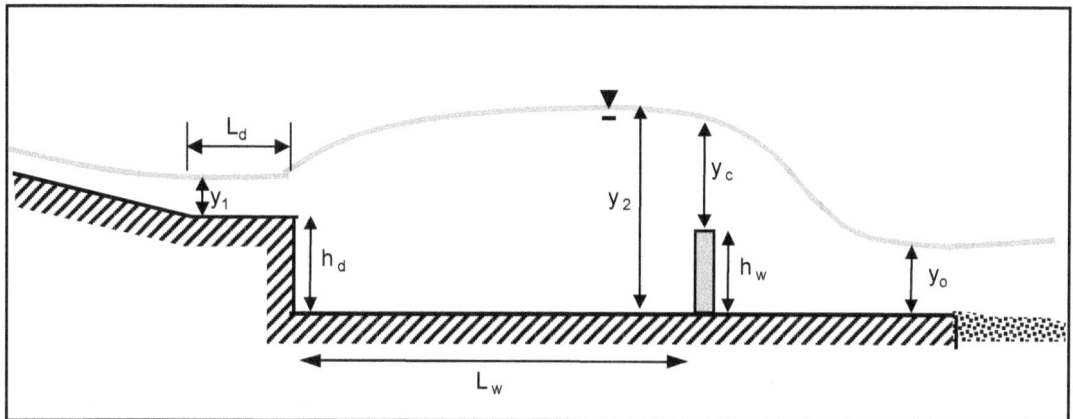

Figure 7.12. Drop followed by Weir

The design procedure is based upon Hotchkiss and Larson (2004) and Hotchkiss, et al. (2005). An extensive set of experiments was performed to define reductions in energy, momentum, and velocity at the culvert outlet due to the presence of a drop followed by a weir. Empirical results relate the drop height, h_d, to approach Froude number and weir height. Designers interested in this design may wish to compare the results with a straight drop structure (Section 11.1) at the end of the culvert or chute.

The recommended design is limited to the following conditions:

1. Approach Froude number between 3.5 and 6
2. Weir height to approach depth (h_w/y_1) between 1.3 and 4
3. Drop height between 60 and 65% of the weir height (suggested, not required)
4. Rectangular culverts

The approach hydraulic conditions, at location 1, may be determined for broken-back culverts (See Section 7.4) or for chutes or any other steep approach to a horizontal runout section. However, the design procedure that follows has only been developed for rectangular shapes. Future extensions of the methodology will need to be supported by additional experimental testing.

The procedure makes use of the critical depth and the sequent depth for the hydraulic jump. The critical depth for a rectangular culvert was given earlier by Equation 7.1. The sequent depth was also provided earlier as Equation 7.26.

Design of the weir primarily involves selecting its location and height. The relationship between weir height, approach depth, and Froude number is given by Equation 7.32. The weir height is also related to the drop height.

$$h_w = \frac{h_d}{h_w}\left(0.9326Fr_1^2 - 6.8218Fr_1 + 14.859\right)y_1 \qquad (7.32)$$

where,

h_w = weir height, m (ft)

h_d = drop height, m (ft)

y_1 = depth at the beginning of the runout section, m (ft)

To solve Equation 7.32, the ratio of h_d/h_w is selected to fall within the range of 0.60 and 0.65. The distance from the drop to the weir is calculated as follows:

$$L_w = 6(y_c + h_w)$$ (7.33)

The quantity $(y_c + h_w)$ approximates the sequent depth downstream from a classic hydraulic jump.

To calculate conditions downstream of the weir, near the culvert outlet, it is necessary to solve the energy equation iteratively for the depth downstream from the weir assuming no losses. Equation 7.29, presented earlier is used for this purpose. The theoretical depth calculated from Equation 7.29 is adjusted for energy losses from the experimental data of Hotchkiss, et al. (2005) using previously presented Equation 7.30.

The following design procedure may be used:

Step 1. Find the depth of flow, y_1, velocity, and Froude number at the beginning of the horizontal runout section. (This can be calculated using BCAP (Hotchkiss et al. 2004), HY8 (FHWA culvert analysis software), or other calculation tool to determine the depth entering the runout section.)

Step 2. Find critical depth, y_c, using Equation 7.1.

Step 3. Find the weir height (Equation 7.32), weir location (Equation 7.33), and drop height.

Step 4. Solve the energy equation (Equation 7.29) iteratively for the depth downstream from the weir.

Step 5. Compute the outlet depth (Equation 7.30) and velocity (Equation 7.31). Evaluate if energy dissipation is sufficient. Check culvert height for sufficient clearance.

Design Example: Drop and Outlet Weir in a Box Culvert (SI)

Design an outlet weir in the runout section of a RCB and determine the outlet conditions. The approach depth to the runout section is $y_1 = 0.375$ m. Given:

Q = 14.2 m³/s

D = 2.44 m

B = 4.3 m

Solution

Step 1. Find the depth of flow, y_1, velocity, and Froude number at the beginning of the horizontal runout section. y_1 was given.

$V_1 = Q/(By_1) = 14.2/((4.3)(0.375)) = 8.81$ m/s

$$Fr_1 = \frac{V_1}{\sqrt{gy_1}} = \frac{8.81}{\sqrt{9.81(0.375)}} = 4.6$$

Step 2. Find critical depth, y_c, using Equation 7.1. Unit discharge, q = Q/B =14.2/4.3 = 3.302 m²/s.

$$y_c = \left(\frac{q^2}{g}\right)^{\frac{1}{3}} = \left(\frac{3.302^2}{9.81}\right)^{\frac{1}{3}} = 1.036 \text{ m}$$

Step 3. Find the weir height (Equation 7.32), weir location (Equations 7.27 and 7.33), and drop height. Select the ratio $h_d/h_w = 0.64$.

$$h_w = \frac{h_d}{h_w}\left(0.9326 Fr_1^2 - 6.8218 Fr_1 + 14.859\right)y_1$$

$$= 0.64\left(0.9326(4.6)^2 - 6.8218(4.6) + 14.859\right)0.375 = 0.771 \text{ m}$$

$$y_2 = \frac{y_1}{2}\left(\sqrt{1 + 8Fr_1^2} - 1\right) = \frac{0.375}{2}\left(\sqrt{1 + 8(4.6)^2} - 1\right) = 2.255 \text{ m}$$

$L_W = 6(y_c + h_w) = 6(1.04 + 0.771) = 10.87$ (round to 10.9 m)

$h_d = 0.64(h_w) = 0.64(0.771) = 0.49$ m

Step 4. Solve the energy equation (Equation 7.29) iteratively for the depth downstream from the weir. From this trial and error process, $y_3 = 0.568$ m

Step 5. Compute the outlet depth (Equation 7.30) and velocity (Equation 7.31). Evaluate if energy dissipation is sufficient.

$y_o = 1.23y_3 + \alpha = 1.23(0.568) + 0.015 = 0.714$ m

$$V_o = \frac{Q}{By_o} = \frac{14.2}{4.3(0.714)} = 4.7 \text{ m/s}$$

If this velocity is acceptable, then the weir design is appropriate. Also, verify that the depth inside the culvert does not touch the top of the culvert. In this case, the rise of the culvert (2.44 m) is higher than the jump height less the drop (2.25 –0.49 m) and the design is acceptable.

Design Example: Drop and Outlet Weir in a Box Culvert (CU)

Design an outlet weir in the runout section of a RCB and determine the outlet conditions. The approach depth to the runout section is $y_1 = 1.23$ ft. Given:

Q = 500 ft³/s

D = 8.0 ft

B = 14.0 ft

Solution

Step 1. Find the depth of flow, y_1, velocity, and Froude number at the beginning of the horizontal runout section. y_1 was given.

$$V_1 = Q/(By_1) = 500/((14.0)(1.23)) = 29.0 \text{ ft/s}$$

$$Fr_1 = \frac{V_1}{\sqrt{gy_1}} = \frac{29.0}{\sqrt{32.2(1.23)}} = 4.6$$

Step 2. Find critical depth, y_c, using Equation 7.1. Unit discharge, $q = Q/B = 500/14.0 = 35.71 \text{ ft}^2/\text{s}$.

$$y_c = \left(\frac{q^2}{g}\right)^{\frac{1}{3}} = \left(\frac{35.71^2}{32.2}\right)^{\frac{1}{3}} = 3.41 \text{ ft}$$

Step 3. Find the weir height (Equation 7.32), weir location (Equations 7.27 and 7.33), and drop height. Select the ratio $h_d/h_w = 0.64$.

$$h_w = \frac{h_d}{h_w}\left(0.9326Fr_1^2 - 6.8218Fr_1 + 14.859\right)y_1$$

$$= 0.64\left(0.9326(4.6)^2 - 6.8218(4.6) + 14.859\right)1.23 = 2.53 \text{ ft}$$

$$y_2 = \frac{y_1}{2}\left(\sqrt{1 + 8Fr_1^2} - 1\right) = \frac{1.23}{2}\left(\sqrt{1 + 8(4.6)^2} - 1\right) = 7.43 \text{ ft}$$

$$L_W = 6(y_c + h_w) = 6(3.41 + 2.53) = 35.64 \text{ ft (round to 35.6 ft)}$$

$$h_d = 0.64(h_w) = 0.64(2.53) = 1.62 \text{ ft}$$

Step 4. Solve the energy equation (Equation 7.29) iteratively for the depth downstream from the weir. From this trial and error process, $y_3 = 1.86$ ft

Step 5. Compute the outlet depth (Equation 7.30) and velocity (Equation 7.31). Evaluate if energy dissipation is sufficient.

$$y_o = 1.23y_3 + \alpha = 1.23(1.86) + 0.05 = 2.34 \text{ ft}$$

$$V_o = \frac{Q}{By_o} = \frac{500}{14.0(2.34)} = 15.3 \text{ ft/s}$$

If this velocity is acceptable, then the weir design is appropriate. Also, verify that the depth inside the culvert does not touch the top of the culvert. In this case, the rise of the culvert (8.0 ft) is higher than the jump height less the drop (7.43 − 1.62 ft) and the design is acceptable.

CHAPTER 8: STILLING BASINS

Stilling basins are external energy dissipators placed at the outlet of a culvert, chute, or rundown. These basins are characterized by some combination of chute blocks, baffle blocks, and sills designed to trigger a hydraulic jump in combination with a required tailwater condition. With the required tailwater, velocity leaving a properly designed stilling basin is equal to the velocity in the receiving channel.

Depending on the specific design, they operate over a range of approach flow Froude numbers from 1.7 to 17 as summarized in Table 8.1. This chapter includes the following stilling basins: USBR Type III, USBR Type IV, and SAF. The United States Bureau of Reclamation (USBR) basins were developed based on model studies and evaluation of existing basins (USBR, 1987). The St. Anthony Falls (SAF) stilling basin is based on model studies conducted by the Soil Conservation Service at the St. Anthony Falls Hydraulic Laboratory of the University of Minnesota (Blaisdell, 1959).

Table 8.1. Applicable Froude Number Ranges for Stilling Basins

Stilling Basin	Minimum Approach Froude Number	Maximum Approach Froude Number
USBR Type III	4.5	17
USBR Type IV	2.5	4.5
SAF	1.7	17

The selection of a stilling basin depends on several considerations including hydraulic limitations, constructibility, basin size, and cost. The design examples in this chapter all use the identical site conditions to provide a comparison between the size of basins and a free hydraulic jump basin for one case. Table 8.2 summarizes the results of these examples with the incoming Froude number, the required tailwater at the exit of the basin along with basin length and depth. For this example, the SAF stilling basin results in the shortest and shallowest basin. Details of the design procedures and this design example are found in the following sections.

Table 8.2. Example Comparison of Stilling Basin Dimensions

Basin Type[1]	Froude Number	Required Tailwater[3], m (ft)	Basin Length, m (ft)	Basin Depth, m (ft)
Free jump	7.6	3.1 (10.1)	33.7 (109.2)	4.8 (15.5)
USBR Type III	6.9	3.0 (9.6)	20.6 (67.3)	3.8 (12.5)
USBR Type IV[2]	8.0	3.5 (11.2)	38.1 (121.8)	5.5 (17.4)
SAF	6.1	2.4 (7.9)	12.4 (39.7)	2.7 (8.6)

[1]Based on a 3 m by 1.8 m (10 ft by 6 ft) box culvert at a design discharge of 11.8 m^3/s (417 ft^3/s). All basins have a constant width equal to the culvert width. Detailed description of the example is found in Section 8.1.

[2]The USBR Type IV approach Froude number is outside of the recommended range, but was included for comparison.

[3]Required tailwater influences basin depth. Velocity leaving each of these basins is the same and depends on the tailwater channel.

8.1 EXPANSION AND DEPRESSION FOR STILLING BASINS

As explained in Chapter 4, the higher the Froude number at the entrance to a basin, the more efficient the hydraulic jump and the shorter the resulting basin. To increase the Froude number as the water flows from the culvert to the basin, an expansion and depression is used as is shown in Figure 8.1. The expansion and depression converts depth, or potential energy, into kinetic energy by allowing the flow to expand, drop, or both. The result is that the depth decreases and the velocity and Froude number increase.

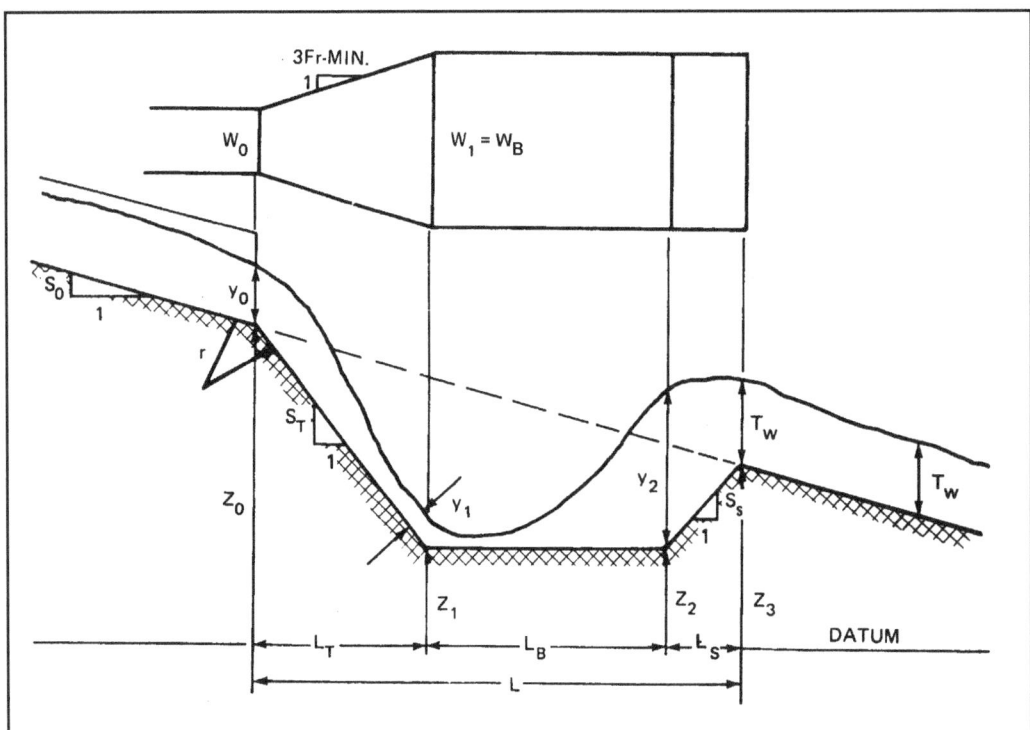

Figure 8.1. Definition Sketch for Stilling Basin

The Froude number used to determine jump efficiency and to evaluate the suitability of alternative stilling basins as described in Table 8.1 is defined in Equation 8.1.

$$Fr_1 = \frac{V_1}{\sqrt{gy_1}} \qquad (8.1)$$

where,

Fr_1 = Froude number at the entrance to the basin
V_1 = velocity entering the basin, m/s (ft/s)
y_1 = depth entering the basin, m (ft)
g = acceleration due to gravity, m/s^2 (ft/s^2)

To solve for the velocity and depth entering the basin, the energy balance is written from the culvert outlet to the basin. Substituting $Q/(y_1W_B)$ for V_1 and solving for Q results in:

$$Q = y_1 W_B \left[2g(z_o - z_1 + y_o - y_1) + V_o^2 \right]^{\frac{1}{2}}$$

(8.2)

where,

W_B = width of the basin, m/s (ft/s)

V_o = culvert outlet velocity, m/s (ft/s)

y_1 = depth entering the basin, m (ft)

y_o = culvert outlet depth, m (ft)

z_1 = ground elevation at the basin entrance, m (ft)

z_o = ground elevation at the culvert outlet, m (ft)

Equation 8.2 has three unknowns y_1, W_B, and z_1. The depth y_1 can be determined by trial and error if W_B and z_1 are assumed. W_B should be limited to the width that a jet would flare naturally in the slope distance L.

$$W_B \leq W_o + \frac{2L_T \sqrt{S_T^2 + 1}}{3Fr_o}$$

(8.3)

where,

L_T = length of transition from culvert outlet to basin, m (ft)

S_T = slope of the transition, m/m (ft/ft)

Fr_o = outlet Froude number

Since the flow is supercritical, the trial y_1 value should start near zero and increase until the design Q is reached. This depth, y_1, is used to find the sequent (conjugate) depth, y_2, using the hydraulic jump equation:

$$y_2 = \frac{Cy_1}{2} \left(\sqrt{1 + 8Fr_1^2} - 1 \right)$$

(8.4)

where,

y_2 = conjugate depth, m (ft)

y_1 = depth approaching the jump, m (ft)

C = ratio of tailwater to conjugate depth, TW/y_2

Fr_1 = approach Froude number

For a free hydraulic jump, C = 1.0. Later sections on the individual stilling basin types provide guidance on the value of C for those basins. For the jump to occur, the value of $y_2 + z_2$ must be equal to or less than $TW + z_3$ as shown in Figure 8.1. If $z_2 + y_2$ is greater than $z_3 + TW$, the basin must be lowered and the trial and error process repeated until sufficient tailwater exists to force the jump.

In order to perform this check, z_3 and the basin lengths must be determined. The length of the transition is calculated from:

$$L_T = \frac{z_o - z_1}{S_T}$$

(8.5)

where,

L_T = length of the transition from the culvert outlet to the bottom of the basin, m (ft)

S_T = slope of the transition entering the basin, m/m (ft/ft)

The length of the basin, L_B, depends on the type of basin, the entrance flow depth, y_1, and the entrance Froude number, Fr_1. Figure 8.2 describes these relationships for the free hydraulic jump as well as several USBR stilling basins.

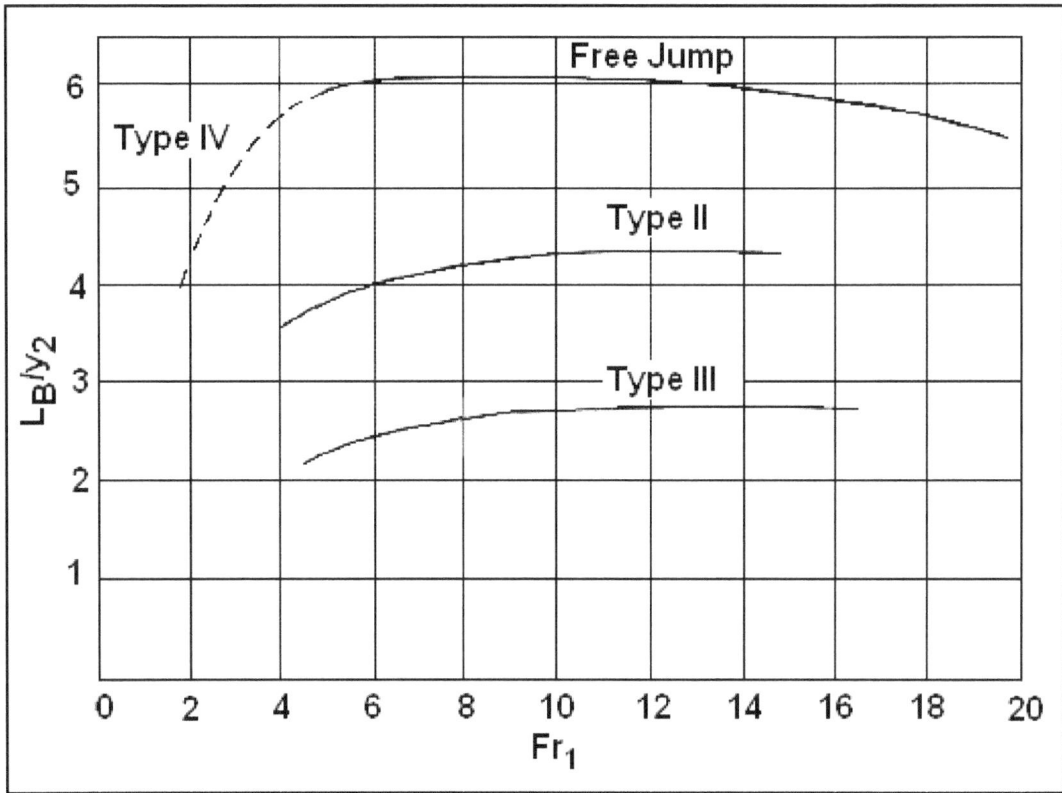

Figure 8.2. Length of Hydraulic Jump on a Horizontal Floor

The length of the basin from the floor to the sill is calculated from:

$$L_S = \frac{L_T(S_T - S_o) - L_B S_o}{S_S + S_o}$$ (8.6)

where,

L_S = length of the basin from the bottom of the basin to the basin exit (sill), m (ft)

S_S = slope leaving the basin, m/m (ft/ft)

The elevation at the entrance to the tailwater channel is then calculated from:

$$z_3 = L_S S_S + z_1$$ (8.7)

where,

z_3 = elevation of basin at basin exit (sill), m (ft)

Figure 8.1 also illustrates a radius of curvature between the culvert outlet and the transition to the stilling basin. If the transition slope is 0.5V:1H or steeper, use a circular curve at the transition with a radius defined by Equation 8.8 (Meshgin and Moore, 1970). It is also advisable to use the same curved transition going from the transition slope to the stilling basin floor.

$$r = \frac{y}{e^{\frac{1.5}{Fr^2}} - 1} \tag{8.8}$$

where,

r   = radius of the curved transition, m (ft)

Fr   = Froude number

y   = depth approaching the curvature, m (ft)

For the curvature between the culvert outlet and the transition, the Froude number and depth are taken at the culvert outlet. For the curvature between the transition and the stilling basin floor, the Froude number and depth are taken as Fr_1 and y_1.

8.2   GENERAL DESIGN PROCEDURE

The design procedure for all of these stilling basins may be summarized in the following steps. Basin specific variations to these steps are discussed in the following sections on each basin.

Step 1. Determine the velocity and depth at the culvert outlet.   For the culvert outlet, calculate culvert brink depth, y_o, velocity, V_o, and Fr_o. For subcritical flow, use Figure 3.3 or Figure 3.4. For supercritical flow, use normal depth in the culvert for y_o. (See HDS 5 (Normann, et al., 2001) for additional information on culvert brink depths.)

Step 2. Determine the velocity and TW depth in the receiving channel downstream of the basin. Normal depth may be determined using Table B.1 or other appropriate technique.

Step 3. Estimate the conjugate depth for the culvert outlet conditions using Equation 8.4 to determine if a basin is needed. Substitute y_o and Fr_o for y_1 and Fr_1, respectively. The value of C is dependent, in part, on the type of stilling basin to be designed. However, in this step the occurrence of a free hydraulic jump without a basin is considered so a value of 1.0 is used. Compare y_2 and TW. If $y_2 <$ TW, there is sufficient tailwater and a jump will form without a basin. The remaining steps are unnecessary.

Step 4. If step 3 indicates a basin is needed ($y_2 >$ TW), make a trial estimate of the basin bottom elevation, z_1, a basin width, W_B, and slopes S_T and S_S. A slope of 0.5 (0.5V:1H) or 0.33 (0.33V:1H) is satisfactory for both S_T and S_s. Confirm that W_B is within acceptable limits using Equation 8.3. Determine the velocity and depth conditions entering the basin and calculate the Froude number. Select candidate basins based on this Froude number.

Step 5. Calculate the conjugate depth for the hydraulic conditions entering the basin using Equation 8.4 and determine the basin length and exit elevation. Basin length and exit elevation are computed using Equations 8.5, 8.6, and 8.7 as well as Figure 8.2. Verify that sufficient tailwater exists to force the hydraulic jump. If the tailwater is insufficient go back to step 4. If excess tailwater exists, the designer may either go on to step 6 or return to step 4 and try a shallower (and smaller) basin.

Step 6. Determine the needed radius of curvature for the slope changes entering the basin using Equation 8.8.

Step 7. Size the basin elements for basin types other than a free hydraulic jump basin. The details for this process differ for each basin and are included in the individual basin sections.

Design Example: Stilling Basin with Free Hydraulic Jump (SI)

Find the dimensions for a stilling basin (see Figure 8.1) with a free hydraulic jump providing energy dissipation for a reinforced concrete box culvert. Given:

Q = 11.8 m^3/s

Culvert

B = 3.0 m

D = 1.8 m

n = 0.015

S_o = 0.065 m/m

z_o = 30.50 m

Downstream channel (trapezoidal)

B = 3.10 m

Z = 1V:2H

n = 0.030

Solution

Step 1. Determine the velocity and depth at the culvert outlet. By trial and error using Manning's Equation, the normal depth is calculated as:

V_o = 8.50 m/s, y_o = 0.463 m

$$Fr_o = \frac{V_o}{\sqrt{gy_o}} = \frac{8.50}{\sqrt{9.81(0.463)}} = 4.0$$

Since the Froude number is greater than 1.0, the normal depth is supercritical and the normal depth is taken as the brink depth.

Step 2. Determine the velocity and depth (TW) in the receiving channel. By trial and error using Manning's Equation or by using Table B.1:

V_n = 4.84 m/s, y_n = TW = 0.574 m

Step 3. Estimate the conjugate depth for the culvert outlet conditions using Equation 8.4. C = 1.0.

$$y_2 = \frac{Cy_0}{2}\left(\sqrt{1+8Fr_0^2}-1\right) = \frac{1.0\,(0.463)}{2}\left(\sqrt{1+8(4.0)^2}-1\right) = 2.4\text{ m}$$

Since y_2 (2.4 m) > TW (0.574 m) a jump will not form and a basin is needed.

Step 4. Since y_2 - TW = 2.64 – 0.574 = 2.07 m, try $z_1 = z_0 - 2.07 = 28.4$ m

Also, choose $W_B = 3.0$ m (no expansion from culvert to basin) and slopes $S_T = 0.5$ and $S_S = 0.5$.

Check W_B using Equation 8.3, but first calculate the transition length from Equation 8.5.

$$L_T = \frac{z_0 - z_1}{S_T} = \frac{30.50 - 28.4}{0.5} = 4.2\text{ m}$$

$$W_B \le W_0 + \frac{2L_T\sqrt{S_T^2+1}}{3Fr_0} = 3.0 + \frac{2(4.2)\sqrt{(0.5)^2+1}}{3(4.0)} = 3.8\text{ m}\,;\ W_B \text{ is OK}$$

By using Equation 8.2 or other appropriate method by trial and error, the velocity and depth conditions entering the basin are:

$V_1 = 10.74$ m/s, $y_1 = 0.366$ m

$$Fr_1 = \frac{V_1}{\sqrt{gy_1}} = \frac{10.74}{\sqrt{9.81(0.366)}} = 5.7$$

Step 5. Calculate the conjugate depth for a free hydraulic jump (C=1) using Equation 8.4.

$$_2 \quad \frac{}{2}\left(\sqrt{1+8Fr_1}-1\right) = \frac{}{2}\left(\sqrt{1+8(5.7)}-1\right) = 2.77\text{ m}$$

From Figure 8.2 basin length, $L_B/y_2 = 6.1$. Therefore, $L_B = 6.1(2.77) = 16.9$ m.

The length of the basin from the floor to the sill is calculated from Equation 8.6:

$$L_S = \frac{L_T(S_T - S_0) - L_B S_0}{S_S + S_0} = \frac{4.2(0.5 - 0.065) - 16.9(0.065)}{0.5 + 0.065} = 1.29\text{ m}$$

The elevation at the entrance to the tailwater channel is from Equation 8.7:

$$z_3 = L_S S_S + z_1 = 1.29(0.5) + 28.4 = 29.05\text{ m}$$

Since $y_2 + z_2$ (2.77+28.4) > z_3 + TW (29.05+ 0.574), tailwater is not sufficient to force a jump in the basin. Go back to step 4.

Step 4 (2nd iteration). Try $z_1 = 25.7$ m. Maintain W_B, S_T, and S_S.

$$L_T = \frac{z_0 - z_1}{S_T} = \frac{30.50 - 25.7}{0.5} = 9.6\text{ m}$$

By using Equation 8.2 or other appropriate method by trial and error, the velocity and depth conditions entering the basin are:

$V_1 = 13.02$ m/s, $y_1 = 0.302$ m

$$Fr_1 = \frac{V_1}{\sqrt{gy_1}} = \frac{13.02}{\sqrt{9.81(0.302)}} = 7.6$$

Step 5 (2nd iteration). Calculate the conjugate depth for a free hydraulic jump (C=1) using Equation 8.4.

$$y_2 = \frac{Cy_1}{2}\left(\sqrt{1+8Fr_1^2}-1\right) = \frac{1.0(0.302)}{2}\left(\sqrt{1+8(7.6)^2}-1\right) = 3.10 \text{ m}$$

From Figure 8.2 basin length, $L_B/y_2 = 6.1$. Therefore, $L_B = 6.1(3.10) = 18.9$ m.

The length of the basin from the floor to the sill is calculated from Equation 8.6:

$$L_S = \frac{L_T(S_T - S_o) - L_B S_o}{S_S + S_o} = \frac{9.6(0.5 - 0.065) - 18.9(0.065)}{0.5 + 0.065} = 5.2 \text{ m}$$

The elevation at the entrance to the tailwater channel is from Equation 8.7:

$$z_3 = L_S S_S + z_1 = 5.2(0.5) + 25.7 = 28.30 \text{ m}$$

Since $y_2 + z_2$ (3.10+25.7) < z_3 + TW (28.30+ 0.574), tailwater is sufficient to force a jump in the basin. Continue on to step 6.

Step 6. For the slope change from the outlet to the transition, determine the needed radius of curvature using Equation 8.8 and the results from step 1.

$$r = \frac{y}{\left(e^{\frac{1.5}{Fr^2}} - 1\right)} = \frac{0.463}{\left(e^{\frac{1.5}{(4.0)^2}} - 1\right)} = 4.71 \text{ m}$$

Step 7. Size the basin elements. Since this is a free hydraulic jump basin, there are no additional elements and the design is complete. The basin is shown in the following sketch.

Total basin length = 9.6 + 18.9 + 5.2 = 33.7 m

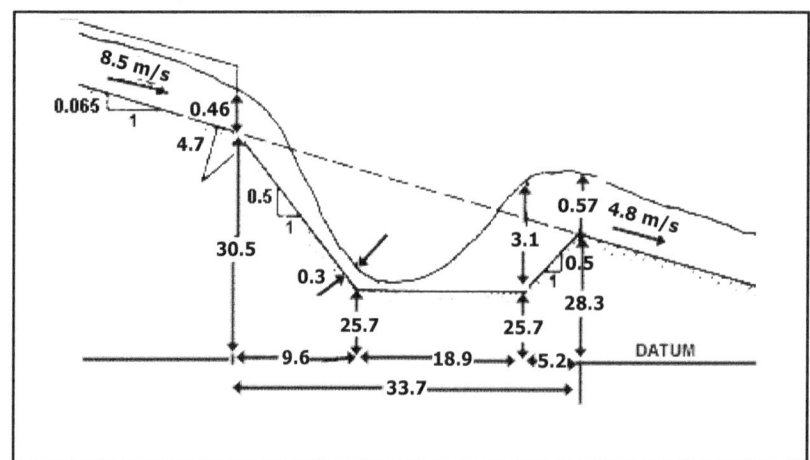

Sketch for Free Hydraulic Jump Stilling Basin Design Example (SI)

Design Example: Stilling Basin with Free Hydraulic Jump (CU)

Find the dimensions for a stilling basin (see Figure 8.1) with a free hydraulic jump providing energy dissipation for a reinforced concrete box culvert. Given:

Q = 417 ft^3/s

Culvert

B = 10.0 ft

D = 6 ft

n = 0.015

S_o = 0.065 ft/ft

z_o = 100 ft

Downstream channel (trapezoidal)

B = 10.2 ft

Z = 1V:2H

n = 0.030

Solution

Step 1.   Determine the velocity and depth at the culvert outlet. By trial and error using Manning's Equation, the normal depth is calculated as:

V_o = 27.8 ft/s, y_o = 1.50 ft

$$Fr_o = \frac{V_o}{\sqrt{gy_o}} = \frac{27.8}{\sqrt{32.2(1.50)}} = 4.0$$

Since the Froude number is greater than 1.0, the normal depth is supercritical and the normal depth is taken as the brink depth.

Step 2.   Determine the velocity and depth (TW) in the receiving channel. By trial and error using Manning's Equation or by using Table B.1:

V_n = 15.9 ft/s, y_n = TW = 1.88 ft

Step 3.   Estimate the conjugate depth for the culvert outlet conditions using Equation 8.4. C = 1.0.

$$y_2 = \frac{Cy_o}{2}\left(\sqrt{1+8Fr_o^2}-1\right) = \frac{1.0(1.50)}{2}\left(\sqrt{1+8(4.0)^2}-1\right) = 7.8 \text{ ft}$$

Since y_2 (7.8 ft) > TW (1.88 ft) a jump will not form and a basin is needed.

Step 4.   Since y_2 - TW = 8.55 – 1.88 = 6.67 ft, try z_1 = z_o –6.67 = 93.3 ft, use 93.

Also, choose W_B = 10.0 ft (no expansion from culvert to basin) and slopes S_T = 0.5 and S_S = 0.5.

Check W_B using Equation 8.3, but first calculate the transition length from Equation 8.5.

$$L_T = \frac{z_o - z_1}{S_T} = \frac{100 - 93}{0.5} = 14 \text{ ft}$$

8-9

$$W_B \leq W_o + \frac{2L_T\sqrt{S_T^2 + 1}}{3Fr_o} = 10.0 + \frac{2(14)\sqrt{(0.5)^2 + 1}}{3(4.0)} = 12.6 \text{ ft}; \; W_B \text{ is OK}$$

By using Equation 8.2 or other appropriate method by trial and error, the velocity and depth conditions entering the basin are:

$V_1 = 35.3$ ft/s, $y_1 = 1.18$ ft

$$Fr_1 = \frac{V_1}{\sqrt{gy_1}} = \frac{35.3}{\sqrt{32.2(1.18)}} = 5.7$$

Step 5. Calculate the conjugate depth for a free hydraulic jump (C=1) using Equation 8.4.

$$y_2 = \frac{Cy_1}{2}\left(\sqrt{1 + 8Fr_1^2} - 1\right) = \frac{1.0(1.18)}{2}\left(\sqrt{1 + 8(5.7)^2} - 1\right) = 8.94 \text{ ft}$$

From Figure 8.2 basin length, $L_B/y_2 = 6.1$. Therefore, $L_B = 6.1(8.94) = 54.5$ ft.

The length of the basin from the floor to the sill is calculated from Equation 8.6:

$$L_S = \frac{L_T(S_T - S_o) - L_B S_o}{S_S + S_o} = \frac{14(0.5 - 0.065) - 54.5(0.065)}{0.5 + 0.065} = 4.5 \text{ ft}$$

The elevation at the entrance to the tailwater channel is from Equation 8.7:

$$z_3 = L_S S_S + z_1 = 4.5(0.5) + 93.0 = 95.25 \text{ ft}$$

Since $y_2 + z_2$ (8.94+93) > z_3 + TW (95.25+1.88), tailwater is not sufficient to force a jump in the basin. Go back to step 4.

Step 4 (2nd iteration). Try $z_1 = 84.5$ ft. Maintain W_B, S_T, and S_S.

$$L_T = \frac{z_o - z_1}{S_T} = \frac{100 - 84.5}{0.5} = 31.0 \text{ ft}$$

By using Equation 8.2 or other appropriate method by trial and error, the velocity and depth conditions entering the basin are:

$V_1 = 42.5$ ft/s, $y_1 = 0.98$ ft

$$Fr_1 = \frac{V_1}{\sqrt{gy_1}} = \frac{42.5}{\sqrt{32.2(0.98)}} = 7.6$$

Step 5 (2nd Iteration). Calculate the conjugate depth for a free hydraulic jump (C=1) using Equation 8.4.

$$y_2 = \frac{Cy_1}{2}\left(\sqrt{1 + 8Fr_1^2} - 1\right) = \frac{1.0(0.98)}{2}\left(\sqrt{1 + 8(7.6)^2} - 1\right) = 10.07 \text{ ft}$$

From Figure 8.2 basin length, $L_B/y_2 = 6.1$. Therefore, $L_B = 6.1(10.07) = 61.4$ ft.

The length of the basin from the floor to the sill is calculated from Equation 8.6:

$$L_S = \frac{L_T(S_T - S_o) - L_B S_o}{S_S + S_o} = \frac{31.0(0.5 - 0.065) - 61.4(0.065)}{0.5 + 0.065} = 16.8 \text{ ft}$$

The elevation at the entrance to the tailwater channel is from Equation 8.7:

$$z_3 = L_S S_S + z_1 = 16.8(0.5) + 84.5 = 92.90 \text{ ft}$$

Since $y_2 + z_2$ (10.1 + 84.5) < z_3 + TW (92.90 + 1.88), tailwater is sufficient to force a jump in the basin. Continue on to step 6.

Step 6. For the slope change from the outlet to the transition, determine the needed radius of curvature using Equation 8.8 and the results from step 1.

$$r = \frac{y}{\left(e^{\frac{1.5}{Fr^2}} - 1\right)} = \frac{1.50}{\left(e^{\frac{1.5}{(4.0)^2}} - 1\right)} = 15.3 \text{ ft}$$

Step 7. Size the basin elements. Since this is a free hydraulic jump basin, there are no additional elements and the design is complete. The basin is shown in the following sketch.

Total basin length = 31.0 + 61.4 + 16.8 = 109.2 ft

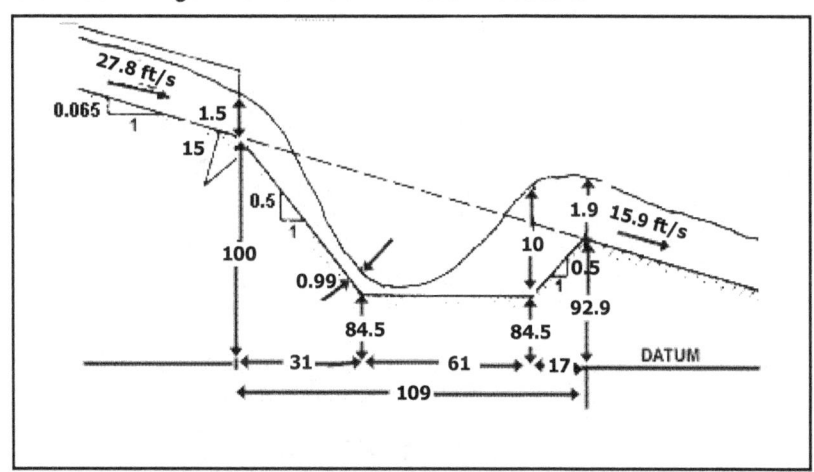

Sketch for Free Hydraulic Jump Stilling Basin Design Example (CU)

8.3 USBR TYPE III STILLING BASIN

The USBR Type III stilling basin (USBR, 1987) employs chute blocks, baffle blocks, and an end sill as shown in Figure 8.3. The basin action is very stable with a steep jump front and less wave action downstream than with the free hydraulic jump. The position, height, and spacing of the baffle blocks as recommended below should be adhered to carefully. If the baffle blocks are too far upstream, wave action in the basin will result; if too far downstream, a longer basin will be required; if too high, waves can be produced; and, if too low, jump sweep out or rough water may result.

The baffle blocks may be shaped as shown in Figure 8.3 or cubes; both are effective. The corners should not be rounded as this reduces energy dissipation.

The recommended design is limited to the following conditions:

1. Maximum unit discharge of 18.6 m³/s/m (200 ft³/s/ft).

2. Velocities up to 18.3 m/s (60 ft/s) entering the basin.

3. Froude number entering the basin between 4.5 and 17.

4. Tailwater elevation equal to or greater than full conjugate depth elevation. This provides a 15 to 18 percent factor of safety.

5. The basin sidewalls should be vertical rather than trapezoidal to insure proper performance of the hydraulic jump.

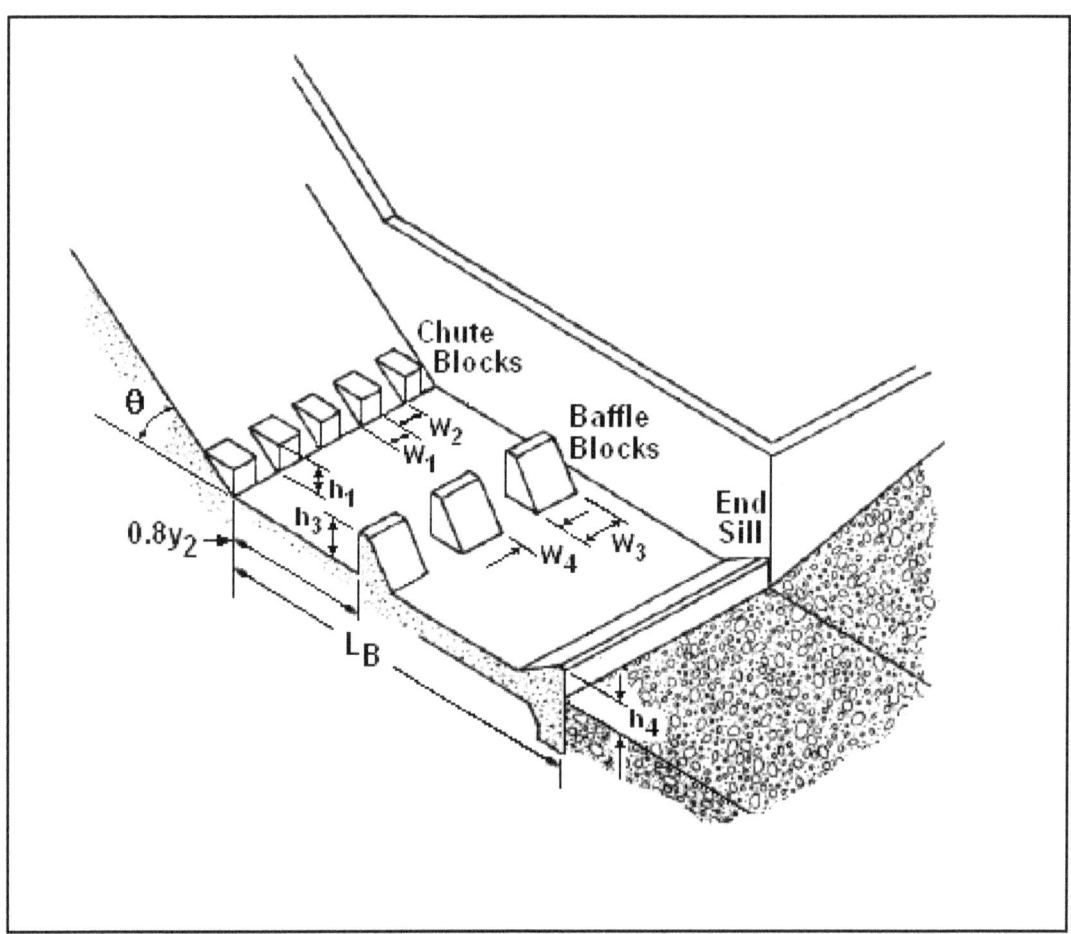

Figure 8.3. USBR Type III Stilling Basin

The general design procedure outlined in Section 8.1 applies to the USBR Type III stilling basin. Steps 1 through 4 and step 6 are applied without modification. For step 5, two adaptations to the general design procedure are made:

1. For computing conjugate depth, C=1.0. (This value is also applicable for the free hydraulic jump.) At a minimum, C=0.85 could be used, but C=1.0 is recommended.

8-12

2. For obtaining the length of the basin, L_B, use Figure 8.2 based on the Froude number calculated in step 4.

For step 7, sizing the basin elements (chute blocks, baffle blocks, and an end sill), the following guidance is recommended. The height of the chute blocks, h_1, is set equal to y_1. If y_1 is less than 0.2 m (0.66 ft), then h_1 = 0.2 m (0.66 ft).

The number of chute blocks is determined by Equation 8.9 rounded to the nearest integer.

$$N_c = \frac{W_B}{2y_1} \qquad (8.9)$$

where,

N_c = number of chute blocks

Block width and block spacing are determined by:

$$W_1 = W_2 = \frac{W_B}{2N_c} \qquad (8.10)$$

where,

W_1 = block width, m (ft)

W_2 = block spacing, m (ft)

Equations 8.9 and 8.10 will provide N_c blocks and N_c-1 spaces between those blocks. The remaining basin width is divided equally for spaces between the outside blocks and the basin sidewalls. With these equations, the height, width, and spacing of chute blocks should approximately equal the depth of flow entering the basin, y_1. The block width and spacing may be reduced as long as W_1 continues to equal W_2.

The height, width, and spacing of the baffle blocks are shown on Figure 8.3. The height of the baffles is computed from the following equation:

$$h_3 = y_1(0.168Fr_1 + 0.58) \qquad (8.11)$$

where,

h_3 = height of the baffle blocks, m (ft)

The top thickness of the baffle blocks should be set at $0.2h_3$ with the back slope of the block on a 1:1 slope. The number of baffle blocks is as follows:

$$N_B = \frac{W_B}{1.5h_3} \qquad (8.12)$$

where,

N_B = number of baffle blocks (rounded to an integer)

Baffle width and spacing are determined by:

$$W_3 = W_4 = \frac{W_B}{2N_B} \qquad (8.13)$$

8-13

where,

W_3 = baffle width, m (ft)

W_4 = baffle spacing, m (ft)

As with the chute blocks, Equations 8.12 and 8.13 will provide N_B baffles and N_B-1 spaces between those baffles. The remaining basin width is divided equally for spaces between the outside baffles and the basin sidewalls. The width and spacing of the baffles may be reduced for narrow structures provided both are reduced by the same amount. The distance from the downstream face of the chute blocks to the upstream face of the baffle block should be $0.8y_2$.

The height of the final basin element, the end sill, is given as:

$$h_4 = y_1(0.0536Fr_1 + 1.04) \tag{8.14}$$

where,

h_4 = height of the end sill, m (ft)

The fore slope of the end sill should be set at 0.5:1 (V:H).

If these recommendations are followed, a short, compact basin with good dissipation action will result. If they cannot be followed closely, a model study is recommended.

Design Example: USBR Type III Stilling Basin (SI)

Design a USBR Type III stilling basin for a reinforced concrete box culvert. Given:

Q = 11.8 m³/s

Culvert

B = 3.0 m

D = 1.8 m

n = 0.015

S_o = 0.065 m/m

z_o = 30.50 m

Downstream channel (trapezoidal)

B = 3.10 m

Z = 1V:2H

n = 0.030

Solution

The culvert, design discharge, and tailwater channel are the same as considered for the free hydraulic jump stilling basin addressed in the design example in Section 8.1. Steps 1 through 3 of the general design process are identical for this example so they are not repeated here. The tailwater depth from the previous design example is TW=0.574 m.

Step 4. Try z_1 = 26.7 m. W_B = 3.0 m, S_T = 0.5 m/m, and S_S = 0.5 m/m. From Equation 8.5:

$$L_T = \frac{z_o - z_1}{S_T} = \frac{30.50 - 26.7}{0.5} = 7.6 \text{ m}$$

By using Equation 8.2 or other appropriate method by trial and error, the velocity and depth conditions entering the basin are:

V_1 = 12.2 m/s, y_1 = 0.322 m

$$Fr_1 = \frac{V_1}{\sqrt{gy_1}} = \frac{12.1}{\sqrt{9.81(0.322)}} = 6.9$$

Step 5. Calculate the conjugate depth in the basin (C=1) using Equation 8.4.

$$y_2 = \frac{Cy_1}{2}\left(\sqrt{1+8Fr_1^2}-1\right) = \frac{1.0(0.322)}{2}\sqrt{1+8(6.9)^2}-1\right) = 2.98 \text{ m}$$

From Figure 8.2 basin length, L_B/y_2 = 2.7. Therefore, L_B = 2.7(2.98) = 8.0 m.

The length of the basin from the floor to the sill is calculated from Equation 8.6:

$$L_S = \frac{L_T(S_T - S_o) - L_B S_o}{S_S + S_o} = \frac{7.6(0.5-0.065)-8.0(0.065)}{0.5+0.065} = 4.9 \text{ m}$$

The elevation at the entrance to the tailwater channel is from Equation 8.7:

$$z_3 = L_S S_S + z_1 = 4.9(0.5) + 26.7 = 29.15 \text{ m}$$

Since $y_2 + z_2$ (2.98+26.7) < z_3 + TW (29.15+ 0.574), tailwater is sufficient to force a jump in the basin. If the tailwater had not been sufficient, repeat step 4 with a lower assumption for z_1.

Step 6. Determine the needed radius of curvature for the slope changes entering the basin. See the design example Section 8.1 for this step. It is unchanged.

Step 7. Size the basin elements. For the USBR Type III basin, the elements include the chute blocks, baffle blocks, and end sill.

For the chute blocks, the height of the chute blocks, $h_1 = y_1 = 0.322$ m (round to 0.32 m). The number of chute blocks is determined by Equation 8.9 and rounded to the nearest integer.

$$N_c = \frac{W_B}{2y_1} = \frac{3.0}{2(0.322)} = 4.6 \approx 5$$

Block width and block spacing are determined by Equation 8.10:

$$W_1 = W_2 = \frac{W_B}{2N_c} = \frac{3.0}{2(5)} = 0.30 \text{ m}$$

With 5 blocks at 0.30 m and 4 spaces at 0.30 m, 0.30 m of space remains. This is divided equally for spaces between the outside blocks and the basin sidewalls.

For the baffle blocks, the height of the baffles is computed from Equation 8.11:

$$h_3 = y_1(0.168Fr_1 + 0.58) = 0.322(0.168(6.9)+0.58) = 0.56 \text{ m}$$

The number of baffles blocks is calculated from Equation 8.12:

$$N_B = \frac{W_B}{1.5h_3} = \frac{3.0}{1.5(0.56)} = 3.6 \approx 4$$

Baffle width and spacing are determined by Equation 8.13:

$$W_3 = W_4 = \frac{W_B}{2N_B} = \frac{3}{2(4)} = 0.38 \text{ m}$$

With 4 baffles at 0.38 m and 3 spaces at 0.38 m, 0.34 m of space remains. This is divided equally for spaces between the outside baffles and the basin sidewalls. The distance from the downstream face of the chute blocks to the upstream face of the baffle block should be $0.8y_2=0.8(2.98)=2.38$ m.

For the end sill, the height of the end sill is given by Equation 8.14:

$$h_4 = y_1(0.0536Fr_1 + 1.04) = 0.322(0.053(6.9) + 1.04) = 0.45 \text{ m}$$

Total basin length = 7.6 + 8.0 + 4.9 = 20.5 m. The basin is shown in the following sketch.

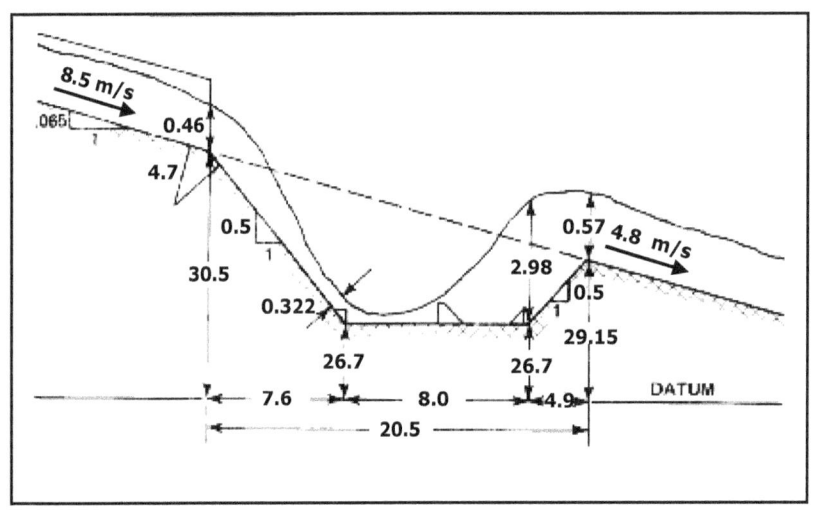

Sketch for USBR Type III Stilling Basin Design Example (SI)

Design Example: USBR Type III Stilling Basin (CU)

Design a USBR Type III stilling basin for a reinforced concrete box culvert. Given:

Q   =   417 ft³/s
Culvert
B   =   10.0 ft
D   =   6 ft
n   =   0.015
S_o =   0.065 ft/ft
z_o =   100 ft

Downstream channel (trapezoidal)

B   = 10.2 ft

Z   = 1V:2H

n   = 0.030

Solution

The culvert, design discharge, and tailwater channel are the same as considered for the free hydraulic jump stilling basin addressed in the design example in Section 8.1. Steps 1 through 3 of the general design process are identical for this example so they are not repeated here. The tailwater depth from the previous design example is TW=1.88 ft.

Step 4. Try z_1 = 87.5 ft. W_B = 10.0 ft, S_T = 0.5 ft/ft, and S_S = 0.5 ft/ft. From Equation 8.5:

$$L_T = \frac{z_o - z_1}{S_T} = \frac{100 - 87.5}{0.5} = 25.0 \text{ ft}$$

By using Equation 8.2 or other appropriate method by trial and error, the velocity and depth conditions entering the basin are:

V_1 = 40.1 ft/s, y_1 = 1.04 ft

$$Fr_1 = \frac{V_1}{\sqrt{gy_1}} = \frac{40.1}{\sqrt{32.2(1.04)}} = 6.9$$

Step 5. Calculate the conjugate depth in the basin (C=1) using Equation 8.4.

$$y_2 = \frac{Cy_1}{2}\left(\sqrt{1 + 8Fr_1^2} - 1\right) = \frac{1.0(1.04)}{2}\left(\sqrt{1 + 8(6.9)^2} - 1\right) = 9.64 \text{ ft}$$

From Figure 8.2 basin length, L_B/y_2 = 2.7. Therefore, L_B = 2.7(9.64) = 26.0 ft.

The length of the basin from the floor to the sill is calculated from Equation 8.6:

$$L_S = \frac{L_T(S_T - S_o) - L_B S_o}{S_S + S_o} = \frac{25.0(0.5 - 0.065) - 26.0(0.065)}{0.5 + 0.065} = 16.3 \text{ ft}$$

The elevation at the entrance to the tailwater channel is from Equation 8.7:

$$z_3 = L_S S_S + z_1 = 16.3(0.5) + 87.5 = 95.65 \text{ ft}$$

Since y_2 +z_2 (9.64+87.5) < z_3 + TW (95.65+1.88), tailwater is sufficient to force a jump in the basin. If the tailwater had not been sufficient, repeat step 4 with a lower assumption for z_1.

Step 6. Determine the needed radius of curvature for the slope changes entering the basin. See the design example Section 8.1 for this step. It is unchanged.

Step 7. Size the basin elements. For the USBR Type III basin, the elements include the chute blocks, baffle blocks, and end sill.

    For the chute blocks, the height of the chute blocks, h_1=y_1=1.04 ft (round to 1.0 ft). The number of chute blocks is determined by Equation 8.9 and rounded to the nearest integer.

$$N_c = \frac{W_B}{2y_1} = \frac{10}{2(1.04)} = 4.8 \approx 5$$

Block width and block spacing are determined by Equation 8.10:

$$W_1 = W_2 = \frac{W_B}{2N_c} = \frac{10.0}{2(5)} = 1.0 \text{ ft}$$

With 5 blocks at 1.0 ft and 4 spaces at 1.0 ft, 1.0 ft of space remains. This is divided equally for spaces between the outside blocks and the basin sidewalls.

For the baffle blocks, the height of the baffles is computed from Equation 8.11:

$$h_3 = y_1(0.168Fr_1 + 0.58) = 1.04(0.168(6.9) + 0.58) = 1.8 \text{ ft}$$

The number of baffles blocks is calculated from Equation 8.12:

$$N_B = \frac{W_B}{1.5h_3} = \frac{10.0}{1.5(1.8)} = 3.7 \approx 4$$

Baffle width and spacing are determined by Equation 8.13:

$$W_3 = W_4 = \frac{W_B}{2N_B} = \frac{10}{2(4)} = 1.3 \text{ ft}$$

With 4 baffles at 1.3 ft and 3 spaces at 1.3 ft, 0.9 ft of space remains. This is divided equally for spaces between the outside baffles and the basin sidewalls. The distance from the downstream face of the chute blocks to the upstream face of the baffle block should be $0.8y_2 = 0.8(9.64) = 7.7$ ft.

For the end sill, the height of the end sill is given by Equation 8.14:

$$h_4 = y_1(0.0536Fr_1 + 1.04) = 1.04(0.053(6.9) + 1.04) = 1.5 \text{ ft}$$

Total basin length = 25.0 + 26.0 + 16.3 = 67.3 ft. The basin is shown in the following sketch.

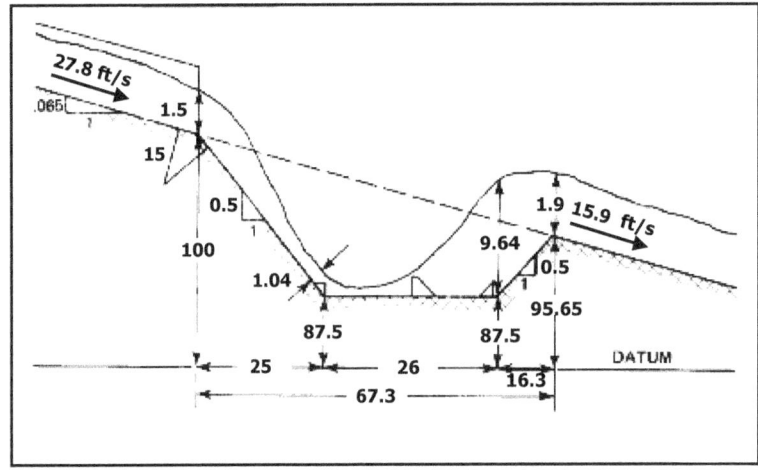

Sketch for USBR Type III Stilling Basin Design Example (CU)

8.4   USBR TYPE IV STILLING BASIN

The USBR Type IV stilling basin (USBR, 1987) is intended for use in the Froude number range of 2.5 to 4.5. In this low Froude number range, the jump is not fully developed and downstream wave action may be a problem as discussed in Chapter 4. For the intermittent flow encountered at most highway culverts, wave action is not judged to be a severe limitation. The basin, illustrated in Figure 8.4, employs chute blocks and an end sill.

The recommended design is limited to the following conditions:

1. The basin sidewalls should be vertical rather than trapezoidal to insure proper performance of the hydraulic jump.

2. Tailwater elevation should be equal to or greater than 110 percent of the full conjugate depth elevation. The hydraulic jump is very sensitive to tailwater depth at the low Froude numbers for which the basin is applicable. The additional tailwater improves jump performance and reduces wave action.

The general design procedure outlined in Section 8.1 applies to the USBR Type IV basin. Steps 1 through 4 and step 6 are applied without modification. For step 5, two adaptations to the general design procedure are made:

1. For computing conjugate depth, C = 1.1.

2. For obtaining the length of the basin, L_B, use Figure 8.2 (dashed portion of the free jump curve) based on the Froude number calculated in step 4.

For step 7, sizing the basin elements (chute blocks and an end sill), the following guidance is recommended. The height of the chute blocks, h_1, is set equal to $2y_1$. The top surface of the chute blocks should be sloped downstream at a 5 degree angle.

The number of chute blocks is determined by Equation 8.15a and rounded to the nearest integer.

$$N_c = \frac{W_B}{2.625y_1} \qquad (8.15a)$$

where,

N_c = number of chute blocks

Block width and block spacing are determined by:

$$W_1 = \frac{W_B}{3.5N_c} \qquad (8.15b)$$

$$W_2 = 2.5W_1 \qquad (8.15c)$$

where,

W_1 = block width, m (ft)
W_2 = block spacing, m (ft)

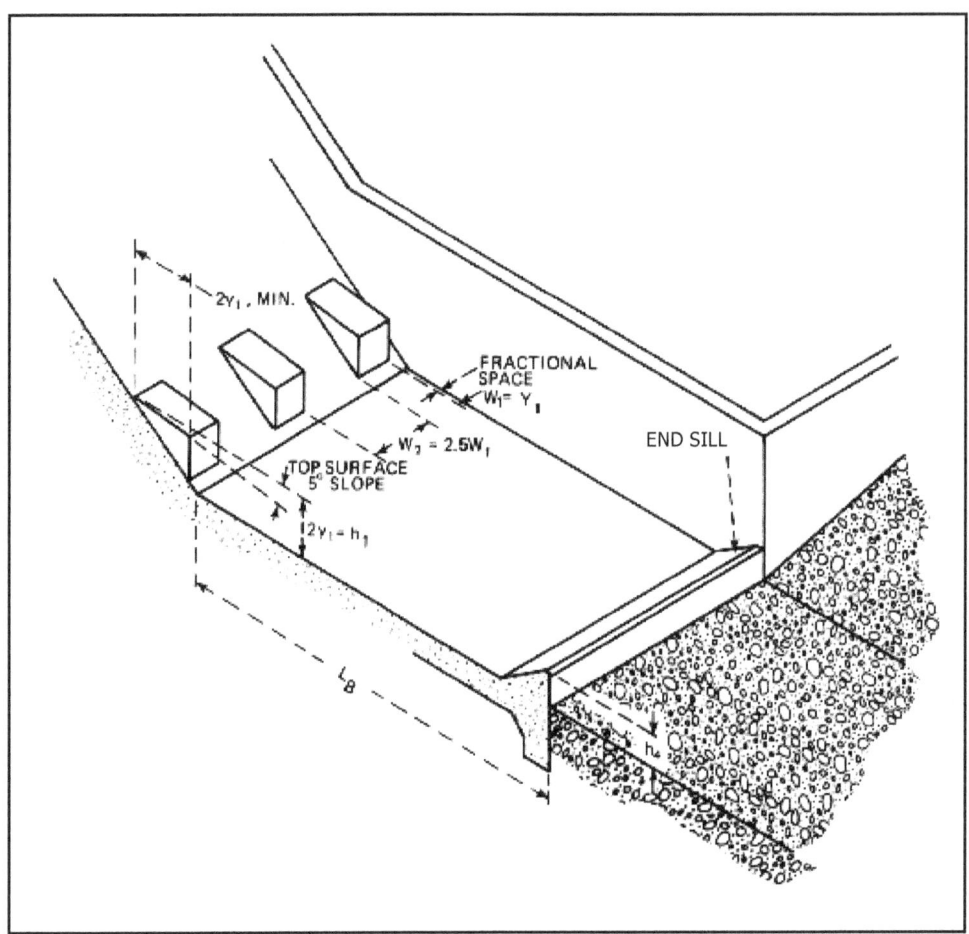

Figure 8.4. USBR Type IV Stilling Basin

With Equation 8.15b, the block width, W_1, should be less than or equal to the depth of the incoming flow, y_1. Equations 8.15a, 8.15b, and 8.15c will provide N_c blocks and N_c-1 spaces between those blocks. The remaining basin width is divided equally for spaces between the outside blocks and the basin sidewalls.

The height of the end sill, is given as:

$$h_4 = y_1(0.0536Fr_1 + 1.04) \qquad (8.16)$$

where,

    h_4 = height of the end sill, m (ft)

The fore slope of the end sill should be set at 0.5:1 (V:H).

Design Example: USBR Type IV Stilling Basin (SI)

Design a USBR Type IV stilling basin for a reinforced concrete box culvert. Given:

Q $\quad$ = 11.8 m³/s

Culvert

B $\quad$ = 3.0 m

D $\quad$ = 1.8 m

n $\quad$ = 0.015

S_o $\quad$ = 0.065 m/m

z_o $\quad$ = 30.50 m

Downstream channel (trapezoidal)

B $\quad$ = 3.10 m

Z $\quad$ = 1V:2H

n $\quad$ = 0.030

Solution

The culvert, design discharge, and tailwater channel are the same as considered for the free hydraulic jump stilling basin addressed in the design example in Section 8.1. Steps 1 through 3 of the general design process are identical for this example so they are not repeated here. The tailwater depth from the previous design example is TW=0.574 m.

Step 4. Try z_1 = 25.00 m. W_B=3.0 m, S_T=0.5 m/m, and S_S=0.5 m/m. From Equation 8.5:

$$L_T = \frac{z_o - z_1}{S_T} = \frac{30.50 - 25.75}{0.5} = 11.0 \text{ m}$$

By using Equation 8.2 or other appropriate method by trial and error, the velocity and depth conditions entering the basin are:

V_1 = 13.55 m/s, y_1 = 0.290 m

$$Fr_1 = \frac{V_1}{\sqrt{gy_1}} = \frac{13.55}{\sqrt{9.81(0.290)}} = 8.0$$

It should be noted that this Froude number is outside the applicability range for the Type IV basin, therefore the Type IV is not appropriate for this situation. However, we will proceed with the calculations in order to compare basin dimensions with the other basin options.

Step 5. Calculate the conjugate depth in the basin (C=1.1) using Equation 8.4.

$$y_2 = \frac{Cy_1}{2}\left(\sqrt{1 + 8Fr_1^2} - 1\right) = \frac{1.1(0.303)}{2}\left(\sqrt{1 + 8(8.0)^2} - 1\right) = 3.46 \text{ m}$$

From Figure 8.2 basin length, L_B/y_2 = 6.1. Therefore, L_B = 6.1(3.46) = 21.1 m.

The length of the basin from the floor to the sill is calculated from Equation 8.6:

$$L_S = \frac{L_T(S_T - S_o) - L_B S_o}{S_S + S_o} = \frac{11.0(0.5 - 0.065) - 21.1(0.065)}{0.5 + 0.065} = 6.0 \text{ m}$$

The elevation at the entrance to the tailwater channel is from Equation 8.7:

$$z_3 = L_S S_S + z_1 = 6.0(0.5) + 25.00 = 28.00 \text{ m}$$

Since $y_2 + z_2$ (3.46+25.00) < z_3 + TW (28.00+0.574), tailwater is sufficient to force a jump in the basin. If this had not been the case, repeat step 4 with a lower assumption for z_1.

Step 6. Determine the needed radius of curvature for the slope changes entering the basin. See the design example Section 8.1 for this step. It is unchanged.

Step 7. Size the basin elements. For the USBR Type IV basin, the elements include the chute blocks and end sill.

For the chute blocks:

The height of the chute blocks, $h_1 = 2y_1 = 2(0.290) = 0.58$ m. The number of chute blocks is determined by Equation 8.15a and rounded to the nearest integer.

$$N_c = \frac{W_B}{2.625y_1} = \frac{3.0}{2.625(0.290)} = 3.9 \approx 4$$

Block width and block spacing are determined by Equations 8.15b and 8.15c:

$$W_1 = \frac{W_B}{3.5N_c} = \frac{3.0}{3.5(4)} = 0.21 \text{ m}$$

$$W_2 = 2.5W_1 = 2.5(0.21) = 0.53 \text{ m}$$

With 4 blocks at 0.21 m and 3 spaces at 0.53 m, 0.57 m of space remains. This is divided equally for spaces between the outside blocks and the basin sidewalls.

For the sill:

The height of the end sill, is given by Equation 8.16:

$$h_4 = y_1(0.0536Fr_1 + 1.04) = 0.290(0.0536(8.0) + 1.04) = 0.43 \text{ m}$$

Total basin length = 11.0 + 21.1 + 6.0 = 38.1 m. The basin is shown in the following sketch.

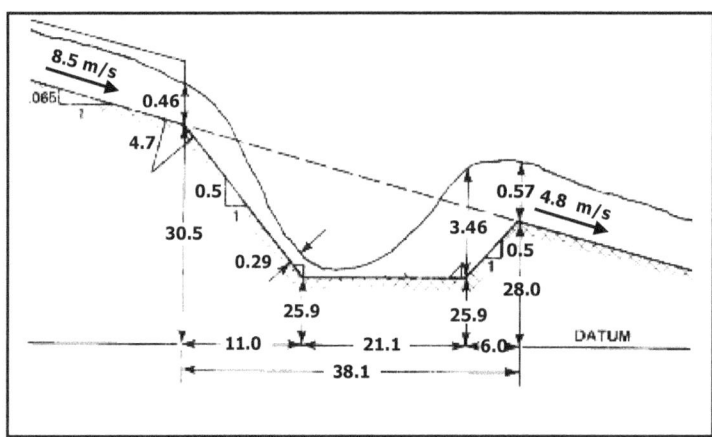

Sketch for USBR Type IV Stilling Basin Design Example (SI)

8-22

Design Example: USBR Type IV Stilling Basin (CU)

Design a USBR Type IV stilling basin for a reinforced concrete box culvert. Given:

Q = 417 ft³/s

Culvert

B = 10 ft

D = 6 ft

n = 0.015

S_o = 0.065 ft/ft

z_o = 100.0 ft

Downstream channel (trapezoidal)

B = 10.2 ft

Z = 1V:2H

n = 0.030

Solution

The culvert, design discharge, and tailwater channel are the same as considered for the free hydraulic jump stilling basin addressed in the design example in Section 8.1. Steps 1 through 3 of the general design process are identical for this example so they are not repeated here. The tailwater depth from the previous design example is TW=1.88 ft.

Step 4. Try z_1 = 82.6 ft. W_B=10.0 ft, S_T=0.5 ft/ft, and S_S=0.5 ft/ft. From Equation 8.5:

$$L_T = \frac{z_o - z_1}{S_T} = \frac{100 - 82.6}{0.5} = 34.8 \text{ ft}$$

By using Equation 8.2 or other appropriate method by trial and error, the velocity and depth conditions entering the basin are:

V_1 = 43.9 ft/s, y_1 = 0.95 ft

$$Fr_1 = \frac{V_1}{\sqrt{gy_1}} = \frac{43.9}{\sqrt{32.2(0.95)}} = 7.9$$

It should be noted that this Froude number is outside the applicability range for the Type IV basin, therefore the Type IV is not appropriate for this situation. However, we will proceed with the calculations in order to compare basin dimensions with the other basin options.

Step 5. Calculate the conjugate depth in the basin (C=1.1) using Equation 8.4.

$$y_2 = \frac{Cy_1}{2}\left(\sqrt{1+8Fr_1^2} - 1\right) = \frac{1.1(0.95)}{2}\left(\sqrt{1+8(7.9)^2} - 1\right) = 11.15 \text{ ft}$$

From Figure 8.2 basin length, L_B/y_2 = 6.1. Therefore, L_B = 6.1(11.15) = 68.0 ft.

The length of the basin from the floor to the sill is calculated from Equation 8.6:

$$L_S = \frac{L_T(S_T - S_o) - L_B S_o}{S_S + S_o} = \frac{34.8(0.5 - 0.065) - 68.0(0.065)}{0.5 + 0.065} = 19.0 \text{ ft}$$

8-23

The elevation at the entrance to the tailwater channel is from Equation 8.7:

$$z_3 = L_S S_S + z_1 = 19.0(0.5) + 82.60 = 92.10 \text{ ft}$$

Since $y_2 + z_2$ (11.15+82.60) < z_3 + TW (92.10+1.88), tailwater is sufficient to force a jump in the basin. If this had not been the case, repeat step 4 with a lower assumption for z_1.

Step 6. Determine the needed radius of curvature for the slope changes entering the basin. See the design example Section 8.1 for this step. It is unchanged.

Step 7. Size the basin elements. For the USBR Type IV basin, the elements include the chute blocks and end sill.

For the chute blocks:

The height of the chute blocks, $h_1 = 2y_1 = 2(0.95) = 1.9$ ft. The number of chute blocks is determined by Equation 8.15a and rounded to the nearest integer.

$$N_c = \frac{W_B}{2.625 y_1} = \frac{10.0}{2.625(0.95)} = 4.01 \approx 4$$

Block width and block spacing are determined by Equations 8.15b and 8.15c:

$$W_1 = \frac{W_B}{3.5 N_c} = \frac{10.0}{3.5(4)} = 0.7 \text{ ft}$$

$$W_2 = 2.5 W_1 = 2.5(0.7) = 1.8 \text{ ft}$$

With 4 blocks at 0.7 ft and 3 spaces at 1.8 ft, 1.8 ft of space remains. This is divided equally for spaces between the outside blocks and the basin sidewalls.

For the sill:

The height of the end sill, is given by Equation 8.16:

$$h_4 = y_1(0.0536 Fr_1 + 1.04) = 0.95(0.0536(7.9) + 1.04) = 1.4 \text{ ft}$$

Total basin length = 34.8 + 68.0 + 19.0 = 121.8 ft. The basin is shown in the following sketch.

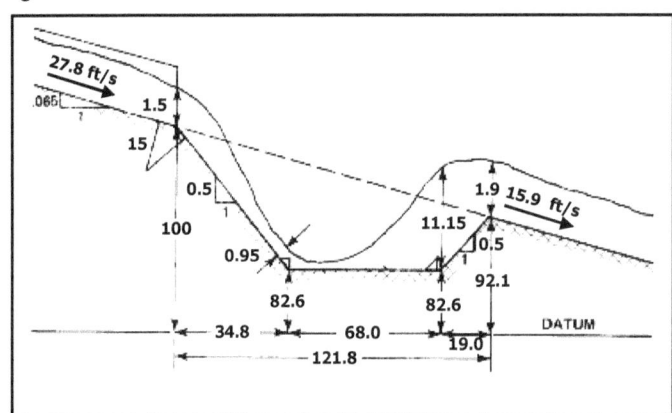

Sketch for USBR Type IV Stilling Basin Design Example (CU)

8.5   SAF STILLING BASIN

The Saint Anthony Falls (SAF) stilling basin, shown in Figure 8.5, provides chute blocks, baffle blocks, and an end sill that allows the basin to be shorter than a free hydraulic jump basin. It is recommended for use at small structures such as spillways, outlet works, and canals where the Froude number at the dissipator entrance is between 1.7 and 17. The reduction in basin length achieved through the use of appurtenances is about 80 percent of the free hydraulic jump length.   The SAF stilling basin provides an economical method of dissipating energy and preventing stream bed erosion.

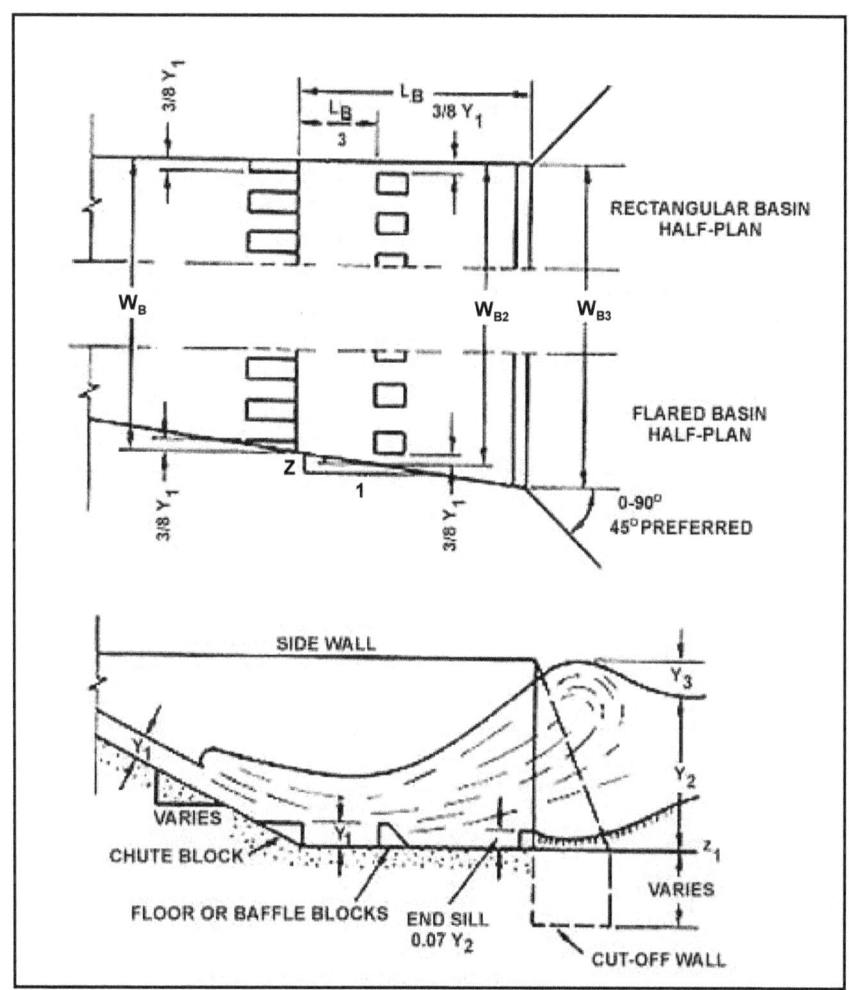

Figure 8.5. SAF Stilling Basin (Blaisdell, 1959)

The general design procedure outlined in Section 8.1 applies to the SAF stilling basin. Steps 1 through 3 and step 6 are applied without modification. As part of step 4, the designer selects a basin width, W_B. For box culverts, W_B must equal the culvert width, W_o.   For circular culverts, the basin width is taken as the larger of the culvert diameter and the value calculated according to the following equation:

$$W_B = 1.7D_o\left(\frac{Q}{g^{0.5}D_o^{2.5}}\right) \tag{8.17}$$

where,

W_B = basin width, m (ft)

Q = design discharge, m³/s (ft³/s)

D_o = culvert diameter, m (ft)

The basin can be flared to fit an existing channel as indicated on Figure 8.5. The sidewall flare dimension z should not be greater than 0.5, i.e., 0.5:1, 0.33:1, or flatter.

For step 5, two adaptations to the general design procedure are made. First, for computing conjugate depth, C is a function of Froude number as given by the following set of equations. Depending on the Froude number, C ranges from 0.64 to 1.08 implying that the SAF basin may operate with less tailwater than the USBR basins, though tailwater is still required.

$$C = 1.1 - \frac{Fr_1^2}{120} \qquad \text{when } 1.7 < Fr_1 < 5.5 \tag{8.18a}$$

$$C = 0.85 \qquad \text{when } 5.5 < Fr_1 < 11 \tag{8.18b}$$

$$C = 1.0 - \frac{Fr_1^2}{800} \qquad \text{when } 11 < Fr_1 < 17 \tag{8.18c}$$

The second adaptation is the determination of the basin length, L_B, using Equation 8.19.

$$L_B = \frac{4.5y_2}{CFr_1^{0.76}} \tag{8.19}$$

For step 7, sizing the basin elements (chute blocks, baffle blocks, and an end sill), the following guidance is recommended. The height of the chute blocks, h_1, is set equal to y_1.

The number of chute blocks is determined by Equation 8.20 rounded to the nearest integer.

$$N_c = \frac{W_B}{1.5y_1} \tag{8.20}$$

where,

N_c = number of chute blocks

Block width and block spacing are determined by:

$$W_1 = W_2 = \frac{W_B}{2N_c} \tag{8.21}$$

where,

W_1 = block width, m (ft)

W_2 = block spacing, m (ft)

Equations 8.20 and 8.21 will provide N_c blocks and N_c spaces between those blocks. A half block is placed at the basin wall so there is no space at the wall.

The height, width, and spacing of the baffle blocks are shown on Figure 8.5. The height of the baffles, h_3, is set equal to the entering flow depth, y_1.

The width and spacing of the baffle blocks must account for any basin flare. If the basin is flared as shown in Figure 8.5, the width of the basin at the baffle row is computed according to the following:

$$W_{B2} = W_B + \left(\frac{2zL_B}{3}\right) \tag{8.22}$$

where,

W_{B2} = basin width at the baffle row, m (ft)
L_B = basin length, m (ft)
z = basin flare, z:1 as defined in Figure 8.5 (z=0.0 for no flare)

The top thickness of the baffle blocks should be set at $0.2h_3$ with the back slope of the block on a 1:1 slope. The number of baffles blocks is as follows:

$$N_B = \frac{W_{B2}}{1.5y_1} \tag{8.23}$$

where,

N_B = number of baffle blocks (rounded to an integer)

Baffle width and spacing are determined by:

$$W_3 = W_4 = \frac{W_{B2}}{2N_B} \tag{8.24}$$

where,

W_3 = baffle width, m (ft)
W_4 = baffle spacing, m (ft)

Equations 8.23 and 8.24 will provide N_B baffles and N_B-1 spaces between those baffles. The remaining basin width is divided equally for spaces between the outside baffles and the basin sidewalls. No baffle block should be placed closer to the sidewall than $3y_1/8$. Verify that the percentage of W_{B2} obstructed by baffles is between 40 and 55 percent. The distance from the downstream face of the chute blocks to the upstream face of the baffle block should be $L_B/3$.

The height of the final basin element, the end sill, is given as:

$$h_4 = \frac{0.07y_2}{C} \tag{8.25}$$

where,

h_4 = height of the end sill, m (ft)

The fore slope of the end sill should be set at 0.5:1 (V:H). If the basin is flared the length of sill (width of the basin at the sill) is:

$$W_{B3} = W_B + 2zL_B \tag{8.26}$$

where,

  W_{B3} = basin width at the sill, m (ft)
  L_B = basin length, m (ft)
  z = basin flare, z:1 as defined in Figure 8.5 (z=0.0 for no flare)

Wingwalls should be equal in height and length to the stilling basin sidewalls. The top of the wingwall should have a 1H:1V slope. Flaring wingwalls are preferred to perpendicular or parallel wingwalls. The best overall conditions are obtained if the triangular wingwalls are located at an angle of 45° to the outlet centerline.

The stilling basin sidewalls may be parallel (rectangular stilling basin) or diverge as an extension of the transition sidewalls (flared stilling basin). The height of the sidewall above the floor of the basin is given by:

$$h_5 \geq y_2 \left(1 + \frac{1}{3C} \right) \tag{8.27}$$

where,

  h_5 = height of the sidewall, m (ft)

A cut-off wall should be used at the end of the stilling basin to prevent undermining. The depth of the cut-off wall must be greater than the maximum depth of anticipated erosion at the end of the stilling basin.

Design Example: SAF Stilling Basin (SI)

Design a SAF stilling basin with no flare for a reinforced concrete box culvert. Given:

  Q   = 11.8 m³/s
  Culvert
  B   = 3.0 m
  D   = 1.8 m
  n   = 0.015
  S_o = 0.065 m/m
  z_o = 30.50 m
  Downstream channel (trapezoidal)
  B   = 3.10 m
  Z   = 1V:2H
  n   = 0.030

Solution

The culvert, design discharge, and tailwater channel are the same as considered for the free hydraulic jump stilling basin addressed in the design example in Section 8.1. Steps 1 through 3 of the general design process are identical for this example so they are not repeated here. The tailwater depth from the previous design example is TW=0.574 m.

Step 4. Try z_1 = 27.80 m. W_B=3.0 m (no flare), S_T=0.5 m/m, and S_S=0.5 m/m. From Equation 8.5:

$$L_T = \frac{z_0 - z_1}{S_T} = \frac{30.50 - 27.80}{0.5} = 5.4 \, m$$

By using Equation 8.2 or other appropriate method by trial and error, the velocity and depth conditions entering the basin are:

V_1 = 11.29 m/s, y_1 = 0.348 m

$$Fr_1 = \frac{V_1}{\sqrt{gy_1}} = \frac{11.29}{\sqrt{9.81(0.348)}} = 6.1$$

Step 5. Calculate the conjugate depth in the basin using Equation 8.4. First estimate C using Equation 8.18. For the calculated Froude number, C=0.85.

$$y_2 = \frac{Cy_1}{2}\left(\sqrt{1 + 8Fr_1^2} - 1\right) = \frac{0.85(0.348)}{2}\left(\sqrt{1 + 8(6.1)^2} - 1\right) = 2.41 \, m$$

From Equation 8.19 basin length is calculated:

$$L_B = \frac{4.5y_2}{CFr_1^{0.76}} = \frac{4.5(2.41)}{0.85(6.1)^{0.76}} = 3.2 \, m$$

The length of the basin from the floor to the sill is calculated from Equation 8.6:

$$L_S = \frac{L_T(S_T - S_o) - L_BS_o}{S_S + S_o} = \frac{5.4(0.5 - 0.065) - 3.2(0.065)}{0.5 + 0.065} = 3.8 \, m$$

The elevation at the entrance to the tailwater channel is from Equation 8.7:

$Z_3 = L_SS_S + Z_1 = 3.8(0.5) + 27.80 = 29.70 \, m$

Since $y_2 + z_2$ (2.41+27.80) < z_3 + TW (29.70+0.574), tailwater is sufficient to force a jump in the basin. If tailwater had not been sufficient, repeat step 4 with a lower assumption for z_1.

Step 6. Determine the needed radius of curvature for the slope changes entering the basin. See the design example Section 8.1 for this step. It is unchanged.

Step 7. Size the basin elements. For the SAF basin, the elements include the chute blocks, baffle blocks, and an end sill.

For the chute blocks:

The height of the chute blocks, h_1=y_1=0.348 (round to 0.35 m).

The number of chute blocks is determined by Equation 8.20:

$$N_c = \frac{W_B}{1.5y_1} = \frac{3.0}{1.5(0.348)} = 5.7 \approx 6$$

Block width and block spacing are determined by Equation 8.21:

$$W_1 = W_2 = \frac{W_B}{2N_c} = \frac{3.0}{2(6)} = 0.25 \text{ m}$$

A half block is placed at each basin wall so there is no space at the wall.

For the baffle blocks:

The height of the baffles, $h_3 = y_1 = 0.348$ m. (round to 0.35 m)

The basin has no flare so the width in the basin is constant and equal to W_B.

The number of baffles blocks is from Equation 8.23:

$$N_B = \frac{W_{B2}}{1.5y_1} = \frac{3.0}{1.5(0.348)} = 5.7 \approx 6$$

Baffle width and spacing are determined from Equation 8.24. In this case $W_{B2} = W_B$.

$$W_3 = W_4 = \frac{W_{B2}}{2N_B} = \frac{3.0}{2(6)} = 0.25 \text{ m}$$

For this design, we have 6 baffles at 0.25 m and 5 spaces between them at 0.25 m. The remaining 0.25 m is divided in half and provided as a space between the sidewall and the first baffle.

The total percentage blocked by baffles is 6(0.25)/3.0=50 percent which falls within the acceptable range of between 40 and 55 percent.

The distance from the downstream face of the chute blocks to the upstream face of the baffle block equals $L_B/3 = 3.2/3 = 1.1$ m.

For the sill:

The height of the end sill, is given in Equation 8.25:

$$h_4 = \frac{0.07y_2}{C} = \frac{0.07(2.41)}{0.85} = 0.20 \text{ m}$$

Total basin length = 5.4 + 3.2 + 3.8 = 12.4 m. The basin is shown in the following sketch.

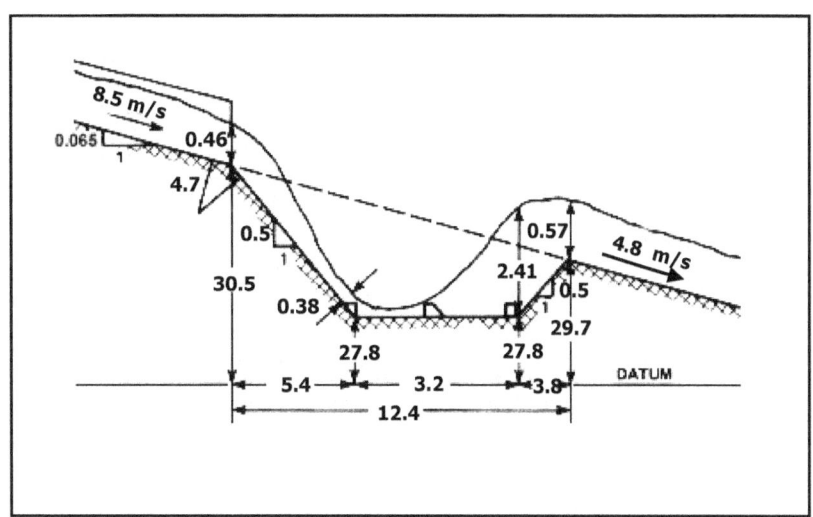

Sketch for SAF Stilling Basin Design Example (SI)

Design Example: SAF Stilling Basin (CU)

Design a SAF stilling basin with no flare for a reinforced concrete box culvert. Given:

Q $\quad$ = 417 ft³/s

Culvert

B $\quad$ = 10.0 ft

D $\quad$ = 6.0 ft

n $\quad$ = 0.015

S_o $\quad$ = 0.065 ft/ft

z_o $\quad$ = 100.0 ft

Downstream channel (trapezoidal)

B $\quad$ = 10.2 ft

Z $\quad$ = 1V:2H

n $\quad$ = 0.030

Solution

The culvert, design discharge, and tailwater channel are the same as considered for the free hydraulic jump stilling basin addressed in the design example in Section 8.1. Steps 1 through 3 of the general design process are identical for this example so they are not repeated here. The tailwater depth from the previous design example is TW=1.88 ft.

Step 4. Try z_1 = 91.40 ft. W_B=10.0 ft (no flare), S_T=0.5 ft/ft, and S_S=0.5 ft/ft. From Equation 8.5:

$$L_T = \frac{z_o - z_1}{S_T} = \frac{100.0 - 91.4}{0.5} = 17.2 \text{ ft}$$

By using Equation 8.2 or other appropriate method by trial and error, the velocity and depth conditions entering the basin are:

8-31

$V_1 = 36.8$ ft/s, $y_1 = 1.13$ ft

$$Fr_1 = \frac{V_1}{\sqrt{gy_1}} = \frac{36.8}{\sqrt{32.2(1.13)}} = 6.1$$

Step 5. Calculate the conjugate depth in the basin using Equation 8.4. First estimate C using Equation 8.18. For the calculated Froude number, C=0.85.

$$y_2 = \frac{Cy_1}{2}\left(\sqrt{1 + 8Fr_1^2} - 1\right) = \frac{0.85(1.13)}{2}\left(\sqrt{1 + 8(6.1)^2} - 1\right) = 7.85 \text{ ft}$$

From Equation 8.19 basin length is calculated:

$$L_B = \frac{4.5y_2}{CFr_1^{0.76}} = \frac{4.5(7.85)}{0.85(6.1)^{0.76}} = 10.5 \text{ ft}$$

The length of the basin from the floor to the sill is calculated from Equation 8.6:

$$L_S = \frac{L_T(S_T - S_o) - L_B S_o}{S_S + S_o} = \frac{17.2(0.5 - 0.065) - 10.5(0.065)}{0.5 + 0.065} = 12.0 \text{ ft}$$

The elevation at the entrance to the tailwater channel is from Equation 8.7:

$Z_3 = L_S S_S + Z_1 = 12.0(0.5) + 91.40 = 97.40$ ft

Since $y_2 + z_2$ (7.85+91.40) < z_3 + TW (97.40+1.88), tailwater is sufficient to force a jump in the basin. If tailwater had not been sufficient, repeat step 4 with a lower assumption for z_1.

Step 6. Determine the needed radius of curvature for the slope changes entering the basin. See the design example Section 8.1 for this step. It is unchanged.

Step 7. Size the basin elements. For the SAF basin, the elements include the chute blocks, baffle blocks, and an end sill.

For the chute blocks:

The height of the chute blocks, $h_1 = y_1 = 1.13$ (round to 1.1 ft).

The number of chute blocks is determined by Equation 8.20:

$$N_c = \frac{W_B}{1.5y_1} = \frac{10.0}{1.5(1.13)} = 5.9 \approx 6$$

Block width and block spacing are determined by Equation 8.21:

$$W_1 = W_2 = \frac{W_B}{2N_c} = \frac{10.0}{2(6)} = 0.8 \text{ ft}$$

A half block is placed at each basin wall so there is no space at the wall.

For the baffle blocks:

The height of the baffles, $h_3 = y_1 = 1.13$ ft. (round to 1.1 ft)

The basin has no flare so the width in the basin is constant and equal to W_B.

The number of baffles blocks is from Equation 8.23:

$$N_B = \frac{W_{B2}}{1.5y_1} = \frac{10.0}{1.5(1.13)} = 5.9 \approx 6$$

Baffle width and spacing are determined from Equation 8.24. In this case $W_{B2}=W_B$.

$$W_3 = W_4 = \frac{W_{B2}}{2N_B} = \frac{10.0}{2(6)} = 0.8 \text{ ft}$$

For this design, we have 6 baffles at 0.8 ft and 5 spaces between them at 0.8 ft. The remaining 1.2 ft is divided in half and provided as a space between the sidewall and the first baffle.

The total percentage blocked by baffles is 6(0.8)/10.0=48 percent which falls within the acceptable range of between 40 and 55 percent.

The distance from the downstream face of the chute blocks to the upstream face of the baffle block equals $L_B/3=10.5/3=3.5$ ft.

For the sill:

The height of the end sill, is given in Equation 8.25:

$$h_4 = \frac{0.07y_2}{C} = \frac{0.07(7.85)}{0.85} = 0.6 \text{ ft}$$

Total basin length = 17.2 + 10.5 + 12.0 = 39.7 ft. The basin is shown in the following sketch.

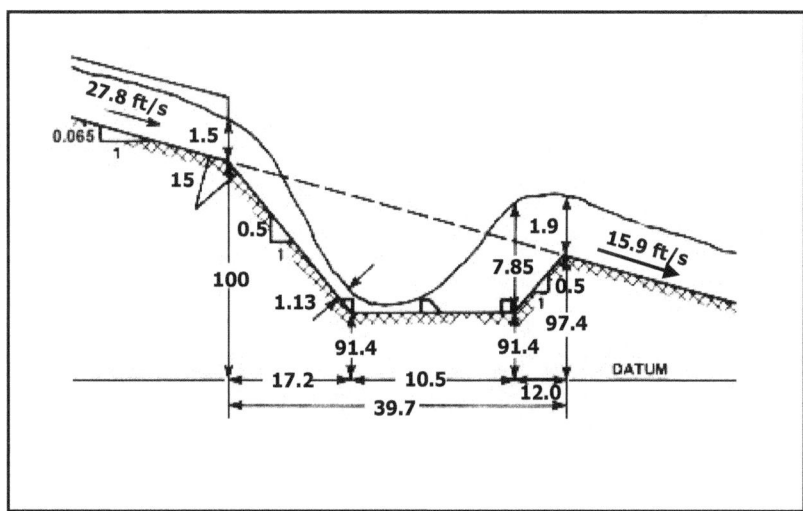

Sketch for SAF Stilling Basin Design Example (CU)

This page intentionally left blank.

CHAPTER 9: STREAMBED LEVEL DISSIPATORS

This chapter contains energy dissipators for culvert outlets that are designed to operate at the streambed level and reestablish natural flow conditions downstream from the culvert outlet. They are also intended to drain by gravity when not in operation. The following sections contain limitations, design guidance, and design examples for the following energy dissipators:

- Colorado State University (CSU) rigid boundary basin

- Contra Costa basin

- Hook basin

- U.S. Bureau of Reclamation (USBR) Type VI impact basin

9.1 CSU RIGID BOUNDARY BASIN

The Colorado State University (CSU) rigid boundary basin, illustrated in Figure 9.1, uses staggered rows of roughness elements to initiate a hydraulic jump (Simons, 1970). CSU tested a number of basins with different roughness configurations to determine the average drag coefficient over the roughened portion of the basins. The effects of the roughness elements are reflected in a drag coefficient that was derived empirically for each roughness configuration. The experimental procedure was to measure depths and velocities at each end of the control volume illustrated in Figure 9.2, and compute the basin drag coefficient, C_B, from the momentum equation by balancing the forces acting on the volume of fluid.

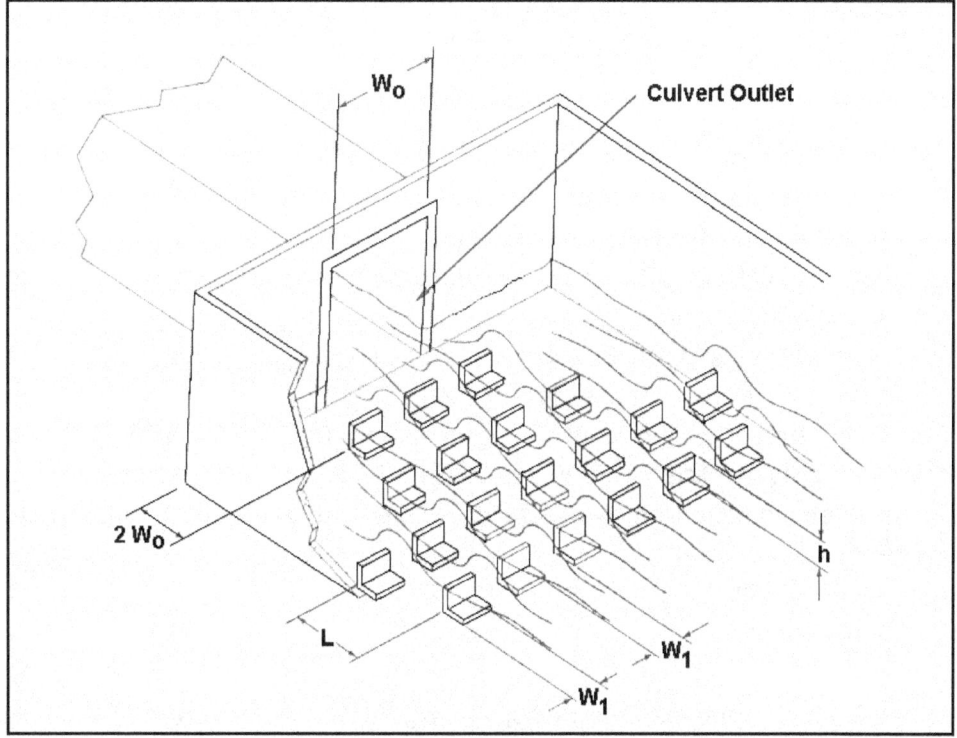

Figure 9.1. CSU Rigid Boundary Basin

The CSU test results indicate several design limitations. The height of the roughness elements, h, must be between 0.31 and 0.91 of the approach flow average depth, y_A, and, the relative spacing, L/h, between rows of elements, must be either 6 or 12. The latter is not a severe restriction since relative spacing is normally a fixed parameter in a design procedure and other tests (Morris, 1968) have shown that the best range for energy dissipation is from 6 to 12.

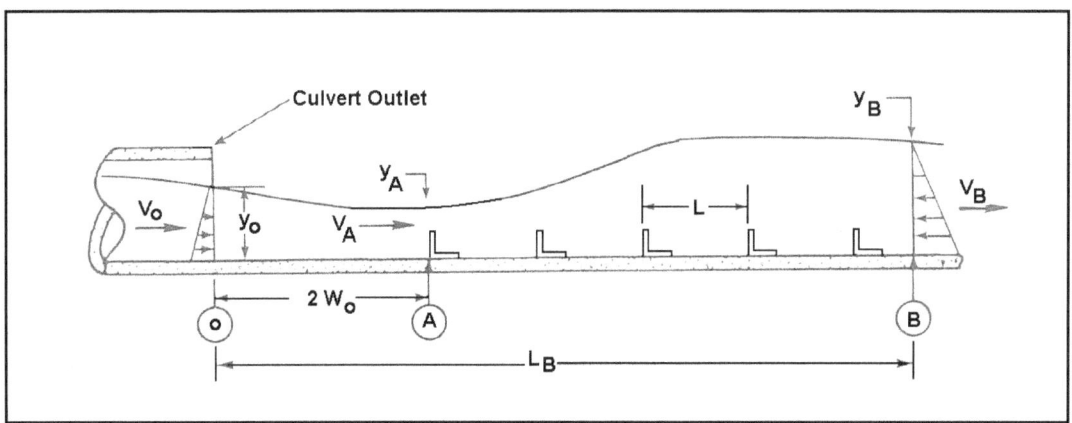

Figure 9.2. Definition Sketch for the Momentum Equation

The roughness configurations tested and the corresponding test results for C_B are shown in Figure 9.3 and Table 9.1, respectively. To design a basin, the designer selects a basin from Figure 9.3 and uses the C_B value from Table 9.1 in the following momentum equation to determine the velocity from the basin (V_B) if the slope is less than 10%:

$$\rho V_o Q + C_p \gamma (y_o^2 /2)W_o = C_B A_F N \rho V_A^2 /2 + \rho V_B Q + \gamma Q^2 /(2V_B^2 W_B) \qquad (9.1)$$

where,

y_o = depth at the culvert outlet, m (ft)

V_o = velocity at the culvert outlet, m/s (ft/s)

W_o = culvert width at the culvert outlet, m (ft)

V_A = approach velocity at two culvert widths downstream of the culvert outlet, m/s (ft/s)

V_B = exit velocity, just downstream of the last row of roughness elements, m/s (ft/s)

W_B = basin width, just downstream of the last row of roughness elements, m/s (ft/s)

N = total number of roughness elements in the basin

A_F = frontal area of one full roughness element, m^2 (ft^2)

C_B = basin drag coefficient (see Figure 9.3)

C_p = momentum correction coefficient for the pressure at the culvert outlet (see Figure 9.4)

γ = unit weight of water, 9810 N/m^3 (62.4 lbs/ft^3)

ρ = density of water, 1000 kg/m^3 (1.94 $slugs/ft^3$)

The C_B values listed are for expansion ratios, W_B /W_o, from 4 to 8 based on the configurations tested. C_B values developed for the $W_B/W_o = 4$ configuration are also valid for expansion ratios less than 4, but greater than or equal to 2, as long as the same number of roughness elements, N, are placed in the basin. For these smaller expansion ratios, this may require increasing the number of rows, N_r, to achieve the required N shown in Figure 9.3. The elements for all basins are arranged symmetrical about the basin centerline. All basins are flared to the width W_B of the corresponding abrupt expansion basin.

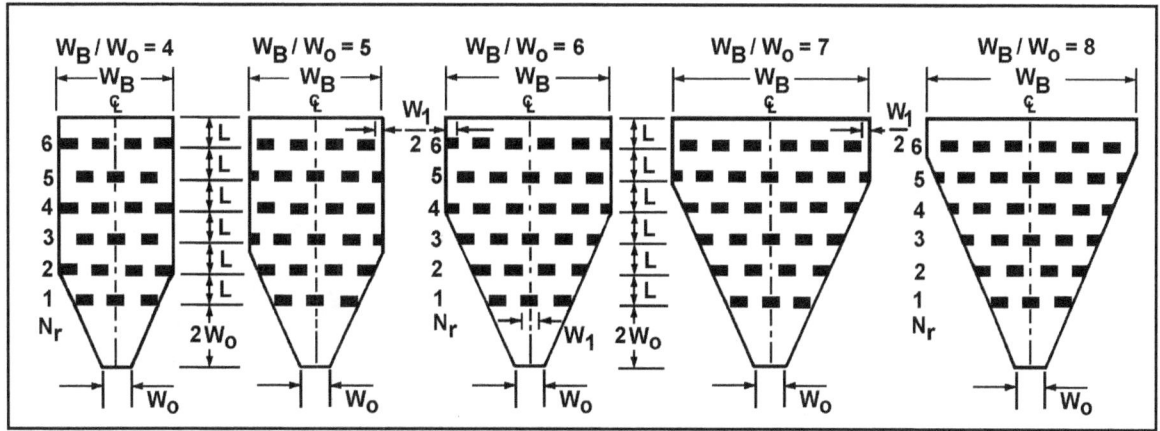

Figure 9.3. Roughness Configurations Tested

Table 9.1. Design Values for Roughness Elements

W_B/W_o		2 to 4			5			6			7	8
W_1/W_o		0.57			0.63			0.6			0.58	0.62
Rows (N_r)		4	5	6	4	5	6	4	5	6	5	6	6
Elements (N)		14	17	21	15	19	23	17	22	27	24	30	30
	h/y_A	L/h					Basin Drag Coefficient, C_B						
RECTANGULAR	0.91	6	0.32	0.28	0.24	0.32	0.28	0.24	0.31	0.27	0.23	0.26	0.22	0.22
	0.71	6	0.44	0.40	0.37	0.42	0.38	0.35	0.40	0.36	0.33	0.34	0.31	0.29
	0.48	12	0.60	0.55	0.51	0.56	0.51	0.47	0.53	0.48	0.43	0.46	0.39	0.35
	0.37	12	0.68	0.66	0.65	0.65	0.62	0.60	0.62	0.58	0.55	0.54	0.50	0.45
CIRCULAR	0.91	6	0.21	0.20	0.48	0.21	0.19	0.17	0.21	0.19	0.17	0.18	0.16	
	0.71	6	0.29	0.27	0.40	0.27	0.25	0.23	0.25	0.23	0.22	0.22	0.20	
	0.31	6	0.38	0.36	0.34	0.36	0.34	0.32	0.34	0.32	0.30	0.30	0.28	
	0.48	12	0.45	0.42	0.25	0.40	0.38	0.36	0.36	0.34	0.32	0.30	0.28	
	0.37	12	0.52	0.50	0.18	0.48	0.46	0.44	0.44	0.42	0.40	0.38	0.36	

Equation 9.1 is applicable for basins on less than 10 percent slopes. For basins with greater slopes, the weight of the water within the hydraulic jump must be considered in the expression. Equation 9.2 includes the weight component by assuming a straight-line water surface profile across the jump:

$$C_P \gamma y_o^2 W_o /2 + \rho V_o Q + w (\sin\theta) = C_B A_F N \rho V_A^2 /2 + \gamma Q^2 /(2V_B^2 W_B) + \rho V_B Q \quad (9.2)$$

where,

w   = weight of water within the basin

Volume = $(y_o W_o + y_A W_A) W_o + (0.75LQ/ V_B) [(N_r -1) - (W_B/W_o - 3) (1 - W_A /W_B)/2]$

Weight = (Volume) γ

θ = arc tan of the channel slope, S_o

N_r = number of rows of roughness elements

L   = longitudinal spacing between rows of elements.

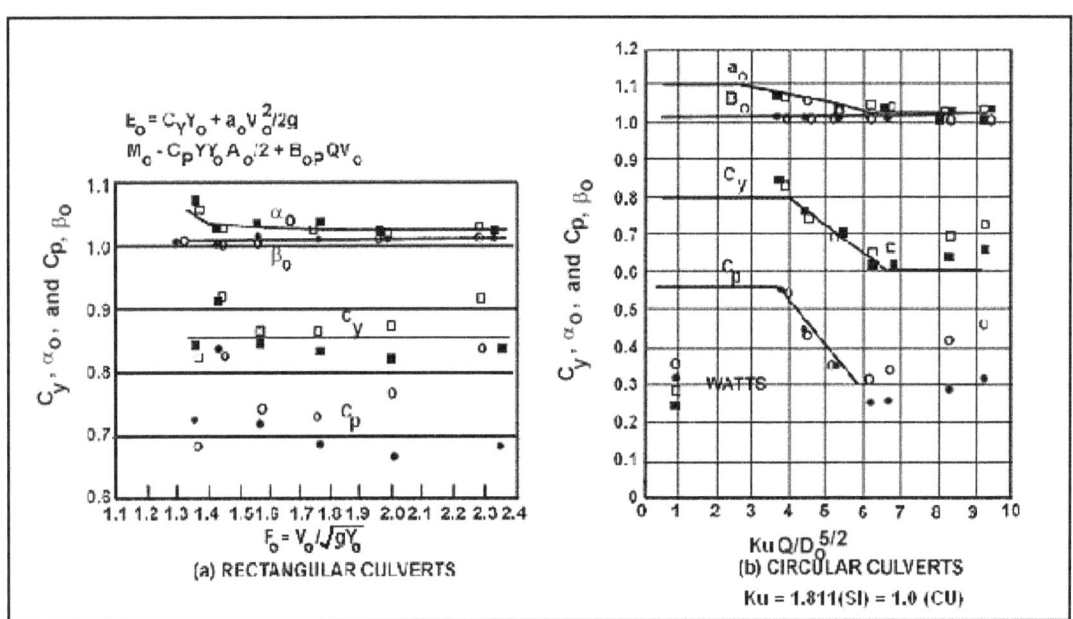

Figure 9.4. Energy and Momentum Coefficients (Simons, 1970)

The depth y_A at the beginning of the roughness elements can be determined from Figure 4.3 and Figure 4.4. These figures are based on slopes less than 10 percent. The velocity V_A can be computed using Equations 4.1 or 4.2. Where slopes are greater than 10 percent, V_A and y_A can be computed using the following energy equation written between the end of the culvert (section o) and two culvert widths downstream (section A).

$$2W_o S_o + y_A + (0.25) (Q/(W_A y_A))^2 /2g = y_o + 0.25(V_o^2 /(2g)) \tag{9.3}$$

where,

$W_A = W_o [4/(3Fr) + 1]$ which is adapted from Equations 4.3 and 4.4

Substantial splashing over the first row of roughness elements will occur if the elements are large and if the approach velocity is high. This problem can be addressed by locating the dissipator partially or totally within the culvert barrel, providing sufficient freeboard in the splash area, or providing some type of splash plate. If feasible both structurally and hydraulically, locating the dissipator partially or totally within the culvert barrel may result in economic, safety, and aesthetic advantages.

The necessary freeboard can be obtained from:

$$FB = h + y_A + 0.5(V_A \sin\phi)^2 /g \qquad (9.4)$$

where,

FB = necessary freeboard, m (ft)

h = roughness element height, m (ft)

y_A = depth approaching first row of roughness elements, m (ft)

g = 9.81 m/s^2 (32.2 ft/s^2)

ϕ = 45° (function of y_A/h and the Froude number but no relationship has been derived)

ϕ is believed to be a function of y_A/h and the Froude number, but no relationship has been derived. 45 degrees is a reasonable approximation.

Another solution is a splash shield, which has been investigated in the FHWA Hydraulics Laboratory by J.S. Jones (unpublished research). A splash shield is a plate with a stiffener suspended between the first two rows of roughness elements as shown in Figure 9.5. The height to the plate was selected rather arbitrarily as a function of the critical depth since flow usually passed through critical in the vicinity of the large roughness elements.

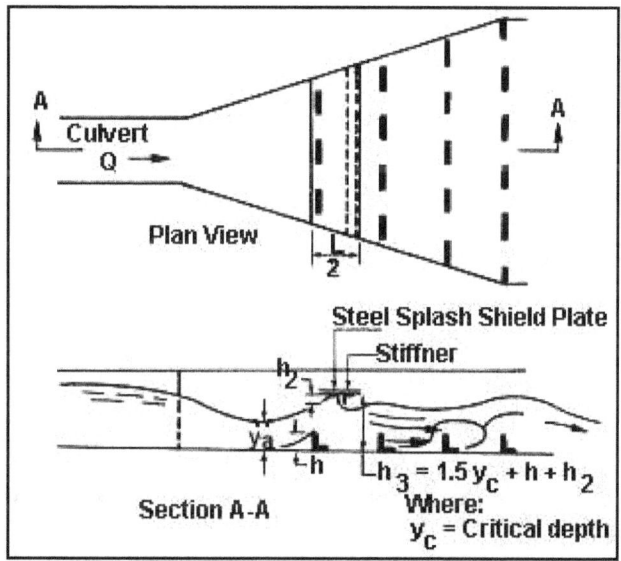

Figure 9.5. Splash Shield

Although the tests were made with abrupt expansions, the configurations recommended for use are the combination flared-abrupt expansion basins shown in Figure 9.3 and above. These basins contain the same number of roughness elements as the abrupt expansion basin. The flare divergence, u_e, is a function of the longitudinal spacing between rows of elements, L, and the culvert barrel width, W_o:

$$u_e = 4/7 + (10/7)L/W_o \qquad (9.5)$$

The design procedure for the CSU rigid boundary basin may be summarized as follows:

Step 1. Compute the velocity, V_o, depth, y_o, and Froude number, Fr, at the culvert outlet or, if the basin is partially or totally located within the culvert barrel, at the beginning of the flared portion of the barrel.

Step 2. Select a trial basin from Table 9.1 based on the W_B/W_o expansion ratio that best matches the site geometry or satisfies other constraints. Choose W_B/W_o, number of rows, N_r, number of elements, N, and ratios h/y_A and L/h.

Step 3. Determine the flow condition V_A and y_A at the approach to the roughness element field (two culvert widths downstream).

Calculate V_A using Equations 4.1 or 4.2.

For $4 < W_B/W_o < 8$, read y_A from Figure 4.3 and Figure 4.4.

For $W_B/W_o < 4$, compute y_A using Equation 9.3.

For slopes > 10 percent, use Equation 9.3 to find both V_A and y_A

Step 4. For the trial roughness height to depth ratio h/y_A and length to height ratio determine dissipator parameters from Figure 9.3:

    a.   roughness element height, h

    b.   longitudinal spacing between rows of elements, L

    c.   width of basin, W_B

    d.   element width, W_1, which equals element spacing

    e.   divergence, u_e

    f.   basin drag, C_B

    g.   roughness element frontal area, $A_F = W_1 h$

    h. C_p from Figure 9.4

    i.   Total basin length, $L_B = 2W_o + LN_r$. This provides a length downstream of the last row of elements equal to the length between rows, L.

Step 5. Confirm that the trial basin produces an exit velocity, V_B, and depth, y_B, that matches the downstream conditions. If W_B matches the downstream channel width or tailwater controls follow option 1. If W_B is less than the channel width follow option 2.

Option 1. Use the downstream depth, y_n, or tailwater if higher, to solve Equation 9.1 or Equation 9.2 for the quantity $C_B A_F N$. Using the C_B, A_F, and N values found in steps 2 and 4 compute $C_B A_F N$ (for basin). The basin value should be greater than or equal to the $C_B A_F N$ value from the equation. If not, select a new roughness configuration.

Option 2. Use the C_B, A_F, and N values found in steps 2 and 4 to solve for V_B in Equation 9.1 or 9.2. Three solutions for V_B are determined by trial and error: two positive roots and a negative root. The negative root may be discarded. The larger positive root is normally used for V_B. If V_B does not match the downstream velocity, select a new roughness configuration. If V_B is satisfactory, calculate y_B. Compare y_B to y_n. If $y_B < y_n$, use the smaller positive root for V_B and calculate y_B.

If tailwater is greater than y_B, V_B should be calculated using the tailwater depth and the trial basin checked using option 1.

Step 6. Sketch the basin. The basin layout is shown on Figure 9.3. The elements are symmetrical about the basin centerline and the spacing between elements is approximately equal to the element width. In no case, should this spacing be made less than 75 percent of the element width. The W_1/h ratio must be between 2 and 8 and at least half the rows of elements should have an element near the wall to prevent high velocity jets from traversing the entire basin length. Alternate rows are staggered so that all streamlines are disrupted.

Step 7. Consider erosion protection downstream of the basin. If option 1 (step 5) is applicable, the flow conditions leaving the basin match the downstream conditions and additional riprap downstream of the basin is not required unless site-specific concern regarding localized turbulence is identified. If, however, option 2 (step 5) is applicable, riprap is likely to be required until flow conditions fully transition to downstream conditions. Chapter 10 contains a section on riprap protection that may be used to size the required riprap.

Design Example: CSU Rigid Boundary Basin (SI)

Design a CSU rigid boundary basin to provide a transition from a RCB culvert to the natural channel. The basin should reduce velocities to approximately the downstream level. Given:

RCB = 2438 x 2438 mm culvert:

L   = 71.6 m

S   = 0.02 m/m

Q   = 39.64 m³/s

n   = 0.013

y_C = 2.987 m

y_n = 1.829 m

Downstream natural channel:

W   = 12.5 m (width)

TW   = 1.00 m (from downstream control)

Solution

Step 1. Compute the velocity, V_o, depth, y_o, and Froude number, Fr, at the culvert outlet

$y_o = y_n = 1.829$ m (from HDS 3)

$V_o = V_n = 8.87$ m/s

$Fr = V_o/(g\, y_o)^{1/2} = 8.87\, / \,[9.81(1.829)]^{1/2} = 2.1$

Step 2. Select a trial basin from Table 9.1 based on the $W_B\,/W_o$ expansion ratio which best matches the site geometry or satisfies other constraints. Choose $W_B\,/W_o$, number of rows, N_r, number of elements, N, and ratios h/y_A and L/h.

Channel Width/Culvert Width = 12.5/2.438 = 5.1

Try the following rectangular basin:

$W_B\,/W_o = 5$ and $W_1/W_o = 0.63$

$N_r = 4$ and $N = 15$

$h/y_A = 0.71$ and $L/h = 6$

Step 3. Determine the flow condition V_A and y_A at the approach to the roughness element field (two culvert widths downstream) = $2W_o$ or $2(2.438) = 4.876$ m.

Calculate V_A using Equations 4.1 or 4.2.

$V_A / V_o = 1.65 - 0.3Fr = 1.65 - 0.3(2.1) = 1.02$ from Equation 4.1

$V_A = 8.870(1.02) = 9.047$ m/s

For $4 < W_B / W_o < 8$, read y_A from Figure 4.3 or Figure 4.4.

$y_A / y_o = 0.33$ from Figure 4.3 for Fr = 2.1 and L = 2B

$y_A = 1.829(0.33) = 0.604$ m

Step 4. For the trial roughness height to depth ratio h/y_A and length to height ratio determine dissipator parameters from Table 9.1:

    a.   roughness element height, $h = (h/y_A)y_A = 0.71(0.604) = 0.429$ m

    b.   spacing between rows of elements, $L = (L/h)h = 6(0.429) = 2.574$ m

    c.   width of basin, $W_B = (W_B / W_o)W_o = 5(2.438) = 12.190$ m

    d.   element width, $W_1 = (W_1/W_o) W_o = 0.63(2.438) = 1.536$ m; use 1.524 m

    e.   divergence, $u_e = 4/7 + 10L/(7W_o) = 4/7 + 10(2.574)/(7(2.438)) = 2.07$ use 2

    f.   basin drag, $C_B = 0.42$

    g.   roughness element frontal area, $A_F = W_1 h = 1.524(0.429) = 0.65$ m^2

    h. C_p from Figure 9.4 = 0.7

    i.   Total basin length, $L_B = 2W_o + LN_r = 2(2.438) + 4(2.574) = 15.172$ m

Step 5. Since the width of the basin ($W_B = 12.190$ m) matches the downstream channel width (12.5 m), confirm trial basin using option 1. Use the normal flow conditions (V_n and y_n) and solve Equation 9.1 for $C_B A_F N$, which will be compared to $C_B A_F N$ for basin:

Calculate $C_B A_F N$ from Equation 9.1

y_n Downstream = 1.001 m

$V_B = Q/ (W_B y_n) = 39.64/ [12.190(1.001)] = 39.64/12.178 = 3.255$ m/s

$\rho V_o Q + C_p \gamma Y_o^2 W_o /2 = C_B A_F N \rho V_A^2/2 + \rho V_B Q + \gamma Q^2 /(2V_B^2 W_B)$ (Equation 9.1)

Terms with V_o and y_o: $1000(8.870) (39.64) + 0.7(9810) (1.829)^2 (2.438)/2 = 379609$

Terms with V_B: $1000(3.255) (39.64) + 9810(39.64)^2 /(2 (3.225)^2 (12.190)) = 189820$

Term with $C_B A_F N$ is $C_B A_F N (1000)(9.047)^2 /2 = 40924 \, C_B A_F N$

$(379609 - 189820) = 40924 \, C_B A_F N$

$C_B A_F N = 4.63$

Calculate $C_B A_F N$ for basin based on parameters determined in steps 2 and 4 (N =15, C_B = 0.42, A_F = 0.65 m^2). Using these values $C_B A_F N$ = 4.12. Since 4.12 is less than the 4.40 calculated from Equation 9.1, try a basin with more resistance (5 rows).

Step 4 (2nd iteration). For the trial roughness height to depth ratio h/y_A and length to height ratio determine dissipator parameters from Table 9.1: W_B / W_o = 5 that had N_r = 5, N =19, h/y_A = 0.71, L/h = 6, and C_B = 0.38.

    a.   roughness element height, h = $(h/y_A)y_A$ = 0.71(0.604) = 0.429 m

    b.   spacing between rows of elements, L = (L/h)h = 6(0.429) = 2.574 m

    c.   width of basin, W_B = (W_B / W_o)W_o = 5(2.438) = 12.190 m

    d.   element width, W_1 = (W_1/W_o) W_o = 0.63(2.438) =1.536 m; use 1.524 m (5 ft)

    e.   divergence, u_e = 4/7+10L/(7W_o) = 4/7 + 10(2.574)/(7(2.438)) = 2.07 use 2

    f.   basin drag, C_B = 0.38

    g.   roughness element frontal area, A_F = W_1 h = 1.524(0.429) = 0.654 m^2

    h. C_p from Figure 9.4 = 0.7

    i.   Total basin length, L_B = 2W_o + LN_r = 2(2.438) + 5(2.574) = 17.746 m

Step 5 (2nd iteration). Calculate $C_B A_F N$ from Equation 9.1.

    $C_B A_F N$ from Equation 9.1 = 4.63 (basin width did not change)

    Calculate $C_B A_F N$ for basin

    $C_B A_F N$ for basin = 0.38(0.654)(19) = 4.72 > 4.63 which is OK

Step 6. Sketch the basin and distribute roughness elements. (See sketch on following page. All dimensions shown in meters.)

    W_1/h = 1.524/0.429 = 3.55 which is between the target range of 2 to 8.

Step 7. Since the design matches the downstream conditions, minimum riprap will be required. See Chapter 10 for guidance on riprap placement.

Design Example: CSU Rigid Boundary Basin (CU)

Design a CSU rigid boundary basin to provide a transition from a RCB culvert to the natural channel. The basin should reduce velocities to approximately the downstream level. Given:

RCB = 8 ft x 8 ft culvert
L   = 235 ft
S   = 0.02 ft/ft
Q   = 1400 ft^3/s
n   = 0.013
y_c = 9.8 ft
y_n = 6.0 ft
Downstream natural channel:
W   = 41 ft (width)
TW = 3.3 ft (from downstream control)

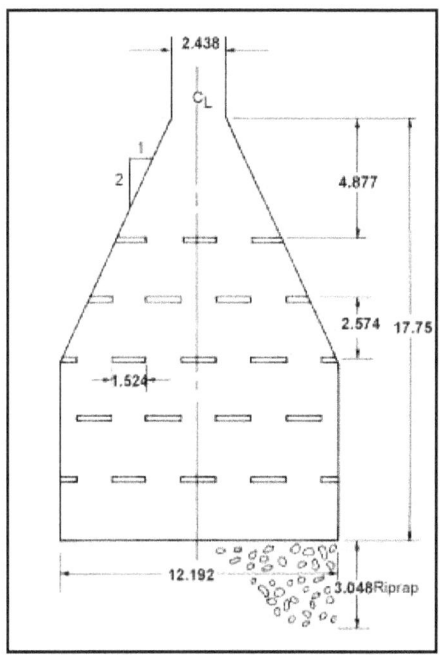

Sketch for the CSU Rigid Boundary Basin Design Example (SI)

Solution

Step 1. Compute the velocity, V_o, depth, y_o, and Froude number, Fr, at the culvert outlet

$y_o = y_n = 6.0$ ft (from HDS 3)

$V_o = V_n = 29.1$ ft/s

Fr $= V_o / (g\ y_o)^{1/2} = 29.1 / [32.2(6.0)]^{1/2} = 2.1$

Step 2. Select a trial basin from Table 9.1 based on the W_B / W_o expansion ratio which best matches the site geometry or satisfies other constraints. Choose W_B / W_o, number of rows, N_r, number of elements, N, and ratios h/y_A and L/h.

Channel Width/Culvert Width = 41/8 = 5.13

Try the following rectangular basin:

$W_B / W_o = 5$ and $W_1/W_o = 0.63$

$N_r = 4$ and N = 15

$h/y_A = 0.71$ and $L/h = 6$

Step 3. Determine the flow condition V_A and y_A at the approach to the roughness element field (two culvert widths downstream) = $2W_o$ or 2(8) = 16 ft.

Calculate V_A using Equations 4.1 or 4.2.

$V_A / V_o = 1.65 - 0.3$Fr $= 1.65 - 0.3(2.1) = 1.02$ from Equation 4.1

$V_A = 29.1(1.02) = 29.7$ ft/s

9-10

For $4 < W_B / W_o < 8$, read y_A from Figure 4.3 or 4.4.

$y_A / y_o = 0.33$ from Figure 4.3 for $Fr = 2.1$ and $L = 2B$

$y_A = 6.0(0.33) = 1.98$ ft

Step 4. For the trial roughness height to depth ratio h/y_A and length to height ratio determine dissipator parameters from Table 9.1:

    a.   roughness element height, $h = (h/y_A)y_A = 0.71(1.98) = 1.4$ ft

    b.   spacing between rows of elements, $L = (L/h)h = 6(1.4) = 8.4$ ft

    c.   width of basin, $W_B = (W_B / W_o)W_o = 5(8) = 40$ ft

    d.   element width, $W_1 = (W_1/W_o) W_o = 0.63(8) = 5.04$ ft; use 5 ft

    e.   divergence, $u_e = 4/7+10L/(7W_o) = 4/7 + 10(8.4)/(7(8)) = 2.07$ use 2

    f.   basin drag, $C_B = 0.42$

    g.   roughness element frontal area, $A_F = W_1 h = 5(1.4) = 7$ ft^2

    h. C_p from Figure 9.4 = 0.7

    i.   Total basin length, $L_B = 2W_o + LN_r = 2(8) + 4(8.4) = 49.6$ ft

Step 5. Since the width of the basin ($W_B = 40$ ft) matches the downstream channel width (41 ft) confirm trial basin using option 1. Use the normal flow conditions (V_n and y_n) and solve Equation 9.1 for $C_B A_F N$, which will be compared to $C_B A_F N$ for basin:

Calculate $C_B A_F N$ from Equation 9.1

y_n Downstream = 3.3 ft

$V_B = Q/(W_B y_n) = 1400/ [40(3.3)] = 1400/132 = 10.6$ ft/s

$\rho V_o Q + C_p \gamma Y_o^2 W_o /2 = C_B A_F N \rho V_A^2/2 + \rho V_B Q + \gamma Q^2 /(2V_B^2 W_B)$ (Equation 9.1)

Terms with V_o and y_o: $1.94(29.1) (1400) + 0.7(62.4) (6)^2 (8)/2 = 85,325.5$

Terms with V_B: $1.94(10.6) (1400) + 62.4(1400)^2 / (2 (10.6)^2 (40)) = 42,395.9$

Term with $C_B A_F N$ is $C_B A_F N (1.94) (30.6)^2 /2 = 908.3(C_B A_F N)$

$(85,325.5 - 42,395.9 = 908.3(C_B A_F N)$

$C_B A_F N = 47.3$

Calculate $C_B A_F N$ for basin based on parameters determined in steps 2 and 4 (($N_r = 4$, $C_B = 0.42$, $A_F = 7$ ft^2). Using these values $C_B A_F N = 44.1$. Since 44.1 is less than the 47.3 calculated from Equation 9.1, try a basin with more resistance (5 rows).

Step 4 (2nd iteration). For the trial roughness height to depth ratio h/y_A and length to height ratio determine dissipator parameters from Table 9.1: $W_B / W_o = 5$ that had $N_r = 5$, $N = 19$, $h/y_A = 0.71$, $L/h = 6$, and $C_B = 0.38$.

    a.   roughness element height, $h = (h/y_A)y_A = 0.71(1.98) = 1.4$ ft

    b.   spacing between rows of elements, $L = (L/h)h = 6(1.4) = 8.4$ ft

    c.   width of basin, $W_B = (W_B / W_o)W_o = 5(8) = 40$ ft

    d.   element width, $W_1 = (W_1/W_o) W_o = 0.63(8) = 5.04$ ft; use 5 ft

e. divergence, $u_e = 4/7 + 10L/(7W_o) = 4/7 + 10(8.4)/(7(8)) = 2.07$ use 2

f. basin drag, $C_B = 0.38$

g. roughness element frontal area, $A_F = W_1 h = 5(1.4) = 7 \text{ ft}^2$

h. C_p from Figure 9.4 = 0.7

i. Total basin length, $L_B = 2W_o + LN_r = 2(8) + 5(8.4) = 58 \text{ ft}$

Step 5 (2nd iteration). Calculate $C_B A_F N$ from Equation 9.1.

$C_B A_F N$ from Equation 9.1 = 47.3 (basin width did not change)

Calculate $C_B A_F N$ for basin

$C_B A_F N$ for basin = 0.38(7) (19) = 50.5 > 47.3 which is OK

Step 6. Sketch basin and distribute roughness elements. (See following figure). All dimensions shown in feet.)

$W_1/h = 5/1.4 = 3.57$ which is between the target range of 2 to 8.

Step 7. Since the design matches the downstream conditions, minimum riprap will be required. See Chapter 10 for guidance on riprap placement.

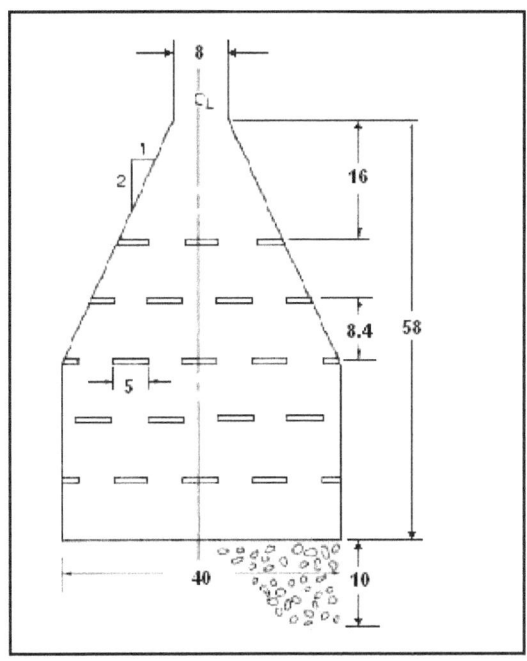

Sketch for the CSU Rigid Boundary Basin Design Example (CU)

9.2   CONTRA COSTA BASIN

The Contra Costa energy dissipator (Keim, 1962) was developed at the University of California, Berkeley, in conjunction with Contra Costa County, California. It is intended for use primarily in urban areas with defined tailwater channels. A sketch of the dissipator is shown in Figure 9.6.

The dissipator was developed to be self-cleaning with minimum maintenance requirements. It is best suited to small and medium size culverts of any cross section where the depth of flow at the outlet is less than or equal to half the culvert height, but is applicable over a wide range of culvert sizes and operating conditions as noted in Table 1.1. The flow leaving the dissipator will be at minimum energy when operating without tailwater. When tailwater is present, the performance will improve. Field experience with this dissipator has been limited. Designers should not extrapolate parameter values in this guidance beyond the ranges cited for the model tests.

Equation 9.6 was obtained by testing model Contra Costa dissipators that had L_2 / h_2 ratios from 2.5 to 7. The equation is in terms of culvert exit velocity, V_o, and depth, y_o, for a circular culvert.

$$\frac{L_2}{h_2} = 1.2 Fr^2 \left(\frac{h_2}{y_o} \right)^{-1.83} \tag{9.6}$$

where,

y_o = outlet depth, m (ft)

V_o = outlet velocity, m/s (ft/s)

Fr = $V_o/(g\, y_o)^{1/2}$

h_2 = height of large baffle, m (ft)

L_2 = length from culvert exit to large baffle, m (ft)

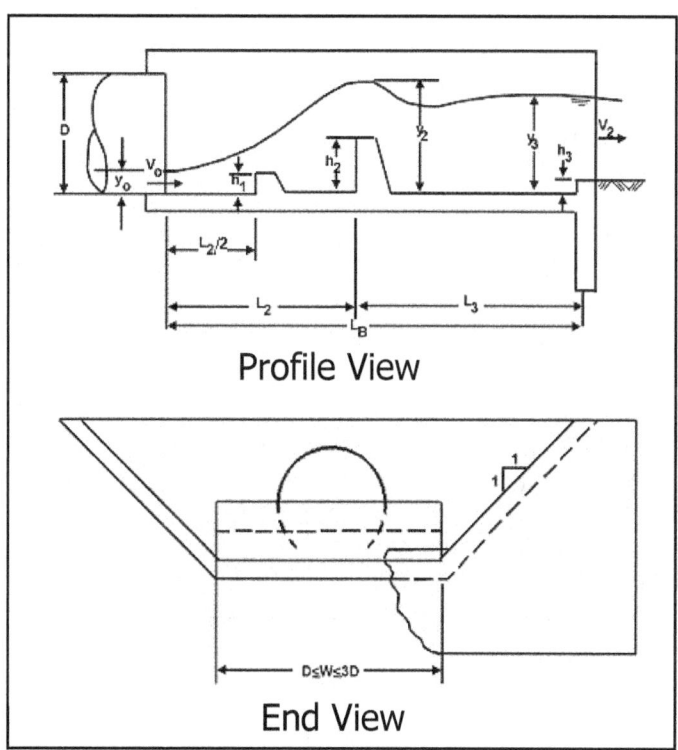

Profile View

End View

Figure 9.6. Contra Costa Basin

Equation 9.6 is generalized for other shapes by substituting the equivalent depth of flow, y_e, for y_o. Equivalent depth is found by converting the area of flow at the culvert outfall to an equivalent rectangular cross section with a width equal to twice the depth of flow. For box culverts, $y_e = y_n$ or y_{brink}.

$$\frac{L_2}{h_2} = 1.35Fr^2\left(\frac{h_2}{y_e}\right)^{-1.83} \tag{9.7a}$$

or:

$$\frac{h_2}{y_e} = \left(\frac{1.35Fr^2}{\dfrac{L_2}{h_2}}\right)^{0.546} \tag{9.7b}$$

where,

y_e = equivalent depth, $(A/2)^{1/2}$, m (ft)
A = outlet flow area, m^2 (ft^2)
V_o = outlet velocity, m/s (ft/s)
Fr = $V_o/(g\,y_e)^{1/2}$

Equation 9.7b is solved by assuming a value of L_2/h_2 between 2.5 and 7. A trial height of the second baffle, h_2, can be determined. If the recommended $L_2/h_2 = 3.5$ value is substituted into Equation 9.7, the design equation becomes Equation 9.8. The value of h_2/y_e should always be greater than unity.

$$\frac{h_2}{y_e} = 0.595Fr^{1.092} \tag{9.8}$$

After determining the values of h_2 and L_2, the length from the large baffle to the end sill, L_3, can be obtained using Equation 9.9.

$$\frac{L_3}{L_2} = 3.75\left(\frac{h_2}{L_2}\right)^{0.68} \tag{9.9}$$

The height of the small baffle, h_1, is half the height of the large baffle, h_2. The position of the small baffle is half way between the culvert outlet and the large baffle or $L_2/2$. Side slopes of the trapezoidal basin for all experimental runs were 1:1 (V:H). The width of basin, W, may vary from one to three times the width of the culvert. The floor of the basin should be essentially level. The height of the end sill, h_3, may vary from $0.06y_2$ to $0.10y_2$. After obtaining satisfactory basin dimensions, the approximate maximum water surface depth, y_2, without tailwater, can be obtained from Equation 9.10 which is for basins with $W_B/W_o = 2$. The depth y_3 is equal to y_c for the basin + h_3.

$$\frac{y_2}{h_2} = 1.3\left(\frac{L_2}{h_2}\right)^{0.36} \tag{9.10}$$

The following steps outline the design procedure for the design of the Contra Costa basin:

Step 1. Determine the flow conditions at the outfall of the culvert for the design discharge. If the depth of flow at the outlet, y_o, is $D/2$ or less, the Contra Costa basin is applicable.

Step 2. Compute equivalent depth, y_e, and Froude number, Fr.

$y_e = y_o$ for rectangular culvert

$y_e = (A/2)^{1/2}$ for other shapes

$Fr = V_o / (gy_e)^{1/2}$

Step 3. The width of the basin floor, W_B, is selected to conform to the natural channel, but must be $1W_o$ to $3W_o$. If there is no defined channel, the width should be no greater than 3 times the culvert width. The basin side slopes should be 1:1.

Step 4. Assume a value of L_2/h_2 between 2.5 and 7. If $L_2/h_2 = 3.5$, use Equation 9.8 to determine h_2. Use Equation 9.7 for other values. Calculate $L_2 = 3.5 h_2$. Calculate the first baffle height, $h_1 = 0.5h_2$ and position, $L_1 = 0.5L_2$.

Step 5. Determine the length from the large baffle to the end sill, L_3, using Equation 9.9. Repeat the procedure, if necessary, until a dissipator is defined which optimizes the design requirements.

Step 6. Estimate the approximate maximum water surface depth without tailwater, y_2, using Equation 9.10 which is for $W_B = 2W_o$. Set the end sill height, h_3, between $0.06y_2$ and $0.1y_2$. If the above dimensions are compatible with the topography at the site, the dimensions are final. If not, a different value of L_2 /h_2 is selected and the design procedure repeated.

Step 7. Determine the basin exit depth, $y_3 = y_c$ and exit velocity, $V_2 = V_c$.

$Q^2/g = (A_c)^3/T_c = [y_c(W_B + y_c)]^3/ (W_B + 2y_c)$ (Substituting for A_c and T_c using the properties of a trapezoid.)

$V_c = Q/A_c$

Step 8. Riprap may be necessary downstream especially for the low tailwater cases. See Chapter 10 for design recommendations. Freeboard to prevent overtopping and a cutoff wall to prevent undermining of the basin also should be considered.

Design Example: Contra Costa Basin (SI)

Determine the design dimensions for a Contra Costa basin. Given:

D   = 1.219 m diameter RCP culvert
Q   = 8.49 m³/s
y_o = 0.701 m
V_o = 12.192 m/s
Channel bottom width = 2.438 m

Solution

Step 1. Determine the flow conditions at the outfall of culvert for the design discharge.

$y_o = 0.701$ m is approximately $D/2$, OK.

Step 2. Compute equivalent depth, y_e, and Froude number, Fr.

Using Equations 7.11 and 7.13, flow area in the culvert = 0.696 m².

$y_e = (A/2)^{1/2} = (0.696/2)^{1/2} = 0.590$ m

$Fr = V_o / (gy_e)^{1/2} = 12.192 / [9.81(0.590)]^{1/2} = 5.07$

Step 3. The width of the basin floor, W_B, is selected to conform to the natural channel. The basin side slopes should be 1:1 (V:H).

Set W = 2.438 m (channel bottom width). $1 \leq W/D \leq 3$ OK

Step 4. Assume $L_2/h_2 = 3.5$, use Equation 9.8 to determine h_2. Calculate $L_2 = 3.5\ h_2$. Calculate the first baffle height, $h_1 = 0.5h_2$ and position, $L_1 = 0.5L_2$

$h_2 / y_e = 0.595\ Fr^{1.092} = 0.595\ (5.07)^{1.092} = 3.5$

$h_2 = y_e (h_2 / y_e) = 0.590\ (3.50) = 2.065$ m

$L_2 = 3.5\ h_2 = 3.5\ (2.065) = 7.228$ m

$h_1 = 0.5h_2 = 0.5\ (2.065) = 1.032$ m

$L_1 = 0.5\ L_2 = 0.5\ (7.228) = 3.614$ m

Step 5. Determine the length from the large baffle to the end sill, L_3, using Equation 9.9. Repeat the procedure, if necessary, until a dissipator is defined which optimizes the design requirements.

$L_3 / L_2 = 3.75(h_2 /L_2)^{0.68} = 3.75(1/3.5)^{0.68} = 1.6$

$L_3 = (L_3 / L_2)\ L_2 = 1.6\ (7.228) = 11.56$ m

Step 6. Estimate the approximate maximum water surface depth without tailwater, y_2, using Equation 9.10 which is for $W_B = 2D$. Determine end sill height, $h_3 = 0.1y_2$

$y_2 / h_2 = 1.3(L_2 /h_2)^{0.36} = 1.3(3.5)^{0.36} = 2.04$

$y_2 = (y_2 / h_2)\ h_2 = (2.04)\ 2.065 = 4.21$ m

$h_3 = 0.1(y_2) = 0.1(4.21) = 0.42$ m

A summary of physical dimensions is shown in the following table.

	First Baffle	Second Baffle	End Sill
Distance from exit (m)	3.61	7.23	18.79
Height (m)	1.03	2.07	0.42

Step 7. Determine the basin exit depth, $y_3 = y_c$ and exit velocity, $V_2 = V_c$.

$Q^2/g = (A_c)^3/T_c = [y_c (W_B + y_c)]^3/ (W_B + 2y_c)$

$8.49^2/9.81 = 7.35 = [y_c (2.438 + y_c)]^3/ (2.438 + 2y_c)$

By trial and success, $y_c = 0.938$ m, $T_c = 4.314$ m, $A_c = 3.17$ m²

$V_c = Q/A_c = 8.49/3.17 = 2.68$ m/s

Step 8. Riprap may be necessary downstream especially for the low tailwater cases. See Chapter 10 for design recommendations. Freeboard to prevent overtopping and a cutoff wall to prevent undermining of the basin also should be considered.

Design Example: Contra Costa Basin (CU)

Determine the design dimensions for a Contra Costa basin. Given:

D = 4 ft diameter RCP culvert
Q = 300 ft³/s
y_o = 2.3 ft
V_o = 40 ft/s
Channel bottom width = 8 ft

Solution

Step 1. Determine the flow conditions at the outfall of culvert for the design discharge.

y_o = 2.3 ft is approximately D/2, OK.

Step 2. Compute equivalent depth, y_e, and Froude number, Fr.

Using Equations 7.11 and 7.13, flow area in the culvert = 7.5 ft².

$y_e = (A/2)^{1/2} = (7.5/2)^{1/2} = 1.94$ ft

$Fr = V_o / (gy_e)^{1/2} = 40 / [32.2(1.94)]^{1/2} = 5.06$

Step 3. The width of the basin floor, W_B, is selected to conform to the natural channel. The basin side slopes should be 1H:1V.

Set W = 8 ft (channel bottom width). $1 \le W/D \le 3$ OK

Step 4. Assume L_2/h_2 = 3.5, use equation 9.8 to determine h_2. Calculate L_2 = 3.5 h_2. Calculate the first baffle height, $h_1 = 0.5h_2$ and position, $L_1 = 0.5L_2$

$h_2 / y_e = 0.595\ Fr^{1.092} = 0.595\ (5.06)^{1.092} = 3.5$

$h_2 = y_e (h_2 / y_e) = 1.94 (3.5) = 6.8$ ft

$L_2 = 3.5\ h_2 = 3.5 (6.8) = 23.8$ ft

$h_1 = 0.5h_2 = 0.5 (6.8) = 3.4$ ft

$L_1 = 0.5\ L_2 = 0.5 (23.8) = 11.9$ ft

Step 5. Determine the length from the large baffle to the end sill, L_3, using Equation 9.9. Repeat the procedure, if necessary, until a dissipator is defined which optimizes the design requirements.

$L_3 / L_2 = 3.75(h_2 /L_2)^{0.68} = 3.75(1/3.5)^{0.68} = 1.6$

$L_3 = (L_3 / L_2)\ L_2 = 1.6 (23.8) = 38.1$ ft

Step 6. Estimate the approximate maximum water surface depth without tailwater, y_2, using Equation 9.10 which is for W_B = 2D. Determine end sill height, $h_3 = 0.1y_2$.

$y_2 / h_2 = 1.3(L_2 /h_2)^{0.36} = 1.3(3.5)^{0.36} = 2.04$

$y_2 = (y_2 / h_2)\ h_2 = (2.04)\ 6.8 = 13.9$ ft

$h_3 = 0.1(y_2) = 0.1(13.9) = 1.4$ ft

A summary of physical dimensions is shown in the following table.

	First Baffle	Second Baffle	End Sill
Distance from exit (ft)	11.9	23.8	61.9
Height (ft)	3.4	6.8	1.4

Step 7.   Determine the basin exit depth, $y_3 = y_c$ and exit velocity, $V_2 = V_c$.

$Q^2/g = (A_c)^3/T_c = [y_c(W_B + y_c)]^3/(W_B + 2y_c)$

$300^2/32.2 = 2795 = [y_c(8 + y_c)]^3/(8 + 2y_c) = 34.06^3/14.15 = 2792$

By trial and success, $y_c = 3.075$ ft, $T_c = 14.15$ ft, $A_c = 34.06$ ft^2

$V_c = Q/A_c = 300/34.06 = 8.8$ ft/s

Step 8.   Riprap may be necessary downstream especially for the low tailwater cases. See Chapter 10 for design recommendations. Freeboard to prevent overtopping and a cutoff wall to prevent undermining of the basin also should be considered.

9.3   HOOK BASIN

The Hook basin was developed at the University of California in cooperation with the California Division of Highways and the Bureau of Public Roads (MacDonald, 1967). The basin was originally developed for large arch culverts with low tailwater, but can be used with box or circular conduits. The dissipator can be used for culvert outlet Froude numbers from 1.8 to 3.0. Two hydraulic model studies were conducted: (1) a basin with wingwalls warped from vertical at the culvert outlet to side slopes of 1:1.5 (V:H) at the end sill and a tapered basin floor which is discussed in Section 9.3.1 and (2) a trapezoidal channel of uniform cross section which is discussed which is discussed in Section 9.3.2.

9.3.1   Hook Basin with Warped Wingwalls

The hook basin with warped wingwalls is shown in Figure 9.7. The design procedure is deterministic except for selecting the width of the hooks. Judgment is necessary in choosing this dimension to insure that the width is sufficient for effective energy dissipation, but not so great that flow passage between the hooks is inadequate. A ratio of $W_4/W_o = 0.16$, which was the minimum tested, is recommended. Each design should be checked to see that the spacing between hooks is 1.5 to 2.5 times the hook width.

The height of wingwalls, h_6, should be at least twice the flow depth at the culvert exit or $2y_e$. This height is based on the highest water surface elevations observed in the basin during the study. Therefore, setting $h_6 = 2y_e$ does not provide freeboard to contain splashing. Depending on the site conditions, the designer should provide for additional freeboard.

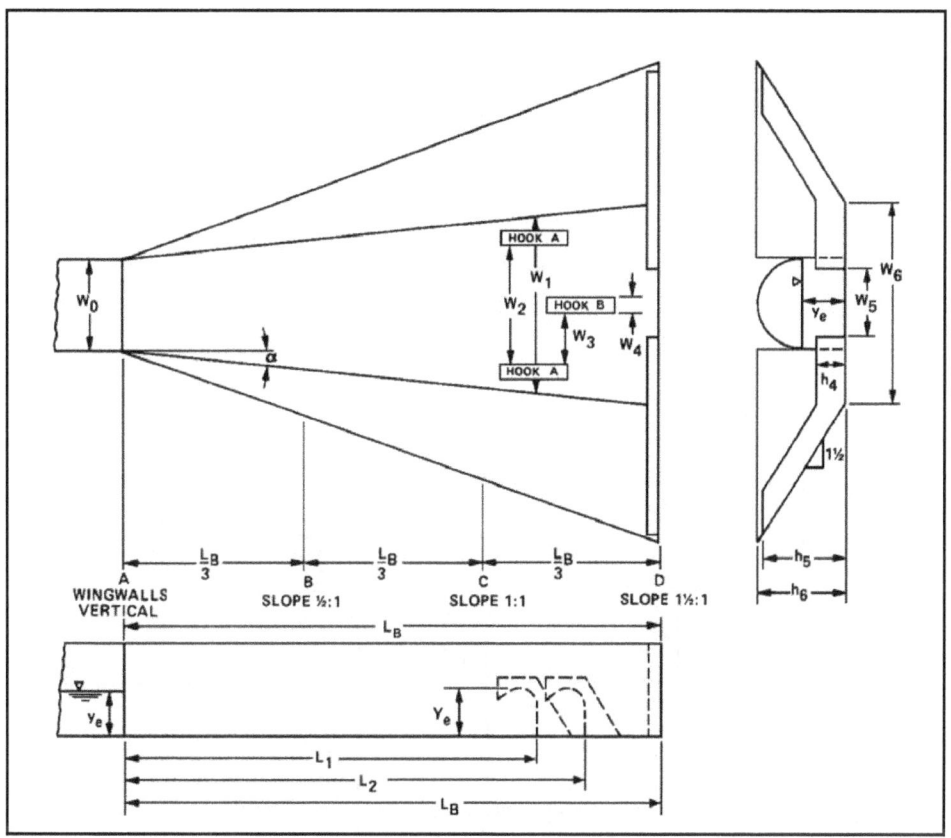

Figure 9.7. Hook Basin with Warped Wingwalls

The best range of design dimensions that were tested are indicated in the design procedure. In most cases, the ratio that will produce the smallest dimension was used. The recommended hook configuration is shown in Figure 9.8. The recommended dimensions are:

1. $h_3 = y_e$
2. $h_2 = 1.28h_1$
3. $h_1 = y_e/1.4$
4. $\beta = 135°$
5. $r = 0.4h_1$

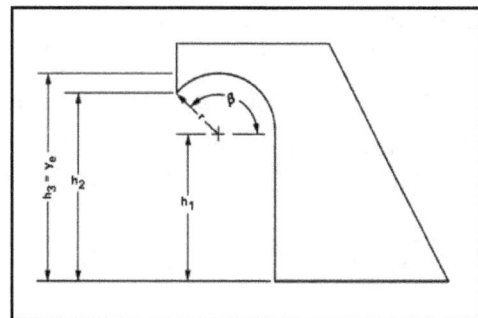

Figure 9.8. Hook for Warped Wingwall Basin

A flare angle, α, of 5.7 degrees per side (tanα = 0.10) is the optimum value for Fr > 2.45. Increasing the length beyond $L_B = 3W_o$ does not improve basin performance. The effectiveness of the dissipator falls off rapidly with increasing Froude number regardless of hook width, for flare angle exceeding 5.7 degrees. The exit velocity of the dissipator, V_B, is estimated from Figure 9.9. The higher the velocity ratio, V_o/V_B, the more effective the basin is in dissipating energy and distributing the flow downstream.

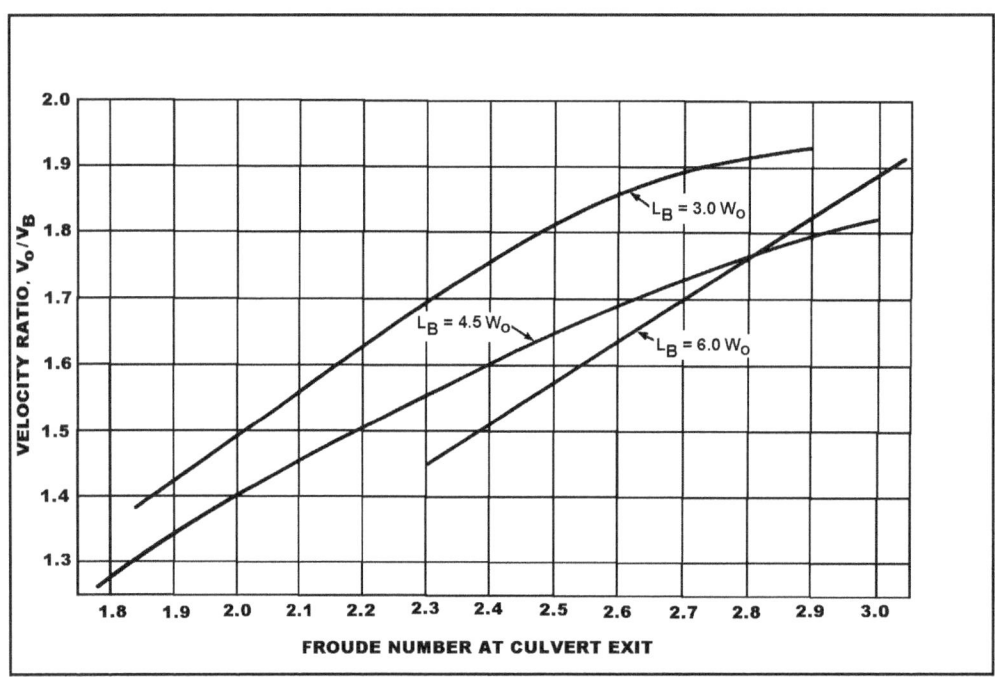

Figure 9.9. Velocity Ratio for Hook Basin With Warped Wingwalls

Depending on final velocity and soil conditions, some scour can be expected downstream of the basin. The designer should, where necessary, provide riprap protection in this area. Chapter 10 contains design guidance for riprap. Where large debris is expected, armor plating the upstream face of the hooks with steel is recommended.

The recommended design procedure for a Hook basin with warped wingwalls is as follows:

Step 1. Compute the culvert outlet velocity, V_o, equivalent depth, y_e, and Froude number, Fr = $V_o/(gy_e)^{1/2}$. If 1.8 < Fr < 3.0, proceed with design.

Step 2. Compute the downstream channel velocity, V_n, and depth, y_n.

Step 3. Select width of the basin at the basin exit, W_B ($W_B = W_6$), and compute L_B. W_6 should be approximately equal to the channel width, if the downstream channel is defined.

$L_B = (W_6 - W_o) / (2\tan\alpha)$, use $\alpha = 5.7°$ (tan = 0.10)

Step 4. Compute the position and spacing of the hooks (see Figure 9.7):

    a. distance to first hooks, $L_1 = 0.75\, L_B$

       (allowable range: $0.75 < L_1/L_B < 0.80$)

    b. width at first hooks, $W_1 = 2L_1(\tan\alpha) + W_o$

    c. distance between first (row A) hooks, $W_2 = 0.66\, W_1$

       (allowable range: $0.66 < W_2/W_1 < 0.70$)

    d. distance to second (row B) hook, $L_2 = 0.83\, L_B$

       (allowable range: $0.83 < L_2/L_B < 0.89$)

    e. width of hooks, $W_4 = 0.16 W_o$

    f. lateral spacing between A and B hook, $W_3 = (W_2 - W_4)/2$

       If spacing does not satisfy $1.5 < W_3/W_4 < 2.5$, adjust W_4.

Step 5. Compute hook dimensions (see Figure 9.8):

    a. height to center of radius, $h_1 = y_e/1.4$

    b. height to point, $h_2 = 1.28 h_1$

    c. height to top of radius, $h_3 = y_e$

    d. angle of radius, $\beta = 135°$

    e. radius, $r = 0.4 h_1$

Step 6. Compute the end sill and wingwall dimensions (see Figure 9.7):

    a. height of end sill, $h_4 = 0.67 y_e$

    b. width of slot in end sill, $W_5 = 0.33 W_6$

    c. height to top of warped wingwall, $h_6 = 2 y_e$ minimum

    d. height to top of end sill, $h_5 = 0.94 h_6$

Step 7. Find V_o/V_B from Figure 9.9 and compute basin exit velocity, V_B. Compare V_B with V_n from step 2. If V_B is unacceptable, adjust basin length. Assess scour potential downstream based on soil condition and outlet velocity. If riprap is needed, see Chapter 10.

Step 8. Where large debris is expected, the upstream face of the hooks should be armored with steel.

Design Example: Hook Basin With Warped Wingwalls (SI)

Determine dimensions for a Hook basin with warped wingwalls (see Figure 9.7) for a long concrete semicircular arch culvert that is 3.658 m wide and 3.658 m from the floor to the crown. Given:

S_o = 0.020 m/m

n   = 0.012

Q   = 76.41 m³/s

y_e =   1.829 m

V_o =   11.43 m/s

The downstream channel has a trapezoidal shape with the following properties:

W_c =   6.096 m

S_o =   0.020 m/m

Z =   1.5

n =   0.030

V_n =   5.27 m/s

y_n =   1.676 m

Solution

Step 1.   Compute the culvert outlet velocity, V_o, equivalent depth, y_e, and Froude number, Fr, $V_o/(gy_e)^{1/2}$. V_o = 11.43 m/s and y_e = 1.829 m were given.

$$Fr_o = V_o / (gy_e)^{1/2} = 11.43/ (9.81 \times 1.829)^{1/2} = 2.70$$

Since 1.8 < 2.70 < 3, proceed to step 2.

Step 2.   Compute the downstream channel velocity, V_n, and depth, y_n. V_n = 5.27 m/s and y_n = 1.676 m were given.

Step 3.   Select W_6 and compute L_B.

Use $W_6 = W_c$ = 6.096 m and $\tan\alpha$ = 0.10.

$$L_B = (W_6 - W_o) / (2\tan\alpha) = (6.096 - 3.658)/ [2(0.10)] = 12.19 \text{ m or } 3.3W_o$$

Step 4.   Compute the position and spacing of the hooks (see Figure 9.7):

     a.   distance to first hooks, $L_1 = 0.75 L_B$ = 0.75(12.19) = 9.143 m

     b.   width at first hooks, $W_1 = 2L_1(\tan\alpha) + W_o$ = 2(9.143)(0.1) + 3.658 = 5.487 m

     c.   distance between first (row A) hooks, $W_2 = 0.66 W_1$ = 0.66(5.487) = 3.621 m

     d.   distance to second (row B) hook, $L_2 = 0.83 L_B$ = 0.83(12.19) = 10.118 m

     e.   width of hooks, $W_4 = 0.16W_o$ = 0.16(3.658) = 0.585 m

     f.   lateral spacing between A and B hook, W_3

        $W_3 = (W_2 - W_4)/2$ = (3.621 - 0.585) /2 = 1.518 m

        W_3/W_4 = 1.518/0.585 = 2.6, which does not satisfy 1.5 < W_3/W_4 < 2.5

        Adjust $W_4 = W_3/2.5$ = 1.518/2.5 = 0.607 m

Step 5.   Compute hook dimensions (see Figure 9.8):

     a.   height to center of radius, $h_1 = y_e /1.4$ = 1.829/1.4 = 1.306 m

     b.   height to point, $h_2 = 1.28h_1$ = 1.28(1.306) = 1.672 m

     c.   height to top of radius, $h_3 = y_e$ = 1.829 m

     d.   angle of radius, β = 135°

     e.   radius, $r = 0.4h_1$ = 0.4(1.306) = 0.522 m

Step 6. Compute the end sill and wingwall dimensions (see Figure 9.7):

    a.   height of end sill, $h_4 = 0.67y_e = 0.67(1.829) = 1.225$ m

    b.   width of slot in end sill, $W_5 = 0.33W_6 = 0.33(6.096) = 2.012$ m

    c.   height to top of warped wingwall, $h_6 = 2y_e = 2(1.829) = 3.658$ m

    d.   height to top of end sill, $h_5 = 0.94h_6 = 0.94(2y_e) = 3.439$ m

Step 7. Find V_o/V_B from Figure 9.9 and compute V_B. Compare with V_n from step 2. Assess scour potential downstream based on soil condition and outlet velocity. If riprap is needed, see Chapter 10.

With Fr = 2.7 and L_B = 3.3 W_o, V_o/V_B will be less than 1.9 making $V_B \cong 11.43/1.9 = 6.016$ m/s. This is somewhat higher than the normal velocity in the downstream channel indicating riprap protection may be desirable. See Chapter 10.

The dissipator design is shown on the sketch below. (All dimensions are in meters.)

Step 8. Since no large debris is expected at this site, the hook face will not be armored with steel.

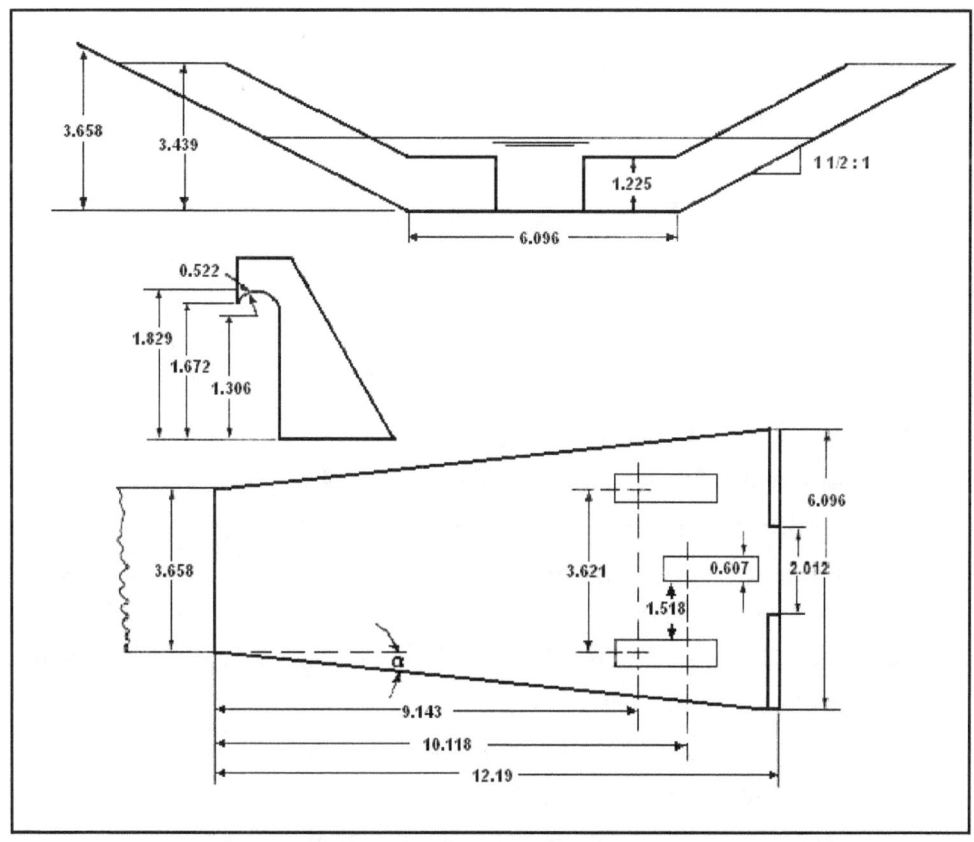

Sketch for the Hook Basin with Warped Wingwalls Design Example (SI)

9-23

Design Example: Hook Basin With Warped Wingwalls (CU)

Determine dimensions for a Hook basin with warped wingwalls (see Figure 9.7) for a long concrete semicircular arch culvert that is 12 ft wide and 12 ft from the floor to the crown. Given:

S_o = 0.020 ft/ft
n = 0.012
Q = 2700 ft³/s
y_e = 6 ft
V_o = 37.5 ft/s

The downstream channel has a trapezoidal shape with the following properties:

W_c = 20 ft
S_o = 0.020 ft/ft
Z = 1.5
n = 0.030
V_n = 16.1 ft/s
y_n = 5.5 ft

Solution

Step 1. Compute the culvert outlet velocity, V_o, equivalent depth, y_e, and Froude number, $Fr\ V_o/(gy_e)^{1/2}$. V_o = 37.5 ft/s and y_e = 6 ft were given.

$$Fr_o = V_o / (gy_e)^{1/2} = 37.5/ (32.2 \times 6)^{1/2} = 2.7$$

Since 1.8 < 2.7 < 3, proceed with step 2.

Step 2. Compute the downstream channel velocity, V_n, and depth, y_n. V_n = 16.1 ft/s and y_n = 5.5 ft were given.

Step 3. Select W_6 and compute L_B.

Use $W_6 = W_c = 20$ and $\tan\alpha = 0.10$.

$$L_B = (W_6 - W_o) / (2\tan\alpha) = (20 - 12)/ [2(0.10)] = 40 \text{ ft or } 3.3W_o$$

Step 4. Compute the position and spacing of the hooks (see Figure 9.7):

    a.   distance to first hooks, $L_1 = 0.75\ L_B = 0.75(40) = 30$ ft

    b.   width at first hooks, $W_1 = 2L_1(\tan\alpha) + W_o = 2(30)(0.1) + 12 = 18$ ft

    c.   distance between first (row A) hooks, $W_2 = 0.66\ W_1 = 0.66(18) = 11.9$ ft, use 12 ft

    d.   distance to second (row B) hook, $L_2 = 0.83\ L_B = 0.83(40) = 33.2$ ft, use 33 ft

    e.   width of hooks, $W_4 = 0.16W_o = 0.16(12) = 1.92$ ft, use 2 ft

    f.   lateral spacing between A and B hook, W_3

        $W_3 = (W_2 - W_4)/2 = (12 - 2) /2 = 5$ ft

        $W_3/W_4 = 5/2 = 2.5$, which satisfies $1.5 < W_3/W_4 < 2.5$

Step 5. Compute hook dimensions (see Figure 9.8):

    a.   height to center of radius, $h_1 = y_e /1.4 = 6/1.4 = 4.3$ ft

b. height to point, $h_2 = 1.28h_1 = 1.28(4.3) = 5.5$ ft

c. height to top of radius, $h_3 = y_e = 6$ ft

d. angle of radius, $\beta = 135°$

e. radius, $r = 0.4h_1 = 0.4(4.3) = 1.72$ ft

Step 6. Compute the end sill and wingwall dimensions (see Figure 9.7):

a. height of end sill, $h_4 = 0.67y_e = 0.67(6) = 4$ ft

b. width of slot in end sill, $W_5 = 0.33W_6 = 0.33(20) = 6.6$ ft

c. height to top of warped wingwall, $h_6 = 2y_e = 2(6) = 12$ ft

d. height to top of end sill, $h_5 = 0.94h_6 = 0.94(2y_e) = 11.3$ ft

Step 7. Find V_o/V_B from Figure 9.9 and compute V_B. Compare with V_n from step 2. Assess scour potential downstream based on soil condition and outlet velocity. If riprap is needed, see Chapter 10.

With Fr = 2.7 and $L_B = 3.3 W_o$, V_o/V_B will be less than 1.9 making $V_B \cong 37.5/1.9 = 19.7$ ft/s. This is somewhat higher than the normal velocity in the downstream channel (16.1 ft/s) indicating riprap protection may be desirable. See Chapter 10.

The dissipator design is shown on the sketch below. (All dimensions are in feet.)

Step 8. Since no large debris is expected at this site, the hook face will not be armored with steel.

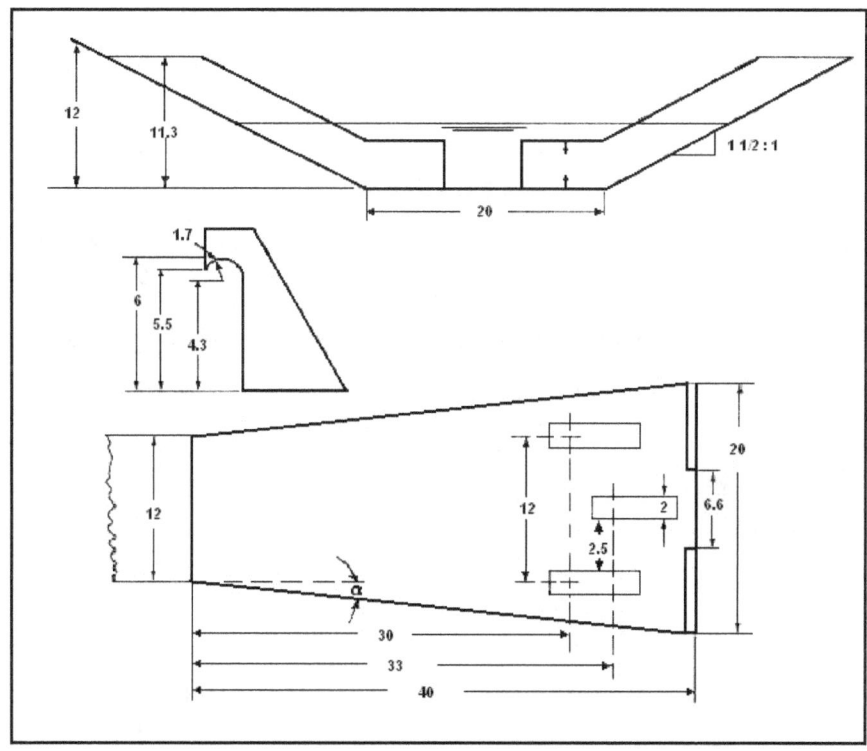

Sketch for the Hook Basin with Warped Wingwalls Design Example (CU)

9.3.2 Hook Basin with Uniform Trapezoidal Channel

The Hook basin with a uniform trapezoidal channel with end sill is shown in Figure 9.10. The hooks and end sill are closer to the outfall of the culvert than the hooks and end sill with warped wingwalls. The research report (MacDonald, 1967) presents several charts depicting the effect of various variables on the performance of the dissipator. These charts show that for a given discharge condition widening the basin produces some reduction in the velocity downstream, and flattening the side slopes improves the performance of the dissipator for values of the Froude number up to 3.0.

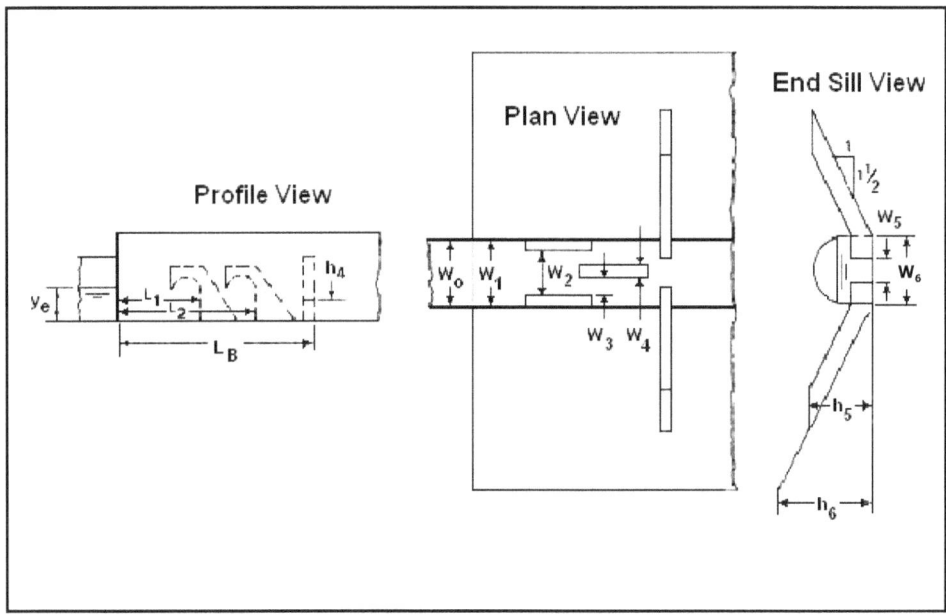

Figure 9.10. Hook Basin with Uniform Trapezoidal Channel

The best range of design dimensions that were tested are indicated in the design procedure. In most cases, the ratio that will produce the smallest dimension was used. The recommended hook configuration is shown in Figure 9.11. The height dimensions are different than those used for the hook energy dissipator with warped wingwalls. The recommended dimensions are:

1. $h_1 = 0.78y_e$

2. $h_2 = y_e$

3. $h_3 = 1.4h_1$

4. $\beta = 135°$

5. $r = 0.4h_1$

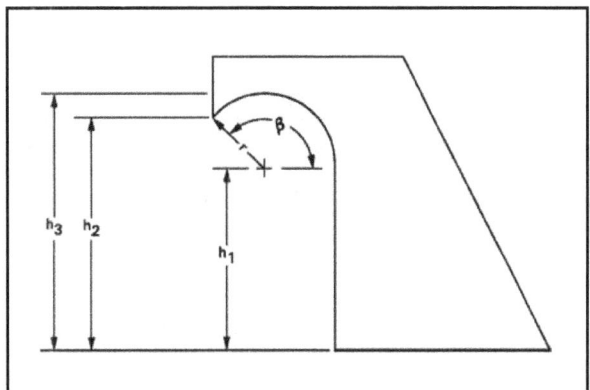

Figure 9.11. Hook for Uniform Trapezoidal Channel Basin

The discharge velocity at the exit of the dissipator, V_B, is estimated from Figure 9.12.

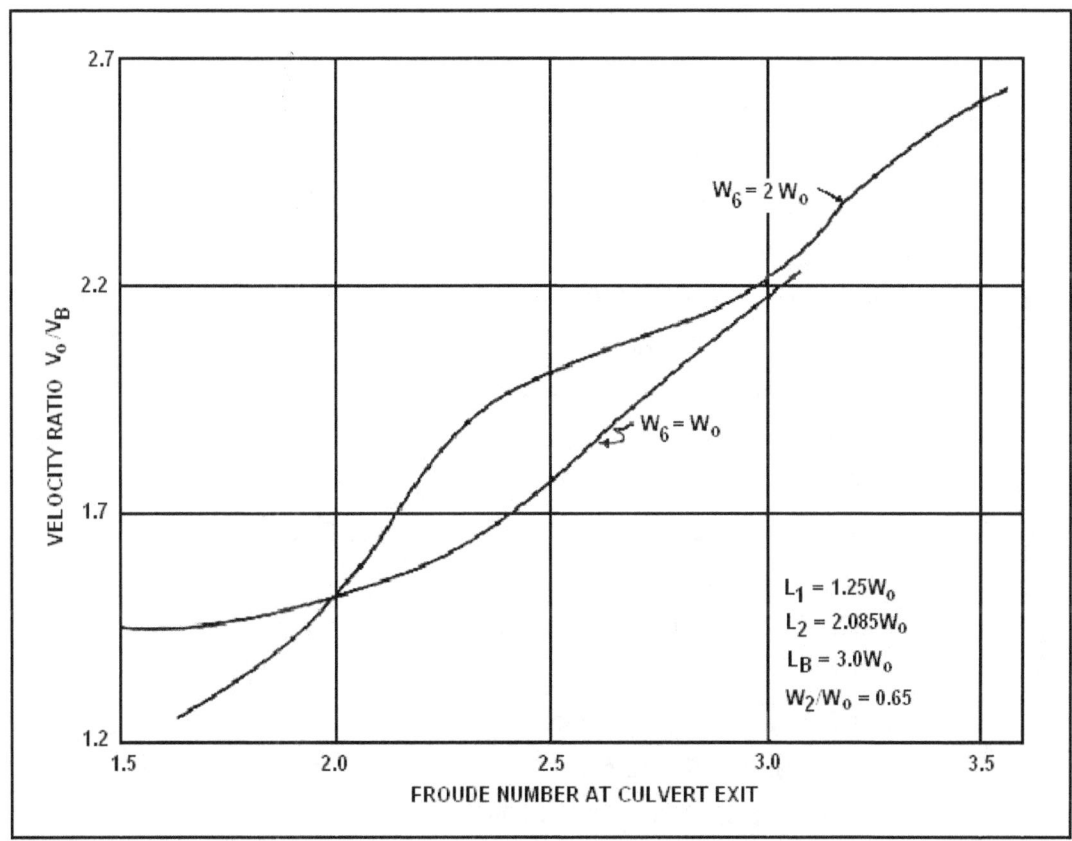

Figure 9.12. Velocity Ratio for Hook Basin With Uniform Trapezoidal Channel

The design procedure for a Hook basin with a uniform trapezoidal channel is as follows:

Step 1. Compute the culvert outlet velocity, V_o, equivalent depth, $y_e = (A/2)^{1/2}$, and Froude number, $Fr = V_o/(gy_e)^{1/2}$.

Culvert width, W_o = width of rectangular culvert or $W_o = 2y_e$ for circular and other shapes.

If $1.8 < Fr < 3.0$, proceed with design.

Step 2. Compute the downstream channel velocity, V_n, and depth, y_n.

Step 3. Select a basin width, W_B ($W_6 = W_B$), side slope, and length, L_B. W_6 should be approximately equal to the channel width if the downstream channel is defined.

$W_6 = W_o$ to $2W_o$

Basin side slope can be from 1:1.5 to 1:2 (V:H).

$L_B = 3.0W_o$

Step 4. Compute the position and spacing of the hooks (see Figure 9.10):

   a.   distance to first hooks, $L_1 = 1.25\ W_o$

   b.   width at first hooks, $W_1 = W_o$

   c.   distance between first hooks, $W_2 = 0.65\ W_o$

   d.   distance to second hook, $L_2 = 2.085\ W_o$

   e.   width of hooks, $W_4 = 0.16W_o$

   f.   spacing between first and second hook, $W_3 = (W_2 - W_4)/2$

   If spacing does not satisfy $W_3/W_4 \geq 1.0$, adjust W_6.

Step 5. Compute hook dimensions (see Figure 9.11):

   a.   height to center of radius, $h_1 = 0.78y_e$

   b.   height to point, $h_2 = y_e$

   c.   height to top of radius, $h_3 = 1.4h_1$

   d.   angle of radius, $\beta = 135°$

   e.   radius, $r = 0.4h_1$

Step 6. Compute the end sill and wingwall dimensions (see Figure 9.10):

   a.   height of end sill, $h_4 = 0.67y_e$

   b.   width of slot in end sill, $W_5 = 0.33W_B$

   c.   height to top of side slope, h_6

   $h_6 = 3.33\ y_e$ for 1:1.5 side slopes

   $h_6 = 2.69y_e$ for 1:2 side slopes

   d.   height to top of end sill, $h_5 = 0.70h_6$

Step 7. Find V_o/V_B from Figure 9.12 and calculate V_B. Compare with V_n from step 2. If V_B is unacceptable, adjust W_6, if feasible. Assess scour potential downstream based on soil condition and outlet velocity. If riprap is needed, see Chapter 10.

Step 8. Where large debris is expected, the upstream face of the hooks should be armored with steel.

Design Example: Hook Basin with a Uniform Trapezoidal Channel (SI)

Determine dimensions for a Hook basin with a uniform trapezoidal channel for a long concrete semicircular arch culvert that is 3.658 m wide and 3.658 m from the floor to the crown. Given:

S_o = 0.020 m/m

n = 0.012

Q = 76.41 m³/s

y_e = 1.829 m

V_o = 11.43 m/s

The downstream channel has a trapezoidal shape with the following properties:

W_c = 6.096 m

S_o = 0.020 m/m

Z = 1.5

n = 0.030

V_n = 5.27 m/s

y_n = 1.676 m

Solution

Step 1. Compute the culvert outlet velocity, V_o, equivalent depth, y_e, and Froude number, $Fr = V_o/(gy_e)^{1/2}$. V_o = 11.43 m/s and y_e = 1.829 m were given.

$Fr_o = V_o / (gy_e)^{1/2} = 11.43/ (9.81 \times 1.829)^{1/2} = 2.70$

Since 1.8 < 2.70 < 3, proceed to step 2.

Step 2. Compute the downstream channel velocity, V_n, and depth, y_n. V_n = 5.273 m/s and y_n = 1.676 m were given.

Step 3. Select a basin width, W_6, side slope, and length, L_B. W_6 should be approximately equal to the channel width, if the downstream channel is defined.

$W_6 = W_c$ = 6.096 m, which is 6.096/3.658 = 1.67 W_o

Basin side slope will be 1:1.5 (V:H)

$L_B = 3.0W_o = 3.0(3.658) = 10.974$ m

Step 4. Compute the position and spacing of the hooks (see Figure 9.10):

a. distance to first hooks, $L_1 = 1.25 W_o = 1.25 (3.658) = 4.573$ m

b. width at first hooks, $W_1 = W_o = 3.658$ m

c. distance between first hooks, $W_2 = 0.65 W_o = 0.65 (3.658) = 2.377$ m

d. distance to second hook, $L_2 = 2.085 W_o = 2.085 (3.658) = 7.627$ m

e. width of hooks, $W_4 = 0.16W_o = 0.16(3.658) = 0.585$ m

f. spacing between first and second hook, W_3

$W_3 = (W_2 - W_4)/2 = (2.377 - 0.585)/2 = 0.896$ m

$W_3 /W_4 = 0.896/0.585 = 1.5$, which satisfies $W_3/W_4 \geq 1.0$.

Step 5. Compute hook dimensions (see Figure 9.11):

    a.   height to center of radius, $h_1 = 0.78y_e = 1.427$ m

    b.   height to point, $h_2 = y_e = 1.829$ m

    c.   height to top of radius, $h_3 = 1.4h_1 = 1.4(1.427) = 1.998$ m

    d.   angle of radius, $\beta = 135°$

    e.   radius, $r = 0.4h_1 = 0.4(1.427) = 0.571$ m

Step 6. Compute the end sill and wingwall dimensions (see Figure 9.10):

    a.   height of end sill, $h_4 = 0.67y_e = 0.67(1.829) = 1.225$ m

    b.   width of slot in end sill, $W_5 = 0.33W_B = 0.33(6.096) = 2.012$ m

    c.   height to top of side slope, h_6 for 1:1.5 (V:H) side slopes

        $h_6 = 3.33 \, y_e = 3.33(1.829) = 6.091$ m

    d.   height to top of end sill, $h_5 = 0.70h_6 = 0.70(6.091) = 4.264$ m

Step 7. Find V_o/V_B from Figure 9.12 and compute V_B. Compare with V_n from step 2. Assess scour potential downstream based on soil condition and outlet velocity. If riprap is needed, see Chapter 10.

From Figure 9.12 with a Froude number of 2.70 and $W_6/W_o = 1.67$, $V_o/V_B \cong 2.0$ making $V_B \cong 11.43/2 = 5.72$ m/s which is slightly higher than the normal channel velocity, $V_n = 5.273$ m/s indicating minimum riprap protection will be necessary. A sketch of this dissipator is shown in the sketch on the next page. (All dimensions are shown in meters.)

Step 8. Where large debris is expected the upstream face of the hooks should be armored. Since no large debris is expected, the hook face will not be armored.

The design example dimensions for both the warped wingwall and the trapezoidal basins are shown in the following table.

Feature	Element	Symbol	Warped Wingwall (m)	Trapezoidal (m)
Basin	Length	L_B	12.19	10.974
	Width	W_6	6.096	6.096
First Hooks	Length	L_1	9.143	4.573
	Spacing	W_2	3.621	2.377
Second Hook	Length	L_2	10.118	7.682
	Spacing	W_3	1.518	0.896
End Wall	Height	h_4	1.225	1.225
	Slot	W_5	2.012	2.012
	Top	h_5	3.439	4.264
Hooks	Height	h_3	1.829	1.998
	Width	W_4	0.607	0.585

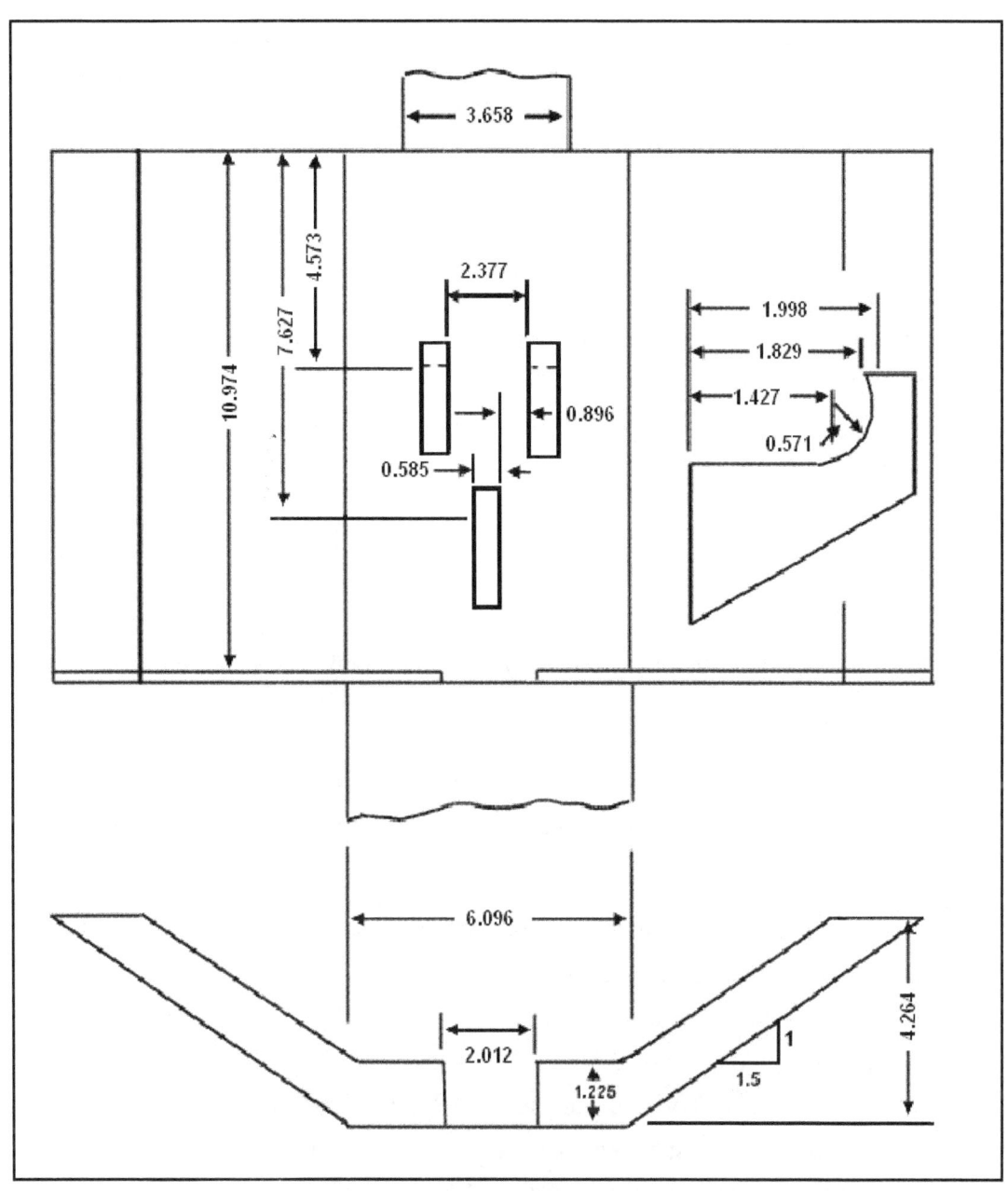

Sketch for the Hook Basin with a Uniform Trapezoidal Channel (SI)

Design Example: Hook Basin with a Uniform Trapezoidal Channel (CU)

Determine dimensions for a Hook basin with a uniform trapezoidal channel for a long concrete semicircular arch culvert that is 12 ft wide and 12 ft from the floor to the crown. Given:

S_o = 0.020 ft/ft
n = 0.012
Q = 2700 ft^3/s
y_e = 6 ft
V_o = 37.5 ft/s

The downstream channel has a trapezoidal shape with the following properties:

W_c = 20 ft
S_o = 0.020 ft/ft
Z = 1.5
n = 0.030
V_n = 16.1 ft/s
y_n = 5.5 ft

Solution

Step 1. Compute the culvert outlet velocity, V_o, equivalent depth, y_e, and Froude number, $Fr = V_o/(gy_e)^{1/2}$. V_o = 37.5 ft/s and y_e = 6 ft were given.

$$Fr_o = V_o / (gy_e)^{1/2} = 37.5/ (32.2 \times 6)^{1/2} = 2.7$$

Since 1.8 < 2.7 < 3, proceed with step 2.

Step 2. Compute the downstream channel velocity, V_n, and depth, y_n. V_n = 16.1 ft/s and y_n = 5.5 ft were given.

Step 3. Select a basin width, W_6, side slope, and length, L_B. W_6 should be approximately equal to the channel width, if the downstream channel is defined.

$W_6 = W_c$ = 20 ft, which is 20/12 = 1.67 W_o

Basin side slope will be 1:1.5 (V:H)

$L_B = 3.0W_o = 3.0(12)$ = 36 ft

Step 4. Compute the position and spacing of the hooks (see Figure 9.10):

a. distance to first hooks, $L_1 = 1.25 W_o = 1.25 (12)$ = 15 ft

b. width at first hooks, $W_1 = W_o$ = 12 ft

c. distance between first hooks, $W_2 = 0.65 W_o = 0.65 (12)$ = 7.8 ft

d. distance to second hook, $L_2 = 2.085 W_o = 2.085 (12)$ = 25 ft

e. width of hooks, $W_4 = 0.16W_o = 0.16(12)$ = 1.92 ft (use 2 ft)

f. spacing between first and second hook, W_3

$W_3 = (W_2 - W_4)/2 = (7.8 - 2)/2$ = 2.9 ft

$W_3 /W_4 = 2.9/2$ = 1.45, which satisfies $W_3/W_4 \geq 1.0$.

Step 5. Compute hook dimensions (see Figure 9.11):

   a.   height to center of radius, $h_1 = 0.78y_e = 0.78(6) = 4.7$ ft

   b.   height to point, $h_2 = y_e = 6$ ft

   c.   height to top of radius, $h_3 = 1.4h_1 = 1.4(4.7) = 6.6$ ft

   d.   angle of radius, $\beta = 135°$

   e.   radius, $r = 0.4h_1 = 0.4(4.7) = 1.9$ ft

Step 6. Compute the end sill and wingwall dimensions (see Figure 9.10):

   a.   height of end sill, $h_4 = 0.67y_e = 0.67(6) = 4$ ft

   b.   width of slot in end sill, $W_5 = 0.33W_B = 0.33(20) = 6.6$ ft

   c.   height to top of side slope, h_6 for 1:1.5 (V:H) side slopes

   $h_6 = 3.33\ y_e = 3.33(6) = 20$ ft

   d.   height to top of end sill, $h_5 = 0.70h_6 = 0.70(20) = 14$ ft

Step 7. Find V_o/V_B from Figure 9.12 and compute V_B. Compare with V_n from step 2. Assess scour potential downstream based on soil condition and outlet velocity. If riprap is needed, see Chapter 10.

From Figure 9.12 with a Froude number of 2.70 and $W_6/W_o = 1.67$, $V_o/V_B \cong 2.0$ making $V_B \cong 37.5/2 = 18.8$ ft/s which is slightly higher than the normal channel velocity, $V_n = 16.1$ ft/s indicating minimum riprap protection will be necessary. A sketch of this dissipator is shown on the next page. (All dimensions are shown in feet.)

Step 8. Where large debris is expected the upstream face of the hooks should be armored. Since no large debris is expected, the hook face will not be armored.

The design example dimensions for both the warped wingwall and the trapezoidal basins are shown in the following table.

Feature	Element	Symbol	Warped Wingwall (ft)	Trapezoidal (ft)
Basin	Length	L_B	36	40
	Width	W_6	20	20
First Hooks	Length	L_1	15	30
	Spacing	W_2	7.8	12
Second Hook	Length	L_2	25	33
	Spacing	W_3	2.9	5
End Wall	Height	h_4	4	4
	Slot	W_5	6.6	6.6
	Top	h_5	14	11.3
Hooks	Height	h_3	6.6	6
	Width	W_4	2	2

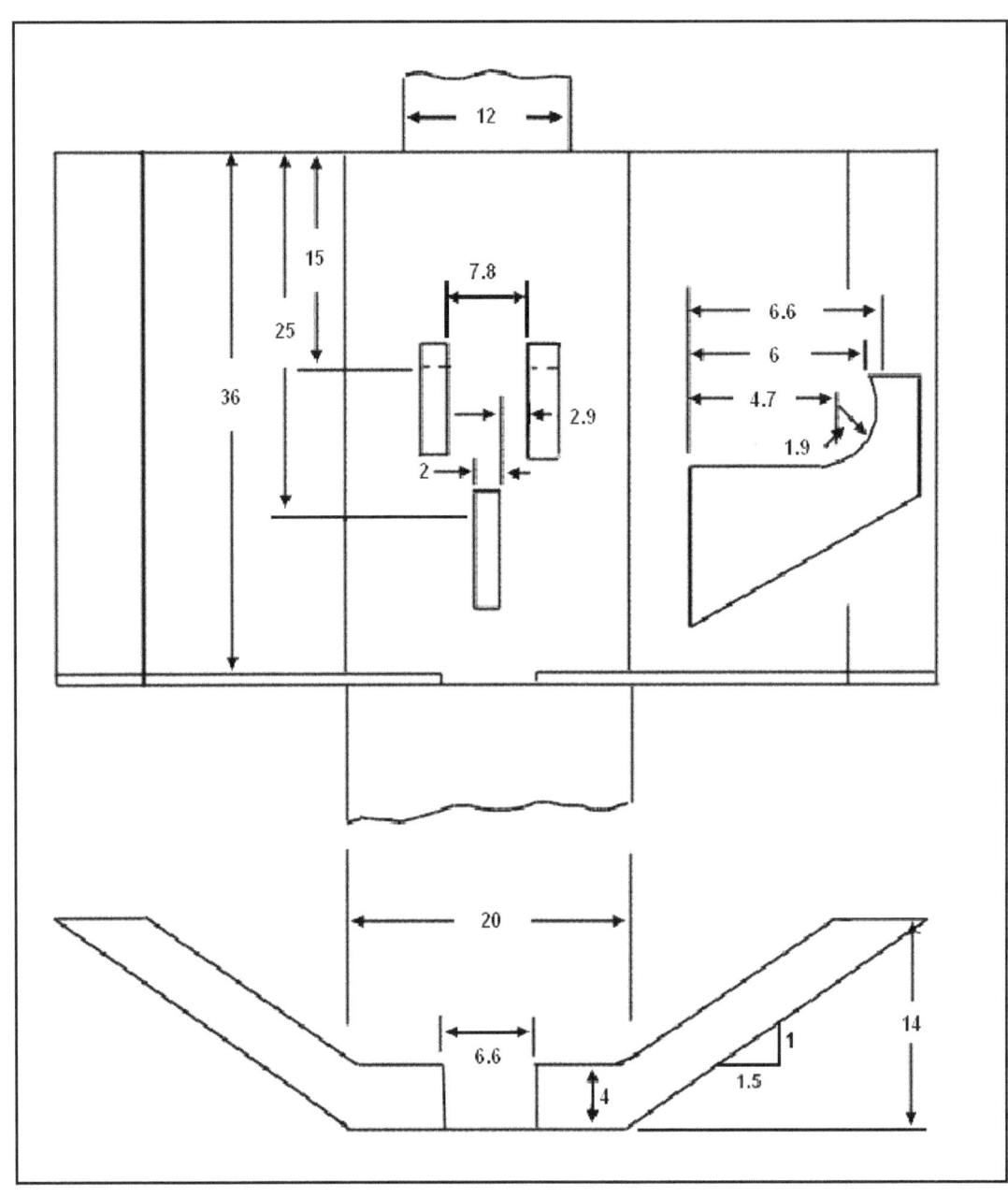

Sketch for Hook Basin with a Uniform Trapezoidal Channel (CU)

9.4   USBR TYPE VI IMPACT BASIN

The U.S. Bureau of Reclamation (USBR) Type VI impact basin was developed at the USBR Laboratory (ASCE, 1957). The dissipator is contained in a relatively small box-like structure that requires no tailwater for successful performance. Although the emphasis in this manual is on its use at culvert outlets, the structure may also be used in open channels.

The shape of the basin has evolved from extensive tests, but these were limited in range by the practical size of field structures required. With the many combinations of discharge, velocity, and depth possible for the incoming flow, it became apparent that some device was needed which would be equally effective over the entire range. The vertical hanging baffle, shown in Figure 9.13, proved to be this device. Energy dissipation is initiated by flow striking the vertical hanging baffle and being deflected upstream by the horizontal portion of the baffle and by the floor, creating horizontal eddies.

Notches in the baffle are provided to aid in cleaning the basin after prolonged periods of low or no flow. If the basin is full of sediment, the notches provide concentrated jets of water for cleaning. The basin is designed to carry the full discharge over the top of the baffle if the space beneath the baffle becomes completely clogged. Although this performance is not good, it is acceptable for short periods of time.

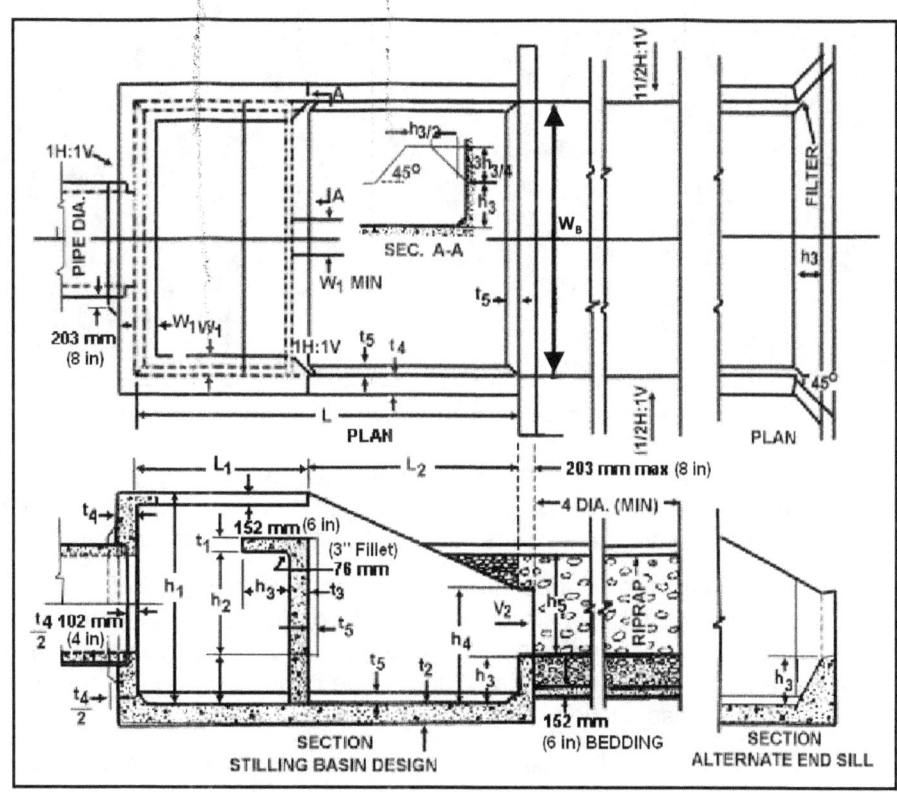

Figure 9.13. USBR Type VI Impact Basin

The design information is presented as a dimensionless curve in Figure 9.14. This curve incorporates the original information contained in ASCE (1957) and the results of additional experimentation performed by the Department of Public Works, City of Los Angeles. The curve shows the relationship of the Froude number to the ratio of the energy entering the dissipator to the width of dissipator required. The Los Angeles tests indicate that limited extrapolation of this curve is permissible.

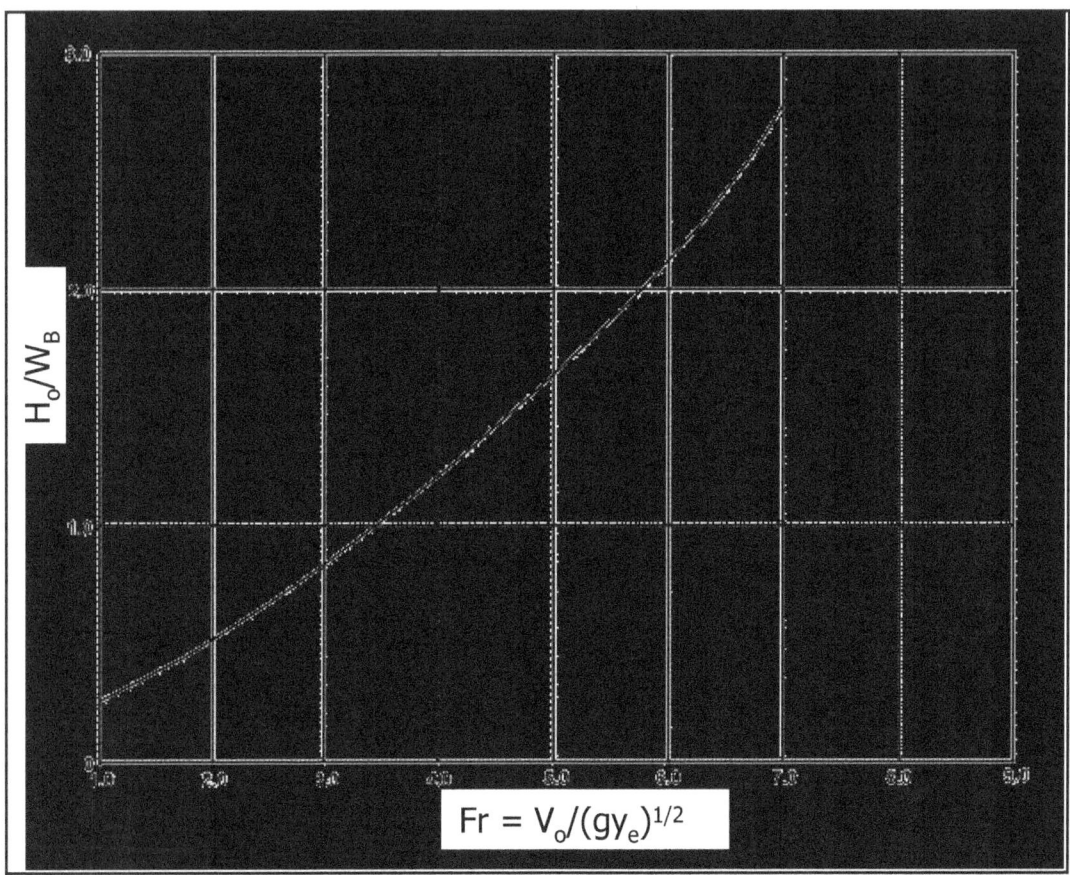

Figure 9.14. Design Curve for USBR Type VI Impact Basin

Once the basin width, W_B, has been determined, many of the other dimensions shown in Figure 9.13 follow according to Table 9.2. To use Table 9.2, round the value of W_B to the nearest entry in the table to determine the other dimensions. Interpolation is not necessary.

In calculating the energy and the Froude number, the equivalent depth of flow, $y_e = (A/2)^{1/2}$, entering the dissipator from a pipe or irregular-shaped conduit must be computed. In other words, the cross section flow area in the pipe is converted into an equivalent rectangular cross section in which the width is twice the depth of flow. The conduit preceding the dissipator can be open, closed, or of any cross section.

The effectiveness of the basin is best illustrated by comparing the energy losses within the structure to those in a natural hydraulic jump, Figure 9.15. The energy loss was computed based on depth and velocity measurements made in the approach pipe and also in the downstream channel with no tailwater. Compared with the natural hydraulic jump, the USBR Type VI impact basin shows a greater capacity for dissipating energy.

Although tailwater is not necessary for successful operation, a moderate depth of tailwater will improve the performance. For best performance set the basin so that maximum tailwater does not exceed $h_3 + (h_2/2)$ which is half of the baffle.

The basin floor should be constructed horizontally and will operate effectively with entrance conduits on slopes up to 15° (27%). For entrance conduits with slopes greater than 15°, a horizontal conduit section of at least four conduit widths long should be provided immediately upstream of the dissipator. Experience has shown that, even for conduits with slopes less than 15 degrees, it is more efficient when the horizontal section of pipe recommended for steeper slopes is used. In every case, the proper position of the entrance invert, as shown in Figure 9.13, should be maintained.

If a horizontal section of pipe is provided before the dissipator, the conduit should be analyzed to determine if a hydraulic jump would form in the conduit. When a hydraulic jump is expected and the pipe outlet is flowing full, a vent about one-sixth the pipe diameter should be installed at a convenient location upstream from the jump.

To provide structural support to the hanging baffle, a short support should be placed under the center of the baffle wall. This support will also provide an additional energy dissipating barrier to the flow.

Table 9.2 (SI). USBR Type VI Impact Basin Dimensions (m) (AASHTO, 1999)

W_B	h_1	h_2	h_3	H_4	L	L_1	L_2
1.0	0.79	0.38	0.17	0.43	1.40	0.59	0.79
1.5	1.16	0.57	0.25	0.62	2.00	0.88	1.16
2.0	1.54	0.75	0.33	0.83	2.68	1.14	1.54
2.5	1.93	0.94	0.42	1.04	3.33	1.43	1.93
3.0	2.30	1.12	0.50	1.25	4.02	1.72	2.30
3.5	2.68	1.32	0.58	1.46	4.65	2.00	2.68
4.0	3.12	1.51	0.67	1.67	5.33	2.28	3.08
4.5	3.46	1.68	0.75	1.88	6.00	2.56	3.46
5.0	3.82	1.87	0.83	2.08	6.52	2.84	3.82
5.5	4.19	2.03	0.91	2.29	7.29	3.12	4.19
6.0	4.60	2.25	1.00	2.50	7.98	3.42	4.60

W_B	W_1	W_2	t_1	t_2	t_3	t_4	t_5
1.0	0.08	0.26	0.15	0.15	0.15	0.15	0.08
1.5	0.13	0.42	0.15	0.15	0.15	0.15	0.08
2.0	0.15	0.55	0.15	0.15	0.15	0.15	0.08
2.5	0.18	0.68	0.16	0.18	0.18	0.16	0.08
3.0	0.22	0.83	0.20	0.20	0.22	0.20	0.08
3.5	0.26	0.91	0.20	0.23	0.23	0.21	0.10
4.0	0.30	0.91	0.20	0.28	0.25	0.25	0.10
4.5	0.36	0.91	0.20	0.30	0.30	0.30	0.13
5.0	0.39	0.91	0.22	0.31	0.30	0.30	0.15
5.5	0.41	0.91	0.22	0.33	0.33	0.33	0.18
6.0	0.45	0.91	0.25	0.36	0.35	0.35	0.19

Table 9.2 (CU). USBR Type VI Impact Basin Dimensions (ft) (AASHTO, 2005)

W_B	h_1	h_2	h_3	h_4	L	L_1	L_2
4.	3.08	1.50	0.67	1.67	5.42	2.33	3.08
5.	3.83	1.92	0.83	2.08	6.67	2.92	3.83
6.	4.58	2.25	1.00	2.50	8.00	3.42	4.58
7.	5.42	2.58	1.17	2.92	9.42	4.00	5.42
8.	6.17	3.00	1.33	3.33	10.67	4.58	6.17
9.	6.92	3.42	1.50	3.75	12.00	5.17	6.92
10.	7.58	3.75	1.67	4.17	13.42	5.75	7.67
11.	8.42	4.17	1.83	4.58	14.58	6.33	8.42
12.	9.17	4.50	2.00	5.00	16.00	6.83	9.17
13.	10.17	4.92	2.17	5.42	17.33	7.42	10.00
14.	10.75	5.25	2.33	5.83	18.67	8.00	10.75
15.	11.50	5.58	2.50	6.25	20.00	8.50	11.50
16.	12.25	6.00	2.67	6.67	21.33	9.08	12.25
17.	13.00	6.33	2.83	7.08	21.50	9.67	13.00
18.	13.75	6.67	3.00	7.50	23.92	10.25	13.75
19.	14.58	7.08	3.17	7.92	25.33	10.83	14.58
20.	15.33	7.50	3.33	8.33	26.58	11.42	15.33

W_B	W_1	W_2	t_1	t_2	t_3	t_4	t_5
4.	0.33	1.08	0.50	0.50	0.50	0.50	0.25
5.	0.42	1.42	0.50	0.50	0.50	0.50	0.25
6.	0.50	1.67	0.50	0.50	0.50	0.50	0.25
7.	0.50	1.92	0.50	0.50	0.50	0.50	0.25
8.	0.58	2.17	0.50	0.58	0.58	0.50	0.25
9.	0.67	2.50	0.58	0.58	0.67	0.58	0.25
10.	0.75	2.75	0.67	0.67	0.75	0.67	0.25
11.	0.83	3.00	0.67	0.75	0.75	0.67	0.33
12.	0.92	3.00	0.67	0.83	0.83	0.75	0.33
13.	1.00	3.00	0.67	0.92	0.83	0.83	0.33
14.	1.08	3.00	0.67	1.00	0.92	0.92	0.42
15.	1.17	3.00	0.67	1.00	1.00	1.00	0.42
16.	1.25	3.00	0.75	1.00	1.00	1.00	0.50
17.	1.33	3.00	0.75	1.08	1.00	1.00	0.50
18.	1.33	3.00	0.75	1.08	1.08	1.08	0.58
19.	1.42	3.00	0.83	1.17	1.08	1.08	0.58
20.	1.50	3.00	0.83	1.17	1.17	1.17	0.67

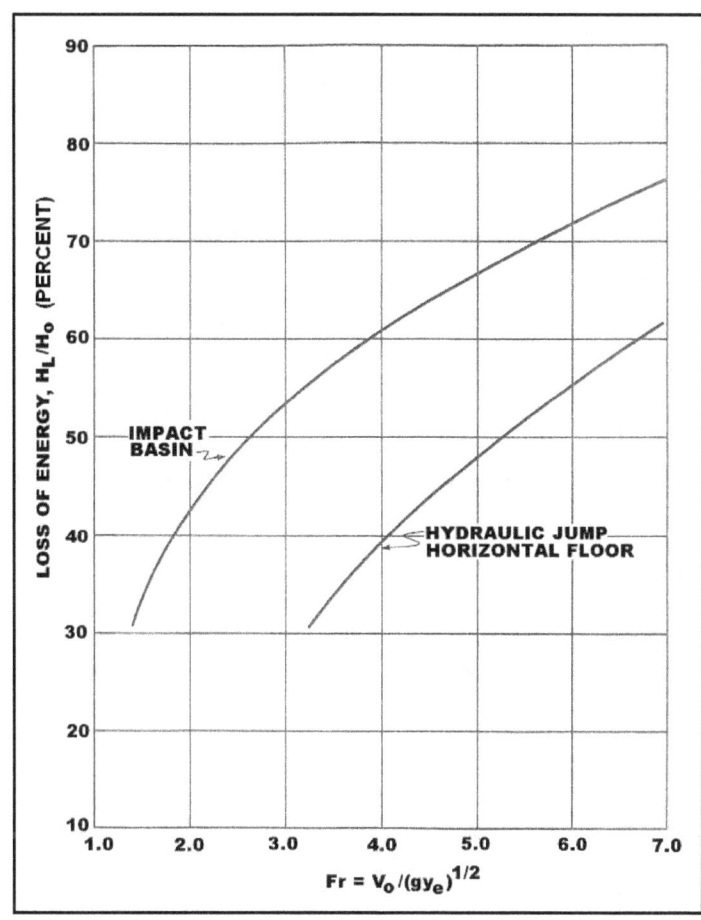

Figure 9.15. Energy Loss of USBR Type VI Impact Basin versus Hydraulic Jump

For erosion reduction and better basin operation, use the alternative end sill and 45° wingwall design as shown in Figure 9.13. The sill should be set as low as possible to prevent degradation downstream. For best performance, the downstream channel should be at the same elevation as the top of the sill. A slot should be placed in the end sill to provide for drainage during periods of low flow. Although the basin is depressed, the slot allows water to drain into the surrounding soil.

For protection against undermining, a cutoff wall should be added at the end of the basin. Its depth will depend on the type of soil present. Riprap should be placed downstream of the basin for a length of at least four conduit widths. For riprap size recommendations see Chapter 10.

The Los Angeles experiments simulated discharges up to 11.3 m^3/s (400 ft^3/s) and entrance velocities as high as 15.2 m/s (50 ft/s). Therefore, use of the basin is limited to installations within these parameters. Velocities up to 15.2 m/s (50 ft/s) can be used without subjecting the structure to damage from cavitation forces. Some structures already constructed have exceeded these thresholds suggesting there may be some design flexibility. For larger installations where discharge is separable, two or more structures may be placed side by side. The USBR Type VI is not recommended where debris or ice buildup may cause substantial clogging.

The recommended design procedure for the USBR Type VI impact basin is as follows:

Step 1. Determine the maximum discharge, Q, and velocity, V_o and check against design limits. Compute the flow area at the end of the approach pipe, A. Compute equivalent depth, $y_e = (A/2)^{1/2}$.

Step 2. Compute the Froude number, Fr, and the energy at the end of the pipe, H_o.

Step 3. Determine H_o/W_B from Figure 9.14. Calculate the required width of basin, W_B.

$$W_B = H_o / (H_o/ W_B)$$

Step 4. Obtain the remaining dimensions of the USBR Type VI impact basin from Table 9.2 using W_B obtained from step 3.

Step 5. Determine exit velocity, $V_B = V_2$, by trial and error using an energy balance between the culvert exit and the basin exit. Determine if this velocity is acceptable and whether or not riprap protection is needed downstream (see Chapter 10.)

$$H_B = Q/(W_B V_B) + V_B^2/(2g) = H_o(1 - H_L/H_o)$$

This equation is a cubic equation yielding 3 solutions, two positive and one negative. The negative solution is discarded. The two positive roots yield a subcritical and supercritical solution. Where low or no tailwater exists, the supercritical solution is taken. Where sufficient tailwater exists, the subcritical solution is taken.

Design Example: USBR Type VI Impact Basin (SI)

Determine the USBR Type VI impact basin dimensions for use at the outlet of a concrete pipe. Compare the design with a dissipator at the end of a rectangular concrete channel. Given:

D	=	1.219 m (pipe diameter and rectangular channel width)
S_o	=	0.15 m/m
Q	=	8.5 m³/s (pipe)
Q	=	10.6 m³/s (channel)
n	=	0.015
V_o	=	12.192 m/s
y_o	=	0.701 m (for both pipe and channel)

Solution

First design dissipator for the pipe conduit.

Step 1. Determine the maximum discharge, Q, and velocity, V_o. Compute the flow area at the end of the approach pipe, A. Compute equivalent depth, $y_e = (A/2)^{1/2}$.

Since Q is less than 11.3 m³/s and V_o less than 15.2 m/s, the dissipator can be tried at this site.

$A = Q/V_o = 8.5/12.192 = 0.697$ m²

$y_e = (A/2)^{1/2} = (0.697/2)^{1/2} = 0.590$ m

Step 2. Compute the Froude number, Fr, and the energy at the end of the pipe, H_o.

$Fr = V_o /(gy_e)^{1/2} = 12.192/ [9.81(0.590)]^{1/2} = 5.07$

$H_o = y_e + V_o^2 /(2g) = 0.590 + (12.192)^2 /19.62 = 8.166$ m

Step 3. Determine H_o/ W_B from Figure 9.14. Calculate the required width of basin, W_B.

$W_B = H_o / (H_o/ W_B) = 8.166 /1.68 = 4.86$ m

Step 4. Obtain the remaining dimensions of the USBR Type VI impact basin from Table 9.2 using $W_B = 5.0$ m obtained from step 3. (The basin width is taken to the nearest 0.5 m.) Results are summarized in the following table.

Step 5. Determine exit velocity, $V_B = V_2$, by trial and error using an energy balance between the culvert exit and the basin exit. Determine if this velocity is acceptable and whether or not riprap protection is needed downstream (see Chapter 10.)

$H_B = Q/(W_B V_B) + V_B^2/(2g) = H_o(1- H_L/H_o)$

$H_B = 8.5/(5.0V_B) + V_B^2/19.62 = 8.166(1 -0.67)$

$H_B = 1.7/V_B + V_B^2/19.62 = 2.695$

No tailwater exists so the supercritical solution is chosen. By trial and error, $V_B = 6.9$ m/s, therefore velocity has been reduced from 12.2 m/s to 6.9 m/s.

Compare the design for the circular pipe with a second USBR Type VI impact basin at the end of a long rectangular concrete channel. The computations and comparison with the pipe are tabulated below. $W_B = 5.5$ m for this case.

Approach Channel	Circular Pipe	Rectangular Channel
Depth of flow y_o (m)	0.701	0.701
Area of flow, A (m^2)	0.697	0.855
Velocity, V_o (m/s)	12.192	12.419
Equivalent depth, y_e (m)	0.590	0.701
Velocity Head, $V_o^2/(2g)$ (m)	7.576	7.861
$H_o = y_e + V_o^2/2g$ (m)	8.166	8.562
$Fr = V_o/(gy_e)^{0.5}$	5.07	4.74
H_o/ W_B from Figure 9.14	1.68	1.55
Width of basin, W_B (m)	5.0	5.5
H_L/H_o from Figure 9.15	67%	65%

Design Example: USBR Type VI Impact Basin (CU)

Determine the USBR Type VI impact basin dimensions for use at the outlet of a concrete pipe. Compare the design with a dissipator at the end of a rectangular concrete channel. Given:

D   = 4 ft (pipe diameter and rectangular channel width)

S_o = 0.15 ft/ft

Q   = 300 ft^3/s (pipe)

Q   = 375 ft^3/s (channel)

$$n \quad = \quad 0.015$$

$$V_o \quad = \quad 40 \text{ ft/s}$$

$$y_o \quad = \quad 2.3 \text{ ft (for both pipe and channel)}$$

Solution

First design dissipator for the pipe conduit.

Step 1. Determine the maximum discharge, Q, and velocity, V_o. Compute the flow area at the end of the approach pipe, A. Compute equivalent depth, $y_e = (A/2)^{1/2}$.

Since Q is less than 400 ft^3/s and V_o less than 50 ft/s, the dissipator can be tried at this site.

$$A = Q/V_o = 300/40 = 7.5 \text{ ft}^2$$

$$y_e = (A/2)^{1/2} = (7.5/2)^{1/2} = 1.94 \text{ ft}$$

Step 2. Compute the Froude number, Fr, and the energy at the end of the pipe, H_o.

$$Fr = V_o / (gy_e)^{1/2} = 40/ [32.2(1.94)]^{1/2} = 5.06$$

$$H_o = y_e + V_o^2 /(2g) = 1.94 + (40)^2 /64.4 = 26.8 \text{ ft}$$

Step 3. Determine H_o/ W_B from Figure 9.14. Calculate the required width of basin, W_B.

$$W_B = H_o / (H_o/ W_B) = 26.8 /1.68 = 16 \text{ ft}$$

Step 4. Obtain the remaining dimensions of the USBR Type VI impact basin from Table 9.2 using W_B = 16 ft obtained from step 3. (The basin width is taken to the nearest 1 ft.) Results are summarized in the following table.

Step 5. Determine exit velocity, $V_B = V_2$, by trial and error using an energy balance between the culvert exit and the basin exit.   Determine if this velocity is acceptable and whether or not riprap protection is needed downstream (see Chapter 10.)

$$H_B = Q/(W_B V_B) + V_B^2/(2g) = H_o(1- H_L/H_o)$$

$$H_B = 300/(16V_B) + V_B^2/64.4 = 26.8(1 -0.67)$$

$$H_B = 18.75/V_B + V_B^2/64.4 = 8.84$$

No tailwater exists so the supercritical solution is chosen. By trial and error, V_B = 22.7 ft/s, therefore velocity has been reduced from 40 ft/s to 22.7 ft/s.

Compare the design for the circular pipe with a second USBR Type VI impact basin at the end of a long rectangular concrete channel. The computations and comparison with the pipe are tabulated below. W_B = 18 ft for this case.

Approach Channel	Circular Pipe	Rectangular Channel
Depth of flow y_o (ft)	2.3	2.3
Area of flow, A (ft^2)	7.5	9.2
Velocity, V_o (ft/s)	40	40.9
Equivalent depth, y_e (ft)	1.9	2.3
Velocity Head, $V_o^2/(2g)$ (ft)	24.9	26
$H_o = y_e + V_o^2/2g$ (ft)	26.8	28.3
$Fr = V_o/(gy_e)^{0.5}$	5.06	4.75
H_o/W_B from Figure 9.14	1.68	1.55
Width of basin, W_B (ft)	16	18
H_L/H_o from Figure 9.15	67%	65%

This page intentionally left blank.

CHAPTER 10: RIPRAP BASINS AND APRONS

Riprap is a material that has long been used to protect against the forces of water. The material can be pit-run (as provided by the supplier) or specified (standard or special). State DOTs have standard specifications for a number of classes (sizes or gradations) of riprap. Suppliers maintain an inventory of frequently used classes. Special gradations of riprap are produced on-demand and are therefore more expensive than both pit-run and standard classes.

This chapter includes discussion of both riprap aprons and riprap basin energy dissipators. Both can be used at the outlet of a culvert or chute (channel) by themselves or at the exit of a stilling basin or other energy dissipator to protect against erosion downstream. Section 10.1 provides a design procedure for the riprap basin energy dissipator that is based on armoring a pre-formed scour hole. The riprap for this basin is a special gradation. Section 10.2 includes discussion of riprap aprons that provide a flat armored surface as the only dissipator or as additional protection at the exit of other dissipators. The riprap for these aprons is generally from State DOT standard classes. Section 10.3 provides additional discussion of riprap placement downstream of energy dissipators.

10.1 RIPRAP BASIN

The design procedure for the riprap basin is based on research conducted at Colorado State University (Simons, et al., 1970; Stevens and Simons, 1971) that was sponsored by the Wyoming Highway Department. The recommended riprap basin that is shown on Figure 10.1 and Figure 10.2 has the following features:

- The basin is pre-shaped and lined with riprap that is at least $2D_{50}$ thick.

- The riprap floor is constructed at the approximate depth of scour, h_s, that would occur in a thick pad of riprap. The h_s/D_{50} of the material should be greater than 2.

- The length of the energy dissipating pool, L_s, is $10h_s$, but no less than $3W_o$; the length of the apron, L_A, is $5h_s$, but no less than W_o. The overall length of the basin (pool plus apron), L_B, is $15h_s$, but no less than $4W_o$.

- A riprap cutoff wall or sloping apron can be constructed if downstream channel degradation is anticipated as shown in Figure 10.1.

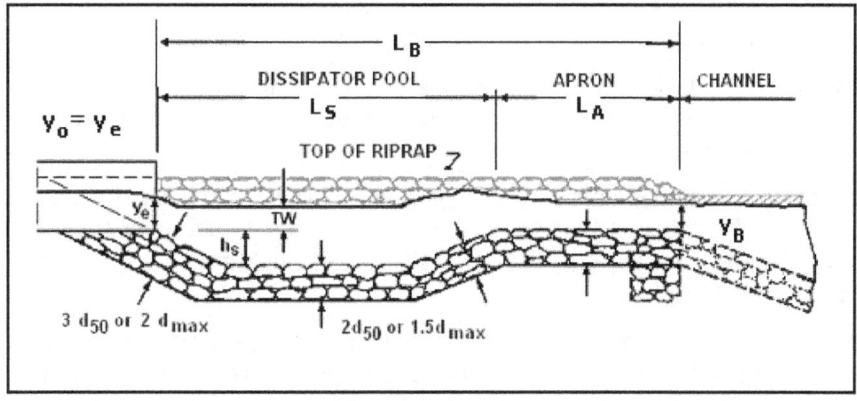

Figure 10.1. Profile of Riprap Basin

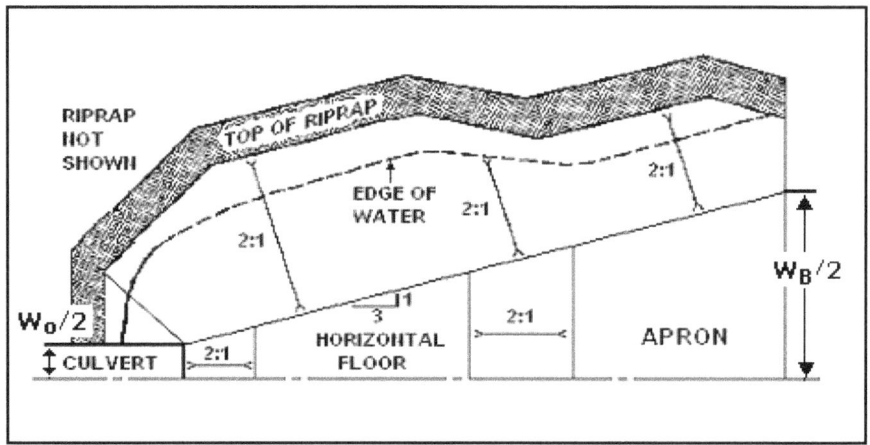

Figure 10.2. Half Plan of Riprap Basin

10.1.1 Design Development

Tests were conducted with pipes from 152 mm (6 in) to 914 mm (24 in) and 152 mm (6 in) high model box culverts from 305 mm (12 in) to 610 mm (24 in) in width. Discharges ranged from 0.003 to 2.8 m³/s (0.1 to 100 ft³/s). Both angular and rounded rock with an average size, D_{50}, ranging from 6 mm (1.4 in) to 177 mm (7 in) and gradation coefficients ranging from 1.05 to 2.66 were tested. Two pipe slopes were considered, 0 and 3.75%. In all, 459 model basins were studied. The following conclusions were drawn from an analysis of the experimental data and observed operating characteristics:

- The scour hole depth, h_s; length, L_s; and width, W_s, are related to the size of riprap, D_{50}; discharge, Q; brink depth, y_o; and tailwater depth, TW.

- Rounded material performs approximately the same as angular rock.

- For low tailwater (TW/y_o < 0.75), the scour hole functions well as an energy dissipator if h_s/D_{50} > 2. The flow at the culvert brink plunges into the hole, a jump forms and flow is generally well dispersed.

- For high tailwater (TW/y_o > 0.75), the high velocity core of water passes through the basin and diffuses downstream. As a result, the scour hole is shallower and longer.

- The mound of material that forms downstream contributes to the dissipation of energy and reduces the size of the scour hole. If the mound is removed, the scour hole enlarges somewhat.

Plots were constructed of h_s/y_e versus $V_o/(gy_e)^{1/2}$ with D_{50}/y_e as the third variable. Equivalent brink depth, y_e, is defined to permit use of the same design relationships for rectangular and circular culverts. For rectangular culverts, $y_e = y_o$ (culvert brink depth). For circular culverts, $y_e = (A/2)^{1/2}$, where A is the brink area.

Anticipating that standard or modified end sections would not likely be used when a riprap basin is located at a culvert outlet, the data with these configurations were not used to develop the design relationships. This assumption reduced the number of applicable runs to 346. A total of 128 runs had a D_{50}/y_e of less than 0.1. These data did not exhibit relationships that appeared useful for design and were eliminated. An additional 69 runs where $h_s/D_{50} < 2$ were also eliminated by the authors of this edition of HEC 14. These runs were not considered reliable for design, especially those with $h_s = 0$. Therefore, the final design development used 149 runs from the study. Of these, 106 were for pipe culverts and 43 were for box culverts. Based on these data, two design relationships are presented here: an envelope design and a best fit design.

To balance the need for avoiding an underdesigned basin against the costs of oversizing a basin, an envelope design relationship in the form of Equation 10.1 and Equation 10.2 was developed. These equations provide a design envelope for the experimental data equivalent to the design figure (Figure XI-2) provided in the previous edition of HEC 14 (Corry, et al., 1983). Equations 10.1 and 10.2, however, improve the fit to the experimental data reducing the root-mean-square (RMS) error from 1.24 to 0.83.

$$\frac{h_s}{y_e} = 0.86 \left(\frac{D_{50}}{y_e} \right)^{-0.55} \left(\frac{V_o}{\sqrt{gy_e}} \right) - C_o \qquad (10.1)$$

where,

h_s = dissipator pool depth, m (ft)
y_e = equivalent brink (outlet) depth, m (ft)
D_{50} = median rock size by weight, m (ft)
C_o = tailwater parameter

The tailwater parameter, C_o, is defined as:

$$
\begin{array}{lll}
C_o = 1.4 & TW/y_e < 0.75 & \\
C_o = 4.0(TW/y_e) -1.6 & 0.75 < TW/y_e < 1.0 & (10.2) \\
C_o = 2.4 & 1.0 < TW/y_e &
\end{array}
$$

A best fit design relationship that minimizes the RMS error when applied to the experimental data was also developed. Equation 10.1 still applies, but the description of the tailwater parameter, C_o, is defined in Equation 10.3. The best fit relationship for Equations 10.1 and 10.3 exhibits a RMS error on the experimental data of 0.56.

$$
\begin{array}{lll}
C_o = 2.0 & TW/y_e < 0.75 & \\
C_o = 4.0(TW/y_e) -1.0 & 0.75 < TW/y_e < 1.0 & (10.3) \\
C_o = 3.0 & 1.0 < TW/y_e &
\end{array}
$$

Use of the envelope design relationship (Equations 10.1 and 10.2) is recommended when the consequences of failure at or near the design flow are severe. Use of the best fit design relationship (Equations 10.1 and 10.3) is recommended when basin failure may easily be addressed as part of routine maintenance. Intermediate risk levels can be adopted by the use of intermediate values of C_o.

10.1.2 Basin Length

Frequency tables for both box culvert data and pipe culvert data of relative length of scour hole ($L_s/h_s < 6$, $6 < L_s/h_s < 7$, $7 < L_s/h_s < 8$ . . . $25 < L_s/h_s < 30$), with relative tailwater depth TW/y_e in increments of 0.03 m (0.1 ft) as a third variable, were constructed using data from 346 experimental runs. For box culvert runs L_s/h_s was less than 10 for 78% of the data and L_s/h_s was less than 15 for 98% of the data. For pipe culverts, L_s/h_s was less than 10 for 91% of the data and, L_s/h_s was less than 15 for all data. A 3:1 flare angle is recommended for the basins walls. This angle will provide a sufficiently wide energy dissipating pool for good basin operation.

10.1.3 High Tailwater

Tailwater influenced formation of the scour hole and performance of the dissipator. For tailwater depths less than 0.75 times the brink depth, scour hole dimensions were unaffected by tailwater. Above this the scour hole became longer and narrower. The tailwater parameter defined in Equations 10.2 and 10.3 captures this observation. In addition, under high tailwater conditions, it is appropriate to estimate the attenuation of the flow velocity downstream of the culvert outlet using Figure 10.3. This attenuation can be used to determine the extent of riprap protection required. HEC 11 (Brown and Clyde, 1989) or the method provided in Section 10.3 can be used for sizing riprap.

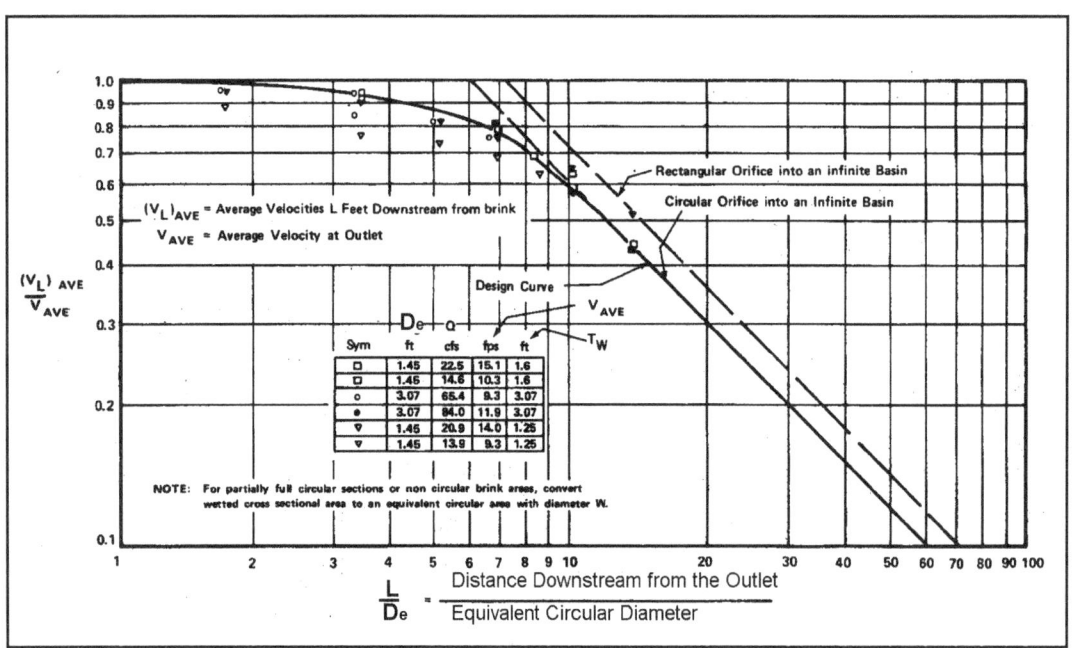

Figure 10.3. Distribution of Centerline Velocity for Flow from Submerged Outlets

10.1.4 Riprap Details

Based on experience with conventional riprap design, the recommended thickness of riprap for the floor and sides of the basin is $2D_{50}$ or $1.50D_{max}$, where D_{max} is the maximum size of rock in the riprap mixture. Thickening of the riprap layer to $3D_{50}$ or $2D_{max}$ on the foreslope of the roadway culvert outlet is warranted because of the severity of attack in the area and the necessity for preventing undermining and consequent collapse of the culvert. Figure 10.1 illustrates these riprap details. The mixture of stone used for riprap and need for a filter should meet the specifications described in HEC 11 (Brown and Clyde, 1989).

10.1.5 Design Procedure

The design procedure for a riprap basin is as follows:

Step 1. Compute the culvert outlet velocity, V_o, and depth, y_o.

For subcritical flow (culvert on mild or horizontal slope), use Figure 3.3 or Figure 3.4 to obtain y_o/D, then obtain V_o by dividing Q by the wetted area associated with y_o. D is the height of a box culvert or diameter of a circular culvert.

For supercritical flow (culvert on a steep slope), V_o will be the normal velocity obtained by using the Manning's Equation for appropriate slope, section, and discharge.

Compute the Froude number, Fr, for brink conditions using brink depth for box culverts ($y_e=y_o$) and equivalent depth ($y_e = (A/2)^{1/2}$) for non-rectangular sections.

Step 2. Select D_{50} appropriate for locally available riprap. Determine C_o from Equation 10.2 or 10.3 and obtain h_s/y_e from Equation 10.1. Check to see that $h_s/D_{50} \geq 2$ and $D_{50}/y_e \geq 0.1$. If h_s/D_{50} or D_{50}/y_e is out of this range, try a different riprap size. (Basins sized where h_s/D_{50} is greater than, but close to, 2 are often the most economical choice.)

Step 3. Determine the length of the dissipation pool (scour hole), L_s, total basin length, L_B, and basin width at the basin exit, W_B, as shown in Figures 10.1 and 10.2. The walls and apron of the basin should be warped (or transitioned) so that the cross section of the basin at the exit conforms to the cross section of the natural channel. Abrupt transition of surfaces should be avoided to minimize separation zones and resultant eddies.

Step 4. Determine the basin exit depth, $y_B = y_c$, and exit velocity, $V_B = V_c$ and compare with the allowable exit velocity, V_{allow}. The allowable exit velocity may be taken as the estimated normal velocity in the tailwater channel or a velocity specified based on stability criteria, whichever is larger. Critical depth at the basin exit may be determined iteratively using Equation 7.14:

$Q^2/g = (A_c)^3/T_c = [y_c(W_B + zy_c)]^3/ (W_B + 2zy_c)$ by trial and success to determine y_B.

$V_c = Q/A_c$

z = basin side slope, z:1 (H:V)

If $V_c \leq V_{allow}$, the basin dimensions developed in step 3 are acceptable. However, it may be possible to reduce the size of the dissipator pool and/or the apron with a larger riprap size. It may also be possible to maintain the dissipator pool, but reduce the flare on the apron to reduce the exit width to better fit the downstream channel. Steps 2 through 4 are repeated to evaluate alternative dissipator designs.

Step 5. Assess need for additional riprap downstream of the dissipator exit. If $TW/y_o \leq 0.75$, no additional riprap is needed. With high tailwater ($TW/y_o \geq 0.75$), estimate centerline velocity at a series of downstream cross sections using Figure 10.3 to determine the size and extent of additional protection. The riprap design details should be in accordance with specifications in HEC 11 (Brown and Clyde, 1989) or similar highway department specifications.

Two design examples are provided. The first features a box culvert on a steep slope while the second shows a pipe culvert on a mild slope.

Design Example: Riprap Basin (Culvert on a Steep Slope) (SI)

Determine riprap basin dimensions using the envelope design (Equations 10.1 and 10.2) for a 2440 mm by 1830 mm reinforced concrete box (RCB) culvert that is in inlet control with supercritical flow in the culvert. Allowable exit velocity from the riprap basin, V_{allow}, is 2.1 m/s. Riprap is available with a D_{50} of 0.50, 0.55, and 0.75 m. Consider two tailwater conditions: 1) TW = 0.85 m and 2) TW = 1.28 m. Given:

Q = 22.7 m³/s

y_o = 1.22 m (normal flow depth) = brink depth

Solution

Step 1. Compute the culvert outlet velocity, V_o, depth, y_o, and Froude number for brink conditions. For supercritical flow (culvert on a steep slope), V_o will be V_n

$y_o = y_e = 1.22$ m

$V_o = Q/A = 22.7/ [1.22 (2.44)] = 7.63$ m/s

$Fr = V_o / (9.81y_e)^{1/2} = 7.63/ [9.81(1.22)]^{1/2} = 2.21$

Step 2. Select a trial D_{50} and obtain h_s/y_e from Equation 10.1. Check to see that $h_s/D_{50} \geq 2$ and $D_{50}/y_e \geq 0.1$.

Try $D_{50} = 0.55$ m; $D_{50}/y_e = 0.55/1.22 = 0.45$ (≥ 0.1 OK)

Two tailwater elevations are given; use the lowest to determine the basin size that will serve the tailwater range, that is, TW = 0.85 m.

$TW/y_e = 0.85/1.22 = 0.7$, which is less than 0.75. Therefore, from Equation 10.2, $C_o = 1.4$

From Equation 10.1,

$$\frac{h_s}{y_e} = 0.86\left(\frac{D_{50}}{y_e}\right)^{-0.55}\left(\frac{V_o}{\sqrt{gy_e}}\right) - C_o = 0.86(0.45)^{-0.55}(2.21) - 1.4 = 1.55$$

$h_S = (h_S /y_e)y_e = 1.55 (1.22) = 1.89$ m

$h_S/D_{50} = 1.89/0.55 = 3.4$ and $h_S/D_{50} \geq 2$ is satisfied

Step 3. Size the basin as shown in Figures 10.1 and 10.2.

$L_S = 10h_S = 10(1.89) = 18.9$ m

L_S min $= 3W_o = 3(2.44) = 7.3$ m, use $L_S = 18.9$ m

$L_B = 15h_S = 15(1.89) = 28.4$ m

L_B min $= 4W_o = 4(2.44) = 9.8$ m, use $L_B = 28.4$ m

$W_B = W_o + 2(L_B/3) = 2.44 + 2(28.4/3) = 21.4$ m

Step 4. Determine the basin exit depth, $y_B = y_c$, and exit velocity, $V_B = V_c$.

$Q^2/g = (A_c)^3/T_c = [y_c(W_B + zy_c)]^3/ (W_B + 2zy_c)$

10-6

$22.7^2/9.81 = 52.5 = [y_c(21.4 + 2y_c)]^3/ (21.4 + 4y_c)$

By trial and success, $y_c = 0.48$ m, $T_c = 23.3$ m, $A_c = 10.7$ m^2

$V_B = V_c = Q/A_c = 22.7/10.7 = 2.1$ m/s (acceptable)

The initial trial of riprap ($D_{50} = 0.55$ m) results in a 28.4 m basin that satisfies all design requirements. Try the next larger riprap size to test if a smaller basin is feasible by repeating steps 2 through 4.

Step 2 (2nd iteration). Select riprap size and compute basin depth.

Try $D_{50} = 0.75$ m; $D_{50}/y_e = 0.75/1.22 = 0.61$ (≥ 0.1 OK)

From Equation 10.1,

$$\frac{h_s}{y_e} = 0.86\left(\frac{D_{50}}{y_e}\right)^{-0.55}\left(\frac{V_o}{\sqrt{gy_e}}\right) - C_o = 0.86(0.61)^{-0.55}(2.21) - 1.4 = 1.09$$

$h_S = (h_S/y_e)y_e = 1.09 (1.22) = 1.34$ m

$h_S/D_{50} = 1.34/0.75 = 1.8$ and $h_S/D_{50} \geq 2$ is not satisfied. Although not available, try a riprap size that will yield h_S/D_{50} close to, but greater than, 2. (A basin sized for smaller riprap may be lined with larger riprap.) Repeat step 2.

Step 2 (3rd iteration). Select riprap size and compute basin depth.

Try $D_{50} = 0.71$ m; $D_{50}/y_e = 0.71/1.22 = 0.58$ (≥ 0.1 OK)

From Equation 10.1,

$$\frac{h_s}{y_e} = 0.86\left(\frac{D_{50}}{y_e}\right)^{-0.55}\left(\frac{V_o}{\sqrt{gy_e}}\right) - C_o = 0.86(0.58)^{-0.55}(2.21) - 1.4 = 1.16$$

$h_S = (h_S/y_e)y_e = 1.16 (1.22) = 1.42$ m

$h_S/D_{50} = 1.42/0.71 = 2.0$ and $h_S/D_{50} \geq 2$ is satisfied.

Step 3 (3rd iteration). Size the basin as shown in Figures 10.1 and 10.2.

$L_S = 10h_S = 10(1.42) = 14.2$ m

L_S min = $3W_o = 3(2.44) = 7.3$ m, use $L_S = 14.2$ m

$L_B = 15h_S = 15(1.42) = 21.3$ m

L_B min = $4W_o = 4(2.44) = 9.8$ m, use $L_B = 21.3$ m

$W_B = W_o + 2(L_B/3) = 2.44 + 2(21.3/3) = 16.6$ m

However, since the trial D_{50} is not available, the next larger riprap size ($D_{50} = 0.75$ m) would be used to line a basin with the given dimensions.

Step 4 (3rd iteration). Determine the basin exit depth, $y_B = y_c$, and exit velocity, $V_B = V_c$.

$Q^2/g = (A_c)^3/T_c = [y_c(W_B + zy_c)]^3/ (W_B + 2zy_c)$

$22.7^2/9.81 = 52.5 = [y_c(16.6 + 2y_c)]^3/ (16.6 + 4y_c)$

By trial and success, $y_c = 0.56$ m, $T_c = 18.8$ m, $A_c = 9.9$ m^2

$V_B = V_c = Q/A_c = 22.7/9.9 = 2.3$ m/s (greater than 2.1 m/s; not acceptable). If the apron were extended (with a continued flare) such that the total basin length was 28.4 m, the velocity would be reduced to the allowable level.

Two feasible options have been identified. First, a 1.89 m deep, 18.9 m long pool, with a 9.5 m apron using $D_{50} = 0.55$ m. Second, a 1.42 m deep, 14.2 m long pool, with a 14.2 m apron using $D_{50} = 0.75$ m. Because the overall length is the same, the first option is likely to be more economical.

Step 5.   For the design discharge, determine if $TW/y_o \leq 0.75$.

For the first tailwater condition, $TW/y_o = 0.85/1.22 = 0.70$, which satisfies $TW/y_o \leq 0.75$. No additional riprap needed downstream.

For the second tailwater condition, $TW/y_o = 1.28/1.22 = 1.05$, which does not satisfy $TW/y_o \leq 0.75$. To determine required riprap, estimate centerline velocity at a series of downstream cross sections using Figure 10.3.

Compute equivalent circular diameter, D_e, for brink area:

$A = \pi D_e^2 /4 = (y_o)(W_o) = (1.22)(2.44) = 3.00$ m^2

$D_e = [3.00(4)/ \pi]^{1/2} = 1.95$ m

Rock size can be determined using the procedures in Section 10.3 (Equation 10.6) or other suitable method. The computations are summarized below.

L/D_e	L (m)	V_L/V_o (Figure 10.3)	V_L (m/s)	Rock size, D_{50} (m)
10	19.5	0.59	4.50	0.43
15	29.3	0.42	3.20	0.22
20	39.0	0.30	2.29	0.11
21	41.0	0.28	2.13	0.10

The calculations above continue until $V_L \leq V_{allow}$. Riprap should be at least the size shown. As a practical consideration, the channel can be lined with the same size rock used for the basin. Protection must extend at least 41.0 m downstream from the culvert brink, which is 12.6 m beyond the basin exit. Riprap should be installed in accordance with details shown in HEC 11.

Design Example: Riprap Basin (Culvert on a Steep Slope) (CU)

Determine riprap basin dimensions using the envelope design (Equations 10.1 and 10.2) for an 8 ft by 6 ft reinforced concrete box (RCB) culvert that is in inlet control with supercritical flow in the culvert. Allowable exit velocity from the riprap basin, V_{allow}, is 7 ft/s. Riprap is available with a D_{50} of 1.67, 1.83, and 2.5 ft. Consider two tailwater conditions: 1) TW = 2.8 ft and 2) TW = 4.2 ft. Given:

Q   = 800 ft^3/s

y_o = 4 ft (normal flow depth) = brink depth

Solution

Step 1. Compute the culvert outlet velocity, V_o, depth, y_o, and Froude number for brink conditions. For supercritical flow (culvert on a steep slope), V_o will be V_n.

$y_o = y_e = 4$ ft

$V_o = Q/A = 800/[4(8)] = 25$ ft/s

$Fr = V_o / (32.2y_e)^{1/2} = 25/[32.2(4)]^{1/2} = 2.2$

Step 2. Select a trial D_{50} and obtain h_s/y_e from Equation 10.1. Check to see that $h_s/D_{50} \geq 2$ and $D_{50}/y_e \geq 0.1$.

Try $D_{50} = 1.83$ ft; $D_{50}/y_e = 1.83/4 = 0.46$ (≥ 0.1 OK)

Two tailwater elevations are given; use the lowest to determine the basin size that will serve the tailwater range, that is, TW = 2.8 ft.

$TW/y_e = 2.8/4 = 0.7$, which is less than 0.75. From Equation 10.2, $C_o = 1.4$

From Equation 10.1,

$$\frac{h_s}{y_e} = 0.86\left(\frac{D_{50}}{y_e}\right)^{-0.55}\left(\frac{V_o}{\sqrt{gy_e}}\right) - C_o = 0.86(0.46)^{-0.55}(2.2) - 1.4 = 1.50$$

$h_s = (h_s/y_e)y_e = 1.50(4) = 6.0$ ft

$h_s/D_{50} = 6.0/1.83 = 3.3$ and $h_s/D_{50} \geq 2$ is satisfied

Step 3. Size the basin as shown in Figures 10.1 and 10.2.

$L_s = 10h_s = 10(6.0) = 60$ ft

L_s min $= 3W_o = 3(8) = 24$ ft, use $L_s = 60$ ft

$L_B = 15h_s = 15(6.0) = 90$ ft

L_B min $= 4W_o = 4(8) = 32$ ft, use $L_B = 90$ ft

$W_B = W_o + 2(L_B/3) = 8 + 2(90/3) = 68$ ft

Step 4. Determine the basin exit depth, $y_B = y_c$, and exit velocity, $V_B = V_c$.

$Q^2/g = (A_c)^3/T_c = [y_c(W_B + zy_c)]^3/(W_B + 2zy_c)$

$800^2/32.2 = 19,876 = [y_c(68 + 2y_c)]^3/(68 + 4y_c)$

By trial and success, $y_c = 1.60$ ft, $T_c = 74.4$ ft, $A_c = 113.9$ ft^2

$V_B = V_c = Q/A_c = 800/113.9 = 7.0$ ft/s (acceptable)

The initial trial of riprap ($D_{50} = 1.83$ ft) results in a 90 ft basin that satisfies all design requirements. Try the next larger riprap size to test if a smaller basin is feasible by repeating steps 2 through 4.

Step 2 (2nd iteration). Select riprap size and compute basin depth.

Try $D_{50} = 2.5$ ft; $D_{50}/y_e = 2.5/4 = 0.63$ (≥ 0.1 OK)

From Equation 10.1,

$$\frac{h_s}{y_e} = 0.86\left(\frac{D_{50}}{y_e}\right)^{-0.55}\left(\frac{V_o}{\sqrt{gy_e}}\right) - C_o = 0.86(0.63)^{-0.55}(2.2) - 1.4 = 1.04$$

$h_S = (h_S/y_e)y_e = 1.04\,(4) = 4.2$ ft

$h_S/D_{50} = 4.2/2.5 = 1.7$ and $h_S/D_{50} \geq 2$ is not satisfied. Although not available, try a riprap size that will yield h_S/D_{50} close to, but greater than, 2. (A basin sized for smaller riprap may be lined with larger riprap.) Repeat step 2.

Step 2 (3rd iteration). Select riprap size and compute basin depth.

Try $D_{50} = 2.3$ ft; $D_{50}/y_e = 2.3/4 = 0.58$ (≥ 0.1 OK)

From Equation 10.1,

$$\frac{h_s}{y_e} = 0.86\left(\frac{D_{50}}{y_e}\right)^{-0.55}\left(\frac{V_o}{\sqrt{gy_e}}\right) - C_o = 0.86(0.58)^{-0.55}(2.2) - 1.4 = 1.15$$

$h_S = (h_S/y_e)y_e = 1.15\,(4) = 4.6$ ft

$h_S/D_{50} = 4.6/2.3 = 2.0$ and $h_S/D_{50} \geq 2$ is satisfied.

Step 3 (3rd iteration). Size the basin as shown in Figures 10.1 and 10.2.

$L_S = 10h_S = 10(4.6) = 46$ ft

L_S min = $3W_o = 3(8) = 24$ ft, use $L_S = 46$ ft

$L_B = 15h_S = 15(4.6) = 69$ ft

L_B min = $4W_o = 4(8) = 32$ ft, use $L_B = 69$ ft

$W_B = W_o + 2(L_B/3) = 8 + 2(69/3) = 54$ ft

However, since the trial D_{50} is not available, the next larger riprap size ($D_{50} = 2.5$ ft) would be used to line a basin with the given dimensions.

Step 4 (3rd iteration). Determine the basin exit depth, $y_B = y_c$, and exit velocity, $V_B = V_c$.

$Q^2/g = (A_c)^3/T_c = [y_c(W_B + zy_c)]^3/(W_B + 2zy_c)$

$800^2/32.2 = 19,876 = [y_c(54 + 2y_c)]^3/(54 + 4y_c)$

By trial and success, $y_c = 1.85$ ft, $T_c = 61.4$ ft, $A_c = 106.9$ ft^2

$V_B = V_c = Q/A_c = 800/106.9 = 7.5$ ft/s (not acceptable). If the apron were extended (with a continued flare) such that the total basin length was 90 ft, the velocity would be reduced to the allowable level.

Two feasible options have been identified. First, a 6-ft-deep, 60-ft-long pool, with a 30-ft-apron using $D_{50} = 1.83$ ft. Second, a 4.6-ft-deep, 46-ft-long pool, with a 44-ft-apron using $D_{50} = 2.5$ ft. Because the overall length is the same, the first option is likely to be more economical.

Step 5. For the design discharge, determine if TW/$y_o \leq 0.75$.

For the first tailwater condition, TW/y_o = 2.8/4.0 = 0.70, which satisfies TW/$y_o \leq 0.75$. No additional riprap needed downstream.

For the second tailwater condition, $TW/y_o = 4.2/4.0 = 1.05$, which does not satisfy $TW/y_o \leq 0.75$. To determine required riprap, estimate centerline velocity at a series of downstream cross sections using Figure 10.3.

Compute equivalent circular diameter, D_e, for brink area:

$A = \pi D_e^2 /4 = (y_o)(W_o) = (4)(8) = 32\ ft^2$

$D_e = [32(4)/ \pi]^{1/2} = 6.4\ ft$

Rock size can be determined using the procedures in Section 10.3 (Equation 10.6) or other suitable method. The computations are summarized below.

L/D_e	L (ft)	V_L/V_o (Figure 10.3)	V_L (ft/s)	Rock size, D_{50} (ft)
10	64	0.59	14.7	1.42
15	96	0.42	10.5	0.72
20	128	0.30	7.5	0.37
21	135	0.28	7.0	0.32

The calculations above continue until $V_L \leq V_{allow}$. Riprap should be at least the size shown. As a practical consideration, the channel can be lined with the same size rock used for the basin. Protection must extend at least 135 ft downstream from the culvert brink, which is 45 ft beyond the basin exit. Riprap should be installed in accordance with details shown in HEC 11.

Design Example: Riprap Basin (Culvert on a Mild Slope) (SI)

Determine riprap basin dimensions using the envelope design (Equations 10.1 and 10.2) for a pipe culvert that is in outlet control with subcritical flow in the culvert. Allowable exit velocity from the riprap basin, V_{allow}, is 2.1 m/s. Riprap is available with a D_{50} of 0.125, 0.150, and 0.250 m. Given:

D = 1.83 m CMP with Manning's n = 0.024
S_o = 0.004 m/m
Q = 3.82 m³/s
y_n = 1.37 m (normal flow depth in the pipe)
V_n = 1.80 m/s (normal velocity in the pipe)
TW = 0.61 m (tailwater depth)

Solution

Step 1. Compute the culvert outlet velocity, V_o, and depth, y_o.

For subcritical flow (culvert on mild slope), use Figure 3.4 to obtain y_o/D, then calculate V_o by dividing Q by the wetted area for y_o.

$K_u Q/D^{2.5} = 1.81 (3.82)/1.83^{2.5} = 1.53$

$TW/D = 0.61/1.83 = 0.33$

From Figure 3.4, $y_o/D = 0.45$

$y_o = (y_o/D)D = 0.45(1.83) = 0.823$ m (brink depth)

From Table B.2, for $y_o /D = 0.45$, the brink area ratio $A/D^2 = 0.343$

$A = (A/D^2)D^2 = 0.343(1.83)^2 = 1.15$ m^2

$V_o = Q/A = 3.82/1.15 = 3.32$ m/s

$y_e = (A/2)^{1/2} = (1.15/2)^{1/2} = 0.76$ m

$Fr = V_o / [9.81(y_e)]^{1/2} = 3.32/ [9.81(0.76)]^{1/2} = 1.22$

Step 2. Select a trial D_{50} and obtain h_s/y_e from Equation 10.1. Check to see that $h_s/D_{50} \geq 2$ and $D_{50}/y_e \geq 0.1$.

Try $D_{50} = 0.15$ m; $D_{50}/y_e = 0.15/0.76 = 0.20$ (≥ 0.1 OK)

$TW/y_e = 0.61/0.76 = 0.80$. Therefore, from Equation 10.2,

$C_o = 4.0(TW/y_e) -1.6 = 4.0(0.80) -1.6 = 1.61$

From Equation 10.1,

$$\frac{h_s}{y_e} = 0.86\left(\frac{D_{50}}{y_e}\right)^{-0.55}\left(\frac{V_o}{\sqrt{gy_e}}\right) - C_o = 0.86(0.20)^{-0.55}(1.22) - 1.61 = 0.933$$

$h_S = (h_S /y_e)y_e = 0.933 (0.76) = 0.71$ m

$h_S/D_{50} = 0.71/0.15 = 4.7$ and $h_S/D_{50} \geq 2$ is satisfied

Step 3. Size the basin as shown in Figures 10.1 and 10.2.

$L_S = 10h_S = 10(0.71) = 7.1$ m

L_S min $= 3W_o = 3(1.83) = 5.5$ m, use $L_S = 7.1$ m

$L_B = 15h_S = 15(0.71) = 10.7$ m

L_B min $= 4W_o = 4(1.83) = 7.3$ m, use $L_B = 10.7$ m

$W_B = W_o + 2(L_B/3) = 1.83 + 2(10.7/3) = 9.0$ m

Step 4. Determine the basin exit depth, $y_B = y_c$ and exit velocity, $V_B = V_c$.

$Q^2/g = (A_c)^3/T_c = [y_c(W_B + zy_c)]^3/ (W_B + 2zy_c)$

$3.82^2/9.81 = 1.49 = [y_c(9.0 + 2y_c)]^3/ (9.0 + 4y_c)$

By trial and success, $y_c = 0.26$ m, $T_c = 10.0$ m, $A_c = 2.48$ m^2

$V_c = Q/A_c = 3.82/2.48 = 1.5$ m/s (acceptable)

The initial trial of riprap ($D_{50} = 0.15$ m) results in a 10.7 m basin that satisfies all design requirements. Try the next larger riprap size to test if a smaller basin is feasible by repeating steps 2 through 4.

Step 2 (2nd iteration). Select a trial D_{50} and obtain h_s/y_e from Equation 10.1.

Try $D_{50} = 0.25$ m; $D_{50}/y_e = 0.25/0.76 = 0.33$ (≥ 0.1 OK)

From Equation 10.1,

$$\frac{h_s}{y_e} = 0.86\left(\frac{D_{50}}{y_e}\right)^{-0.55}\left(\frac{V_o}{\sqrt{gy_e}}\right) - C_o = 0.86(0.33)^{-0.55}(1.22) - 1.61 = 0.320$$

$h_S = (h_S/y_e)y_e = 0.320\,(0.76) = 0.24$ m

$h_S/D_{50} = 0.24/0.25 = 0.96$ and $h_S/D_{50} \geq 2$ is not satisfied. Although not available, try a riprap size that will yield h_S/D_{50} close to, but greater than 2. (A basin sized for smaller riprap may be lined with larger riprap.) Repeat step 2.

Step 2 (3rd iteration). Select a trial D_{50} and obtain h_s/y_e from Equation 10.1.

Try $D_{50} = 0.205$ m; $D_{50}/y_e = 0.205/0.76 = 0.27$ (≥ 0.1 OK)

From Equation 10.1,

$$\frac{h_s}{y_e} = 0.86\left(\frac{D_{50}}{y_e}\right)^{-0.55}\left(\frac{V_o}{\sqrt{gy_e}}\right) - C_o = 0.86(0.27)^{-0.55}(1.22) - 1.61 = 0.545$$

$h_S = (h_S/y_e)y_e = 0.545\,(0.76) = 0.41$ m

$h_S/D_{50} = 0.41/0.205 = 2.0$ and $h_S/D_{50} \geq 2$ is satisfied. Continue to step 3.

Step 3 (3rd iteration). Size the basin as shown in Figures 10.1 and 10.2.

$L_S = 10h_S = 10(0.41) = 4.1$ m

L_S min $= 3W_o = 3(1.83) = 5.5$ m, use $L_S = 5.5$ m

$L_B = 15h_S = 15(0.41) = 6.2$ m

L_B min $= 4W_o = 4(1.83) = 7.3$ m, use $L_B = 7.3$ m

$W_B = W_o + 2(L_B/3) = 1.83 + 2(7.3/3) = 6.7$ m

However, since the trial D_{50} is not available, the next larger riprap size ($D_{50} = 0.25$ m) would be used to line a basin with the given dimensions.

Step 4 (3rd iteration). Determine the basin exit depth, $y_B = y_c$ and exit velocity, $V_B = V_c$.

$Q^2/g = (A_c)^3/T_c = [y_c(W_B + zy_c)]^3/ (W_B + 2zy_c)$

$3.82^2/9.81 = 1.49 = [y_c(6.7 + 2y_c)]^3/ (6.7 + 4y_c)$

By trial and success, $y_c = 0.31$ m, $T_c = 7.94$ m, $A_c = 2.28$ m^2

$V_c = Q/A_c = 3.82/2.28 = 1.7$ m/s (acceptable)

Two feasible options have been identified. First, a 0.71 m deep, 7.1 m long pool, with an 3.6 m apron using $D_{50} = 0.15$ m. Second, a 0.41 m deep, 5.5 m long pool, with a 1.8 m apron using $D_{50} = 0.25$ m. The choice between these two options will likely depend on the available space and the cost of riprap.

Step 5. For the design discharge, determine if $TW/y_o \leq 0.75$

$TW/y_o = 0.61/0.823 = 0.74$, which satisfies $TW/y_o \leq 0.75$. No additional riprap needed.

Design Example: Riprap Basin (Culvert on a Mild Slope) (CU)

Determine riprap basin dimensions using the envelope design (Equations 10.1 and 10.2) for a pipe culvert that is in outlet control with subcritical flow in the culvert. Allowable exit velocity from the riprap basin, V_{allow}, is 7.0 ft/s. Riprap is available with a D_{50} of 0.42, 0.50, and 0.83 ft. Given:

D	=	6 ft CMP with Manning's n = 0.024
S_o	=	0.004 ft/ft
Q	=	135 ft^3/s
y_n	=	4.5 ft (normal flow depth in the pipe)
V_n	=	5.9 ft/s (normal velocity in the pipe)
TW	=	2.0 ft (tailwater depth)

Solution

Step 1. Compute the culvert outlet velocity, V_o, depth, y_o and Froude number.

For subcritical flow (culvert on mild slope), use Figure 3.4 to obtain y_o/D, then calculate V_o by dividing Q by the wetted area for y_o.

$K_u Q/D^{2.5} = 1.0(135)/6^{2.5} = 1.53$

$TW/D = 2.0/6 = 0.33$

From Figure 3.4, $y_o/D = 0.45$

$y_o = (y_o/D)D = 0.45(6) = 2.7$ ft (brink depth)

From Table B.2 for $y_o/D = 0.45$, the brink area ratio $A/D^2 = 0.343$

$A = (A/D^2)D^2 = 0.343(6)^2 = 12.35$ ft^2

$V_o = Q/A = 135/12.35 = 10.9$ ft/s

$y_e = (A/2)^{1/2} = (12.35/2)^{1/2} = 2.48$ ft

$Fr = V_o / [32.2(y_e)]^{1/2} = 10.9/ [32.2(2.48)]^{1/2} = 1.22$

Step 2. Select a trial D_{50} and obtain h_s/y_e from Equation 10.1. Check to see that $h_s/D_{50} \geq 2$ and $D_{50}/y_e \geq 0.1$.

Try $D_{50} = 0.5$ ft; $D_{50}/y_e = 0.5/2.48 = 0.20$ (≥ 0.1 OK)

$TW/y_e = 2.0/2.48 = 0.806$. Therefore, from Equation 10.2,

$C_o = 4.0(TW/y_e) -1.6 = 4.0(0.806) -1.6 = 1.62$

From Equation 10.1,

$$\frac{h_s}{y_e} = 0.86\left(\frac{D_{50}}{y_e}\right)^{-0.55}\left(\frac{V_o}{\sqrt{gy_e}}\right) - C_o = 0.86(0.20)^{-0.55}(1.22) - 1.62 = 0.923$$

$h_S = (h_S/y_e)y_e = 0.923 (2.48) = 2.3$ ft

$h_S/D_{50} = 2.3/0.5 = 4.6$ and $h_S/D_{50} \geq 2$ is satisfied

Step 3. Size the basin as shown in Figures 10.1 and 10.2.

$L_S = 10h_S = 10(2.3) = 23$ ft

L_S min $= 3W_o = 3(6) = 18$ ft, use $L_S = 23$ ft

$L_B = 15h_S = 15(2.3) = 34.5$ ft

L_B min $= 4W_o = 4(6) = 24$ ft, use $L_B = 34.5$ ft

$W_B = W_o + 2(L_B/3) = 6 + 2(34.5/3) = 29$ ft

Step 4. Determine the basin exit depth, $y_B = y_c$ and exit velocity, $V_B = V_c$.

$Q^2/g = (A_c)^3/T_c = [y_c(W_B + zy_c)]^3/ (W_B + 2zy_c)$

$135^2/32.2 = 566 = [y_c(29 + 2y_c)]^3/ (29 + 4y_c)$

By trial and success, $y_c = 0.86$ ft, $T_c = 32.4$ ft, $A_c = 26.4$ ft^2

$V_c = Q/A_c = 135/26.4 = 5.1$ ft/s (acceptable)

The initial trial of riprap ($D_{50} = 0.5$ ft) results in a 34.5 ft basin that satisfies all design requirements. Try the next larger riprap size to test if a smaller basin is feasible by repeating steps 2 through 4.

Step 2 (2nd iteration). Select a trial D_{50} and obtain h_s/y_e from Equation 10.1.

Try $D_{50} = 0.83$ ft; $D_{50}/y_e = 0.83/2.48 = 0.33$ (≥ 0.1 OK)

From Equation 10.1,

$$\frac{h_s}{y_e} = 0.86\left(\frac{D_{50}}{y_e}\right)^{-0.55}\left(\frac{V_o}{\sqrt{gy_e}}\right) - C_o = 0.86(0.33)^{-0.55}(1.22) - 1.62 = 0.311$$

$h_S = (h_S /y_e)y_e = 0.311 (2.48) = 0.8$ ft

$h_S/D_{50} = 0.8/0.83 = 0.96$ and $h_S/D_{50} \geq 2$ is not satisfied. Although not available, try a riprap size that will yield h_S/D_{50} close to, but greater than 2. (A basin sized for smaller riprap may be lined with larger riprap.) Repeat step 2.

Step 2 (3rd iteration). Select a trial D_{50} and obtain h_s/y_e from Equation 10.1.

Try $D_{50} = 0.65$ ft; $D_{50}/y_e = 0.65/2.48 = 0.26$ (≥ 0.1 OK)

From Equation 10.1,

$$\frac{h_s}{y_e} = 0.86\left(\frac{D_{50}}{y_e}\right)^{-0.55}\left(\frac{V_o}{\sqrt{gy_e}}\right) - C_o = 0.86(0.26)^{-0.55}(1.22) - 1.62 = 0.581$$

$h_S = (h_S /y_e)y_e = 0.581 (2.48) = 1.4$ ft

$h_S/D_{50} = 1.4/0.65 = 2.15$ and $h_S/D_{50} \geq 2$ is satisfied. Continue to step 3.

Step 3 (3rd iteration). Size the basin as shown in Figures 10.1 and 10.2.

$L_S = 10h_S = 10(1.4) = 14$ ft

L_S min $= 3W_o = 3(6) = 18$ ft, use $L_S = 18$ ft

$L_B = 15h_S = 15(1.4) = 21$ ft

$L_B \ min = 4W_o = 4(6) = 24 \ ft$, use $L_B = 24 \ ft$

$W_B = W_o + 2(L_B/3) = 6 + 2(24/3) = 22 \ ft$

However, since the trial D_{50} is not available, the next larger riprap size ($D_{50} = 0.83 \ ft$) would be used to line a basin with the given dimensions.

Step 4 (3rd iteration). Determine the basin exit depth, $y_B = y_c$ and exit velocity, $V_B = V_c$.

$Q^2/g = (A_c)^3/T_c = [y_c(W_B + zy_c)]^3/ (W_B + 2zy_c)$

$135^2/32.2 = 566 = [y_c(22 + 2y_c)]^3/ (22 + 4y_c)$

By trial and success, $y_c = 1.02 \ ft$, $T_c = 26.1 \ ft$, $A_c = 24.5 \ ft^2$

$V_c = Q/A_c = 135/24.5 = 5.5 \ ft/s$ (acceptable)

Two feasible options have been identified. First, a 2.3-ft-deep, 23-ft-long pool, with an 11.5-ft-apron using $D_{50} = 0.5 \ ft$. Second, a 1.4-ft-deep, 18-ft-long pool, with a 6-ft-apron using $D_{50} = 0.83 \ ft$. The choice between these two options will likely depend on the available space and the cost of riprap.

Step 5. For the design discharge, determine if $TW/y_o \leq 0.75$

$TW/y_o = 2.0/2.7 = 0.74$, which satisfies $TW/y_o \leq 0.75$. No additional riprap needed.

10.2 RIPRAP APRON

The most commonly used device for outlet protection, primarily for culverts 1500 mm (60 in) or smaller, is a riprap apron. An example schematic of an apron taken from the Federal Lands Division of the Federal Highway Administration is shown in Figure 10.4.

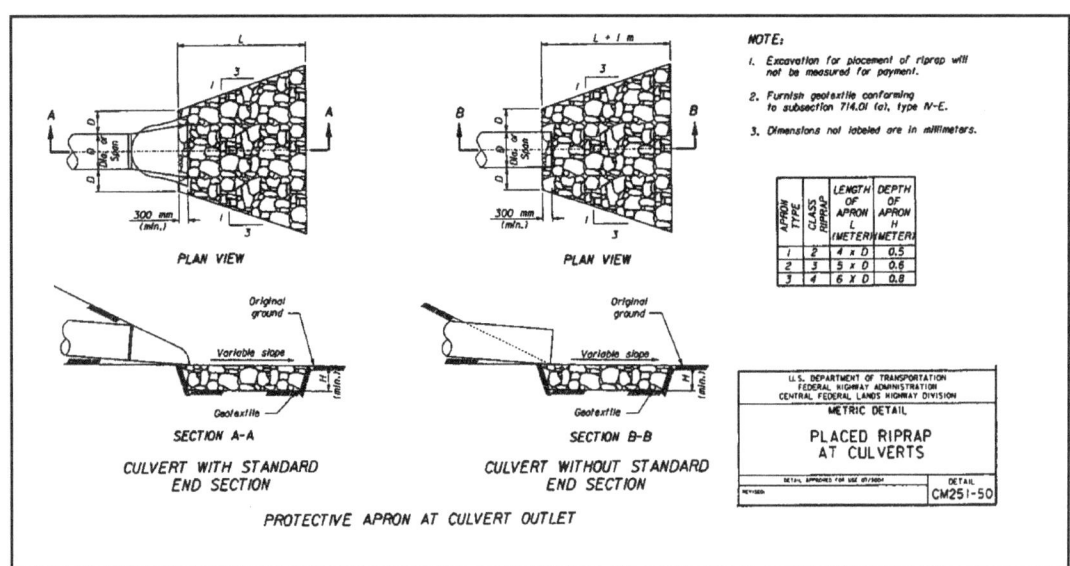

Figure 10.4. Placed Riprap at Culverts (Central Federal Lands Highway Division)

They are constructed of riprap or grouted riprap at a zero grade for a distance that is often related to the outlet pipe diameter. These aprons do not dissipate significant energy except through increased roughness for a short distance. However, they do serve to spread the flow helping to transition to the natural drainage way or to sheet flow where no natural drainage way exists. However, if they are too short, or otherwise ineffective, they simply move the location of potential erosion downstream. The key design elements of the riprap apron are the riprap size as well as the length, width, and depth of the apron.

Several relationships have been proposed for riprap sizing for culvert aprons and several of these are discussed in greater detail in Appendix D. The independent variables in these relationships include one or more of the following variables: outlet velocity, rock specific gravity, pipe dimension (e.g. diameter), outlet Froude number, and tailwater. The following equation (Fletcher and Grace, 1972) is recommended for circular culverts:

$$D_{50} = 0.2\,D \left(\frac{Q}{\sqrt{gD^{2.5}}} \right)^{4/3} \left(\frac{D}{TW} \right) \tag{10.4}$$

where,

D_{50} = riprap size, m (ft)
Q = design discharge, m^3/s (ft^3/s)
D = culvert diameter (circular), m (ft)
TW = tailwater depth, m (ft)
g = acceleration due to gravity, 9.81 m/s^2 (32.2 ft/s^2)

Tailwater depth for Equation 10.4 should be limited to between 0.4D and 1.0D. If tailwater is unknown, use 0.4D.

Whenever the flow is supercritical in the culvert, the culvert diameter is adjusted as follows:

$$D' = \frac{D + y_n}{2} \tag{10.5}$$

where,

D' = adjusted culvert rise, m (ft)
y_n = normal (supercritical) depth in the culvert, m (ft)

Equation 10.4 assumes that the rock specific gravity is 2.65. If the actual specific gravity differs significantly from this value, the D_{50} should be adjusted inversely to specific gravity.

The designer should calculate D_{50} using Equation 10.4 and compare with available riprap classes. A project or design standard can be developed such as the example from the Federal Highway Administration Federal Lands Highway Division (FHWA, 2003) shown in Table 10.1 (first two columns). The class of riprap to be specified is that which has a D_{50} greater than or equal to the required size. For projects with several riprap aprons, it is often cost effective to use fewer riprap classes to simplify acquiring and installing the riprap at multiple locations. In such a case, the designer must evaluate the tradeoffs between over sizing riprap at some locations in order to reduce the number of classes required on a project.

Table 10.1. Example Riprap Classes and Apron Dimensions

Class	D_{50} (mm)	D_{50} (in)	Apron Length[1]	Apron Depth
1	125	5	4D	$3.5D_{50}$
2	150	6	4D	$3.3D_{50}$
3	250	10	5D	$2.4D_{50}$
4	350	14	6D	$2.2D_{50}$
5	500	20	7D	$2.0D_{50}$
6	550	22	8D	$2.0D_{50}$

[1]D is the culvert rise.

The apron dimensions must also be specified. Table 10.1 provides guidance on the apron length and depth. Apron length is given as a function of the culvert rise and the riprap size. Apron depth ranges from $3.5D_{50}$ for the smallest riprap to a limit of $2.0D_{50}$ for the larger riprap sizes. The final dimension, width, may be determined using the 1:3 flare shown in Figure 10.4 and should conform to the dimensions of the downstream channel. A filter blanket should also be provided as described in HEC 11 (Brown and Clyde, 1989).

For tailwater conditions above the acceptable range for Equation 10.4 (TW > 1.0D), Figure 10.3 should be used to determine the velocity downstream of the culvert. The guidance in Section 10.3 may be used for sizing the riprap. The apron length is determined based on the allowable velocity and the location at which it occurs based on Figure 10.3.

Over their service life, riprap aprons experience a wide variety of flow and tailwater conditions. In addition, the relations summarized in Table 10.1 do not fully account for the many variables in culvert design. To ensure continued satisfactory operation, maintenance personnel should inspect them after major flood events. If repeated severe damage occurs, the location may be a candidate for extending the apron or another type of energy dissipator.

Design Example: Riprap Apron (SI)

Design a riprap apron for the following CMP installation. Available riprap classes are provided in Table 10.1. Given:

Q = 2.33 m³/s
D = 1.5 m
TW = 0.5 m

Solution

Step 1. Calculate D_{50} from Equation 10.4. First verify that tailwater is within range.

$TW/D = 0.5/1.5 = 0.33$. This is less than 0.4D, therefore,

use $TW = 0.4D = 0.4(1.5) = 0.6$ m

$$D_{50} = 0.2\,D \left(\frac{Q}{\sqrt{g}D^{2.5}} \right)^{4/3} \left(\frac{D}{TW} \right) = 0.2\,(1.5) \left(\frac{2.33}{\sqrt{9.81}(1.5)^{2.5}} \right)^{4/3} \left(\frac{1.5}{0.6} \right) = 0.13 \text{ m}$$

Step 2. Determine riprap class. From Table 10.1, riprap class 2 ($D_{50} = 0.15$ m) is required.

10-18

Step 3. Estimate apron dimensions.

From Table 10.1 for riprap class 2,

Length, L = 4D = 4(1.5) = 6 m

Depth = $3.3D_{50}$ = 3.3 (0.15) = 0.50 m

Width (at apron end) = 3D + (2/3)L = 3(1.5) + (2/3)(6) = 8.5 m

Design Example: Riprap Apron (CU)

Design a riprap apron for the following CMP installation. Available riprap classes are provided in Table 10.1. Given:

Q   = 85 ft^3/s

D   = 5.0 ft

TW = 1.6 ft

Solution

Step 1. Calculate D_{50} from Equation 10.4. First verify that tailwater is within range.

TW/D = 1.6/5.0 = 0.32.   This is less than 0.4D, therefore,

use TW = 0.4D = 0.4(5) = 2.0 ft

$$D_{50} = 0.2\,D\left(\frac{Q}{\sqrt{gD^{2.5}}}\right)^{4/3}\left(\frac{D}{TW}\right) = 0.2\,(5.0)\left(\frac{85}{\sqrt{32.2}(5.0)^{2.5}}\right)^{4/3}\left(\frac{5.0}{2.0}\right) = 0.43\ \text{ft} = 5.2\ \text{in}$$

Step 2. Determine riprap class. From Table 10.1, riprap class 2 (D_{50} = 6 in) is required.

Step 3. Estimate apron dimensions.

From Table 10.1 for riprap class 2,

Length, L = 4D = 4(5) = 20 ft

Depth = $3.3D_{50}$ = 3.3 (6) = 19.8 in = 1.65 ft

Width (at apron end) = 3D + (2/3)L = 3(5) + (2/3)(20) = 28.3 ft

10.3 RIPRAP APRONS AFTER ENERGY DISSIPATORS

Some energy dissipators provide exit conditions, velocity and depth, near critical. This flow condition rapidly adjusts to the downstream or natural channel regime; however, critical velocity may be sufficient to cause erosion problems requiring protection adjacent to the energy dissipator. Equation 10.6 provides the riprap size recommended for use downstream of energy dissipators. This relationship is from Searcy (1967) and is the same equation used in HEC 11 (Brown and Clyde, 1989) for riprap protection around bridge piers.

$$D_{50} = \frac{0.692}{S-1}\left(\frac{V^2}{2g}\right) \tag{10.6}$$

where,

    D_{50} = median rock size, m (ft)

    V   = velocity at the exit of the dissipator, m/s (ft/s)

    S   = riprap specific gravity

The length of protection can be judged based on the magnitude of the exit velocity compared with the natural channel velocity. The greater this difference, the longer will be the length required for the exit flow to adjust to the natural channel condition. A filter blanket should also be provided as described in HEC 11 (Brown and Clyde, 1989).

CHAPTER 11: DROP STRUCTURES

Drop structures are commonly used for flow control and energy dissipation. Changing the channel slope from steep to mild, by placing drop structures at intervals along the channel reach, changes a continuous steep slope into a series of gentle slopes and vertical drops. Instead of slowing down and transferring high erosion producing velocities into low non-erosive velocities, drop structures control the slope of the channel in such a way that the high, erosive velocities never develop. The kinetic energy or velocity gained by the water as it drops over the crest of each structure is dissipated by a specially designed apron or stilling basin.

The drop structures discussed here (see Figure 11.1) require an aerated nappe and are, in general, for subcritical flow in the upstream as well as downstream channel. The effect of upstream supercritical flow on drop structure design is discussed in a later section. The stilling basin protects the channel against erosion below the drop and dissipates energy. This is accomplished through the impact of the falling water on the floor, redirection of the flow, and turbulence. The stilling basin used to dissipate the excess energy can vary from a simple concrete apron to an apron with flow obstructions such as baffle blocks, sills, or abrupt rises. The length of the concrete apron required can be shortened by the addition of these appurtenances.

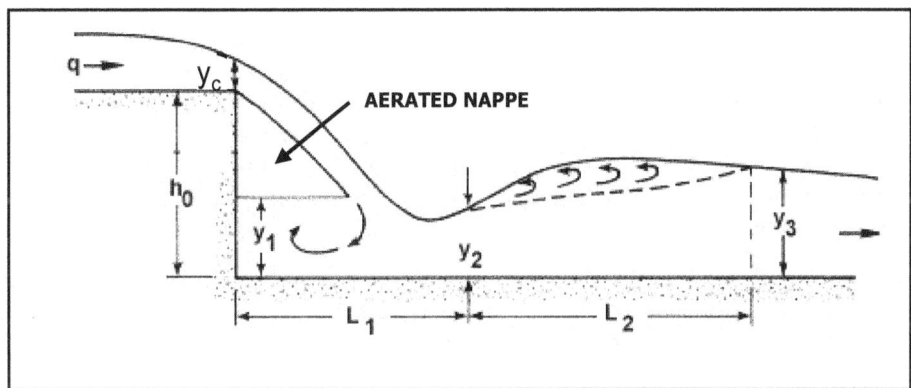

Figure 11.1. Flow Geometry of a Straight Drop Spillway

The drop number gives a quantitative measure for drop:

$$N_d = \frac{q^2}{gh_o^3}$$ (11.1)

where,

- N_d = drop number
- q = unit discharge, $m^3/s/m$ ($ft^3/s/ft$)
- g = acceleration due to gravity, 9.81 m/s^2 (32.2 ft/s^2)
- h_o = drop height, m (ft)

Another commonly used quantitative measure for drop is given by:

$$D_r = \frac{h_o}{y_c}$$ (11.2)

where,

D_r = relative drop

y_c = critical depth at the drop, m (ft)

h_o = drop height, m (ft)

Drop structures may be categorized based on either Equations 11.1 or 11.2. Drops for which N_d is greater than 1 or Drop is less than 1 are considered "low drop" structures. Two dissipators are discussed in this chapter: the straight drop structure and the box inlet drop structure. Neither of these is considered low drop structures.

11.1 STRAIGHT DROP STRUCTURE

A straight drop structure is characterized by flow through a rectangular weir followed by a drop into a stilling basin. The stilling basin may be a flat apron or an apron with various baffles and sills depending on the site conditions. First a simple stilling basin is considered followed by discussion of other features available to modify the drop structure performance.

11.1.1 Simple Straight Drop

The basic flow geometry of a straight drop structure is shown in Figure 11.1. The discharge passes through critical depth as it flows over the drop structure crest. The free-falling nappe reverses its curvature and turns smoothly into supercritical flow on the apron at the distance L_1 from the drop wall. The mean velocity at the distance L is parallel to the apron; the depth y_2 is the smallest depth in the downstream channel, and the pressure is nearly hydrostatic. The depth of supercritical flow in the downstream direction increases due to channel resistance, and at some point will reach a depth sufficient for the formation of a hydraulic jump.

For a given drop height, h_o, and discharge, q, the sequent depth, y_3, in the downstream channel and the drop length, L_1, may be computed. The length of jump L_j, is discussed in Chapter 6. The drop number can be used to estimate the dimensions of a simple straight drop structure.

$$L_1 = 4.30 h_o N_d^{0.27}$$ (11.3a)

$$y_1 = 1.0 h_o N_d^{0.22}$$ (11.3b)

$$y_2 = 0.54 h_o N_d^{0.425}$$ (11.3c)

$$y_3 = 1.66 h_o N_d^{0.27}$$ (11.3d)

where,

L_1 = drop length (the distance from the drop wall to the position of the depth y_2), m (ft)

y_1 = pool depth under the nappe, m (ft)

y_2 = depth of flow at the toe of the nappe or the beginning of the hydraulic jump, m (ft)

y_3 = tailwater depth sequent to y_2, m (ft)

By comparing the channel tailwater depth, TW, with the computed, y_3, one of the following cases will occur. The case will determine design modifications necessary to the structure.

      1. TW > y_3. The hydraulic jump will be submerged and the basin length may need to be increased.

2. $TW = y_3$. The hydraulic jump begins at depth y_2, no supercritical flow exists on the apron and the distance L_1 is a minimum. In this case, the basin will function without additional design modifications.

3. $TW < y_3$. The hydraulic jump will recede downstream and the basin will not function.

For case 3, when the tailwater depth is less than y_3, it is necessary to modify the basin to force the hydraulic jump to stay in the basin. Two alternatives to achieve this are to provide an apron:

1. at the bed level with an end sill or baffles to trigger the jump in the basin, or

2. depressed below the downstream bed level to effectively increase tailwater with an end sill.

The choice of design type and the design dimensions will depend, for a given unit discharge, q, on the drop height, h_o, and on the downstream depth, TW. The apron may be designed to extend to the end of the hydraulic jump. However, including an end sill allows the use of a shorter and more economical stilling basin.

The geometry of the undisturbed flow should be taken into consideration in the design of a straight drop structure. If the overfall crest length is less than the width of the approach channel, it is important that a transition be properly designed by shaping the approach channel to reduce the effect of end contractions. Otherwise the contraction at the ends of the spillway notch may be so pronounced that the jet will land beyond the stilling basin and the concentration of high velocities at the center of the outlet may cause additional scour in the downstream channel (see Chapter 4).

11.1.2 Grate Design

A grate or series of rails forming a "grizzly" may be used in conjunction with drop structures as illustrated in Figure 11.2. The incoming flow is divided into a number of jets as it passes through the grate. These fall almost vertically to the downstream channel resulting in good energy dissipator action. This type of design is also utilized as a debris ejector where the debris rides over the grate and falls into a holding area for later removal and the water passes through the grate.

The Bureau of Reclamation has published design recommendations for grates (USBR, 1987) for use where the incoming flow is subcritical. The length of the grate is calculated from:

$$L_G = \frac{Q}{CWN(2gy_o)^{1/2}} \tag{11.4}$$

where,

C = experimental coefficient equal to 0.245
W = width of the slots, m (ft)
N = number of slots (spaces) between beams
y_o = approach flow depth, m (ft)

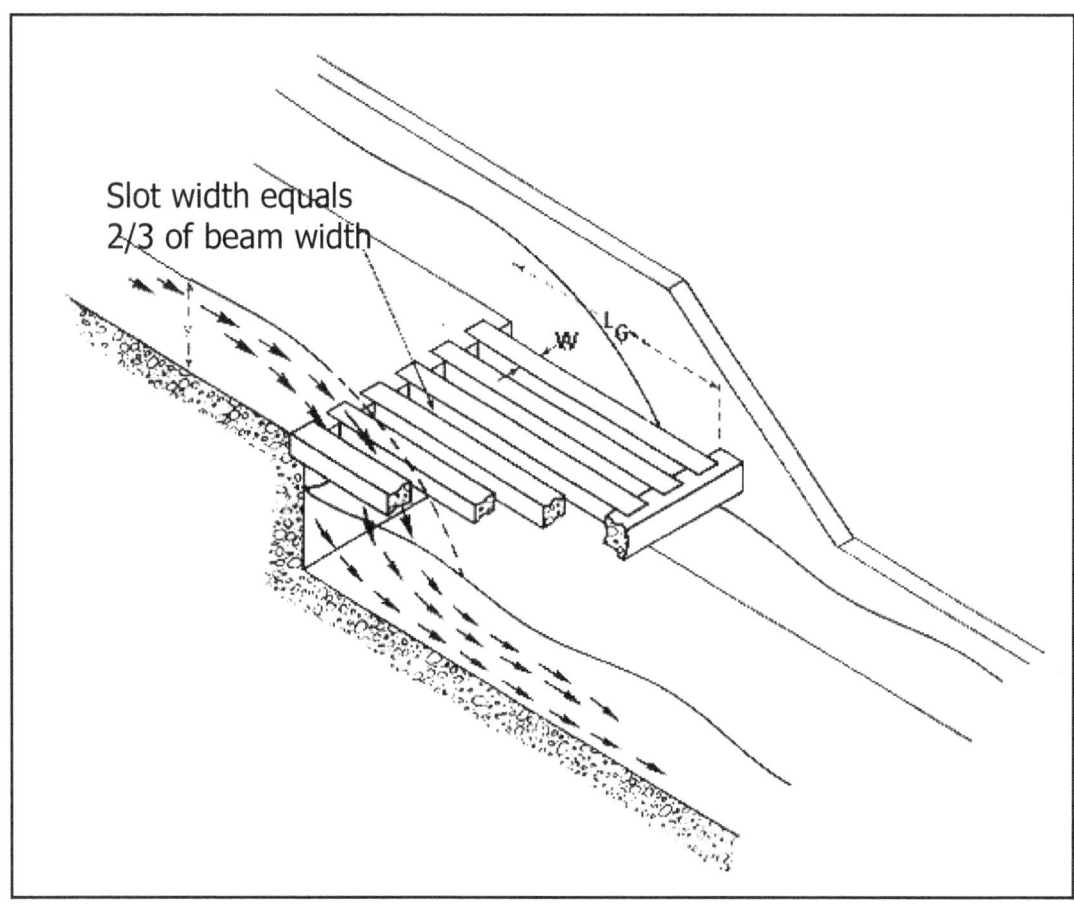

Slot width equals
2/3 of beam width

Figure 11.2. Drop Structure with Grate

The design process is iterative based on the number of slots and the slot width. The USBR (1987) recommends the following guidelines:

1. Select an initial slot width. Provide a full slot width at each wall.

2. Beam width should be approximately 1.5W

3. Estimate the number of slots, N, for use in Equation 11.4

4. Calculate the grate length using Equation 11.4. Adjust the slot width until an acceptable beam length, L_G, is obtained.

5. Tilt the grate about 3° downstream to be self-cleaning.

Examination of the beam length equation indicates the relative effect of higher approach velocities on the design of drop structures. Assuming the slot width, W, approaches the channel width making N equal to 1, and considering a constant flow rate, then the relationship in Equation 11.4 reveals that the grate length is inversely proportional to the square root of the approach depth. For constant Q, as the approach velocity increases the approach depth decreases and the length L_G increases. Therefore, for high velocity flow, above critical velocity, the length of drop structure required, to contain the jet, may very rapidly exceed practical limits.

11.1.3 Straight Drop Structure Design Features

A general design for a stilling basin at the toe of a drop structure was developed by the Agricultural Research Service, St. Anthony Falls Hydraulic Laboratory, University of Minnesota (Donnelly and Blaisdell, 1954). The basin consists of a horizontal apron with blocks and sills to dissipate energy as shown in Figure 11.3. Tailwater also influences the amount of energy dissipated. The stilling basin length computed for the minimum tailwater level required for good performance may be inadequate at high tailwater levels. Scour of the downstream channel may occur if the nappe is supported sufficiently by high tailwater so that it lands beyond the end of the stilling basin. A method for computing the stilling basin length for all tailwater levels is presented.

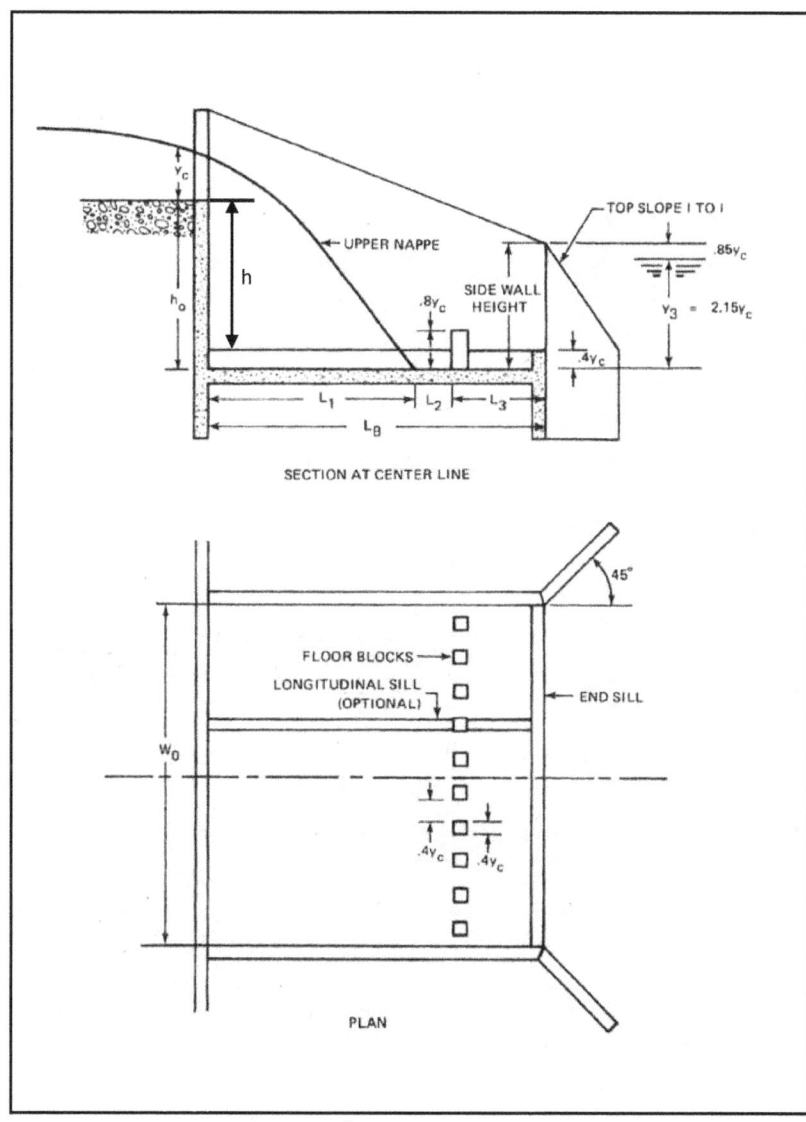

Figure 11.3. Straight Drop Structure (Rand, 1955)

The recommended design is limited to the following conditions:

1. Total drop, h_o, less than 4.6 m (15 ft) with sufficient tailwater.
2. Relative drop, h_o/y_c, between 1.0 and 15.
3. Crest length, W_o, greater than $1.5y_c$.

The elements that must be considered in the design of this stilling basin include the length of basin, the position and size of floor blocks, the position and height of end sill, the position of the wingwalls, and the approach channel geometry. Figure 11.3 illustrates a straight drop structure that provides adequate protection from scour in the downstream channel.

Many of the design parameters for the straight drop structure are based on the critical depth. Critical depth in a rectangular channel or culvert is calculated from the unit discharge (discharge divided by culvert/chute width, B).

$$y_c = \left(\frac{q^2}{g} \right)^{\frac{1}{3}}$$

(11.5)

where,

y_c = critical depth, m (ft)
q = unit discharge (Q/B), m (ft)

Critical flow for an open channel of any shape will occur when:

$$\frac{Q^2 T_c}{g A_c^3} = 1$$

(11.6)

where,

T_c = water surface width at critical flow condition, m (ft)
A_c = flow area at critical flow condition, m (ft)

As discussed earlier, the tailwater must neither be too high nor too low. Therefore, the following relationships must be achieved in the design. First, the tailwater depth above the floor of the stilling basin must be calculated from Equation 11.7.

$$y_3 = 2.15 y_c$$

(11.7)

where,

y_3 = tailwater depth above the floor of the stilling basin, m (ft)

The tailwater also needs to be a distance below the crest to maintain the aerated nappe trajectory as given below. Using the crest as the reference point, this distance is a negative number.

$$h_2 = -(h - y_o)$$

(11.8)

where,

    h_2 = vertical distance of the tailwater below the crest, m (ft)

    h = vertical drop between the approach and tailwater channels, m (ft)

    y_o = normal depth in the tailwater channel (equals normal depth in approach channel assuming same channel characteristics), m (ft)

To achieve sufficient tailwater and to maintain adequate drop from the crest to the tailwater, it is sometimes necessary to depress the floor below the elevation of the tailwater channel. The total drop from the crest to the stilling basin floor is given by:

$$h_o = h_2 - y_3 \tag{11.9}$$

where,

    h_o = drop from crest to stilling basin floor, m (ft)

The horizontal dimensions of the basin must also be established. From Figure 11.3 it can be seen that the total basin length is the sum of three components.

$$L_B = L_1 + L_2 + L_3 \tag{11.10}$$

where,

    L_B = stilling basin length, m (ft)

    L_1 = distance from the headwall to the point where the surface of the upper nappe strikes the stilling basin floor, m (ft)

    L_2 = distance from the point where the surface of the upper nappe strikes the stilling basin floor to the upstream face of the floor blocks, m (ft)

    L_3 = distance from the upstream face of the floor blocks to the end of the stilling basin, m (ft)

L_1 is given by:

$$L_1 = \frac{L_f + L_s}{2} \tag{11.11}$$

where,

    L_f = length given by Equation 11.12, m (ft)

    L_s = length given by Equation 11.13, m (ft)

$$L_f = \left(-0.406 + \sqrt{3.195 - 4.368 \frac{h_o}{y_c}} \right) y_c \tag{11.12}$$

$$L_s = \frac{\left(0.691 + 0.228 \left(\frac{L_t}{y_c} \right)^2 - \left(\frac{h_o}{y_c} \right) \right) y_c}{0.185 + 0.456 \left(\frac{L_t}{y_c} \right)} \tag{11.13}$$

where,

$$L_t = \left(-0.406 + \sqrt{3.195 - 4.368\frac{h_2}{y_c}} \right) y_c \quad (11.14)$$

L_2 and L_3 are determined by:

$$L_2 = 0.8y_c \quad (11.15)$$
$$L_3 \geq 1.75y_c \quad (11.16)$$

In comparison with the simple straight drop structure discussed in Section 11.1.1, the addition of floor blocks and a sill, allows for a shorter basin as given by Equation 11.10. The floor blocks should be proportioned to have a height of $0.8y_c$ with a width and spacing of $0.4y_c$. The basin will perform acceptably if the width and spacing varies within plus or minus $0.15y_c$. The blocks should be square in plan and should occupy between 50 percent and 60 percent of the stilling basin width.

The end sill height should be $0.4y_c$. Longitudinal sills, as shown in Figure 11.3, are optional from a hydraulic perspective. If needed, they reinforce the basin structurally, but should pass through the blocks, not between them.

Final consideration is given to the configuration of the exit of the basin as well as the transition from the approach channel to the basin. With respect to the exit, the sidewall height at the basin exit should be above the tailwater elevation by $0.85y_c$. Wingwalls should be located at an angle of 45° with the outlet centerline and have a top slope of 1 to 1.

With respect to the approach channel, the crest of spillway should be at same elevation as the invert of the approach channel. The bottom width of the approach channel should be equal to the spillway notch length, W_o, at the headwall. Because of the acceleration as the flow approaches the crest, riprap or paving should be provided for a distance upstream from the headwall equal to $3y_c$. (See Section 10.3 for sizing riprap.)

The design procedure for the straight drop structure may be summarized in the following steps.

Step 1. Estimate the elevation difference required between the approach and tailwater channel, h. This may be to address a drop at the outlet of a culvert resulting from erosion or headcutting or it may be to flatten a channel to a series of subcritical slopes and drops.

Step 2. Calculate normal flow conditions approaching the drop to verify subcritical conditions. If not subcritical, repeat step 1.

Step 3. Calculate critical depth over the weir (usually rectangular) into the drop structure. Calculate the vertical dimensions of the stilling basin using Equations 11.7 through 11.9.

Step 4. Estimate the basin length using Equations 11.10 though 11.16.

Step 5. Design the basin floor blocks and end sill.

Step 6. Design the basin exit and entrance transitions.

Design Example: Straight Drop Structure (SI)

Find the dimensions for a straight drop structure with a rectangular weir used to reduce channel slope. Given:

Q   = 7.1 m^3/s

h   = 1.83 m

W_o = 3.10 m

Upstream and downstream channel (trapezoidal)

B   = 3.10 m

Z   = 1V:3H

S_o = 0.002 m/m (after providing for drop)

n   = 0.030

Solution

Step 1. Estimate the required approach and tailwater channel elevation difference, h. This is estimated and given above as 1.83 m. This drop forces the slope of the upstream and downstream channel to 0.002 m/m, as given.

Step 2. Calculate normal flow conditions approaching the drop to verify subcritical conditions. By trial and error,

y_o = 1.025 m, V_o = 1.123 m/s, Fr_o = 0.35; therefore, flow is subcritical. Proceed to step 3.

Step 3. Calculate critical depth over the weir into the drop structure. Calculate the vertical dimensions of the stilling basin. Start by finding the critical depth over the weir using Equation 11.5 based on the unit discharge, q = Q/B = 7.10/3.10 = 2.29 m^2/s.

$$y_c = \left(\frac{q^2}{g}\right)^{\frac{1}{3}} = \left(\frac{2.29^2}{9.81}\right)^{\frac{1}{3}} = 0.812 \text{ m}$$

The required tailwater depth above the floor of the stilling basin is calculated from Equation 11.7.

y_3 = 2.15y_c = 2.15(0.812) = 1.745 m

The distance from the crest down to the tailwater needs to be calculated using Equation 11.8. (The negative indicates the tailwater elevation is below the crest.)

h_2 = -(h – y_o) = -(1.83 – 1.025) = -0.805 m

The total drop from the crest to the stilling basin floor is given by Equation 11.9:

h_o = h_2 – y_3 = -0.805 – 1.745 = -2.55 m

Since the nominal drop, h, is 1.83 m, the floor must be depressed by 0.72 m

Step 4. Estimate the basin length. Start by using Equations 11.12, 11.13, and 11.14.

$$L_f = \left(-0.406 + \sqrt{3.195 - 4.368\frac{h_o}{y_c}}\right)y_c = \left(-0.406 + \sqrt{3.195 - 4.368\frac{-2.55}{0.812}}\right)0.812 = 3.01\,m$$

$$L_t = \left(-0.406 + \sqrt{3.195 - 4.368\frac{h_2}{y_c}}\right)y_c = \left(-0.406 + \sqrt{3.195 - 4.368\frac{-0.805}{0..812}}\right)0.812 = 1.90\,m$$

$$L_s = \frac{\left(0.691 + 0.228\left(\frac{L_t}{y_c}\right)^2 - \left(\frac{h_o}{y_c}\right)\right)y_c}{0.185 + 0.456\left(\frac{L_t}{y_c}\right)} = \frac{\left(0.691 + 0.228\left(\frac{1.90}{0.812}\right)^2 - \left(\frac{-2.55}{0.812}\right)\right)0.812}{0.185 + 0.456\left(\frac{1.90}{0.812}\right)} = 3.30\,m$$

L_1 is given by Equation 11.11:

$$L_1 = \frac{L_f + L_s}{2} = \frac{3.01 + 3.30}{2} = 3.15\,m$$

L_2 and L_3 are determined by Equations 11.15 and 11.16:

$L_2 = 0.8y_c = 0.8(0.812) = 0.65\,m$

$L_3 \geq 1.75y_c = 1.75(0.812) = 1.43\,m$

Total basin length required is given by Equation 11.10:

$L_B = L_1 + L_2 + L_3 = 3.15 + 0.65 + 1.43 = 5.23\,m$

Step 5. Design the basin floor blocks and end sill.

Block height = $0.8y_c = 0.8(0.812) = 0.65\,m$

Block width = block spacing = $0.4y_c = 0.4(0.812) = 0.325\,m$

End sill height = $0.4y_c = 0.4(0.812) = 0.325\,m$

Step 6. Design the basin exit and entrance transitions.

Sidewall height above tailwater elevation = $0.85y_c = 0.85(0.812) = 0.69\,m$

Armour approach channel above headwall to length = $3y_c = 3(0.812) = 2.44\,m$

Design Example: Straight Drop Structure (CU)

Find the dimensions for a straight drop structure with a rectangular weir used to reduce channel slope. Given:

Q   = 250 ft³/s

h   = 6.0 ft

W_o = 10.0 ft

Upstream and downstream channel (trapezoidal)

B = 10.0 ft

Z = 1V:3H

S_o = 0.002 ft/ft (after providing for drop)

n = 0.030

Solution

Step 1. Estimate the required approach and tailwater channel elevation difference, h. This is estimated and given above as 6.0 ft. This drop forces the slope of the upstream and downstream channel to 0.002 ft/ft, as given.

Step 2. Calculate normal flow conditions approaching the drop to verify subcritical conditions. By trial and error,

y_o = 3.36 ft, V_o = 3.71 ft/s, Fr_o = 0.36; therefore, flow is subcritical. Proceed to step 3.

Step 3. Calculate critical depth over the weir into the drop structure. Calculate the vertical dimensions of the stilling basin. Start by finding the critical depth over the weir using Equation 11.5 based on the unit discharge, q = Q/B = 250/10 = 25 ft²/s.

$$y_c = \left(\frac{q^2}{g} \right)^{1/3} = \left(\frac{25^2}{32.2} \right)^{1/3} = 2.69 \text{ ft}$$

The required tailwater depth above the floor of the stilling basin is calculated from Equation 11.7.

y_3 = 2.15y_c = 2.15(2.69) = 5.77 ft

The distance from the crest down to the tailwater needs to be calculated using Equation 11.8. (The negative indicates the tailwater elevation is below the crest.)

h_2 = -(h – y_o) = -(6.0 – 3.36) = -2.64 ft

The total drop from the crest to the stilling basin floor is given by Equation 11.9:

h_o = h_2 –y_3 = -2.64 – 5.77 = -8.41 ft (round to –8.4)

Since the nominal drop, h, is 6.0 ft, the floor must be depressed by 2.4 ft

Step 4. Estimate the basin length. Start by using Equations 11.12, 11.13, and 11.14.

$$L_f = \left(-0.406 + \sqrt{3.195 - 4.368 \frac{h_o}{y_c}} \right) y_c = \left(-0.406 + \sqrt{3.195 - 4.368 \frac{-8.41}{2.69}} \right) 2.69 = 9.94 \text{ ft}$$

$$L_t = \left(-0.406 + \sqrt{3.195 - 4.368 \frac{h_2}{y_c}} \right) y_c = \left(-0.406 + \sqrt{3.195 - 4.368 \frac{-2.64}{2.69}} \right) 2.69 = 6.26 \text{ ft}$$

$$L_s = \frac{\left(0.691 + 0.228\left(\frac{L_t}{y_c}\right)^2 - \left(\frac{h_o}{y_c}\right)\right)y_c}{0.185 + 0.456\left(\frac{L_t}{y_c}\right)} = \frac{\left(0.691 + 0.228\left(\frac{6.26}{2.69}\right)^2 - \left(\frac{-8.41}{2.69}\right)\right)2.69}{0.185 + 0.456\left(\frac{6.26}{2.69}\right)} = 10.89 \text{ ft}$$

L_1 is given by Equation 11.11:

$$L_1 = \frac{L_f + L_s}{2} = \frac{9.94 + 10.89}{2} = 10.4 \text{ ft}$$

L_2 and L_3 are determined by Equations 11.15 and 11.16:

$L_2 = 0.8y_c = 0.8(2.69) = 2.2$ ft

$L_3 \geq 1.75y_c = 1.75(2.69) = 4.7$ ft

Total basin length required is given by Equation 11.10:

$L_B = L_1 + L_2 + L_3 = 10.4 + 2.2 + 4.7 = 17.3$ ft

Step 5.   Design the basin floor blocks and end sill.

Block height = $0.8y_c = 0.8(2.69) = 2.1$ ft

Block width = block spacing = $0.4y_c = 0.4(2.69) = 1.1$ ft

End sill height = $0.4y_c = 0.4(2.69) = 1.1$ ft

Step 6.   Design the basin exit and entrance transitions.

Sidewall height above tailwater elevation = $0.85y_c = 0.85(2.69) = 2.3$ ft

Armour approach channel above headwall to length = $3y_c = 3(2.69) = 8.1$ ft

11.2 BOX INLET DROP STRUCTURE

The box inlet drop structure may be described as a rectangular box open at the top and downstream end (Figure 11.4). Water is directed to the crest of the box inlet by earth dikes and a headwalls. Flow enters over the upstream end and two sides. The long crest of the box inlet permits large flows to pass at relatively low heads.

The outlet structure can be adjusted to fit a wide variety of field conditions. It is possible to lengthen the straight section and cover it to form a highway culvert. The sidewalls of the stilling basin section can be flared if desired, thus permitting use with narrow channels or wide flood plains. Flaring the sidewalls also makes it possible to adjust the outlet depth to that in the natural channel.

The design information is based on an extensive experimental program performed by the Soil Conservation Service, St. Anthony Falls Hydraulic Laboratory, Minneapolis (Blaisdell and Donnelly, 1956). The recommended design is limited to the following conditions:

1. Total drop, h_o, less than 3.7 m (12 ft) and greater than 0.6 m (2 ft). The total drop is that amount required to reduce the channel slope to a desired stable slope.

2. Downstream width of structure should be less than or equal to the width of the tailwater channel.

3. Approach channel is level with the crest of the box inlet

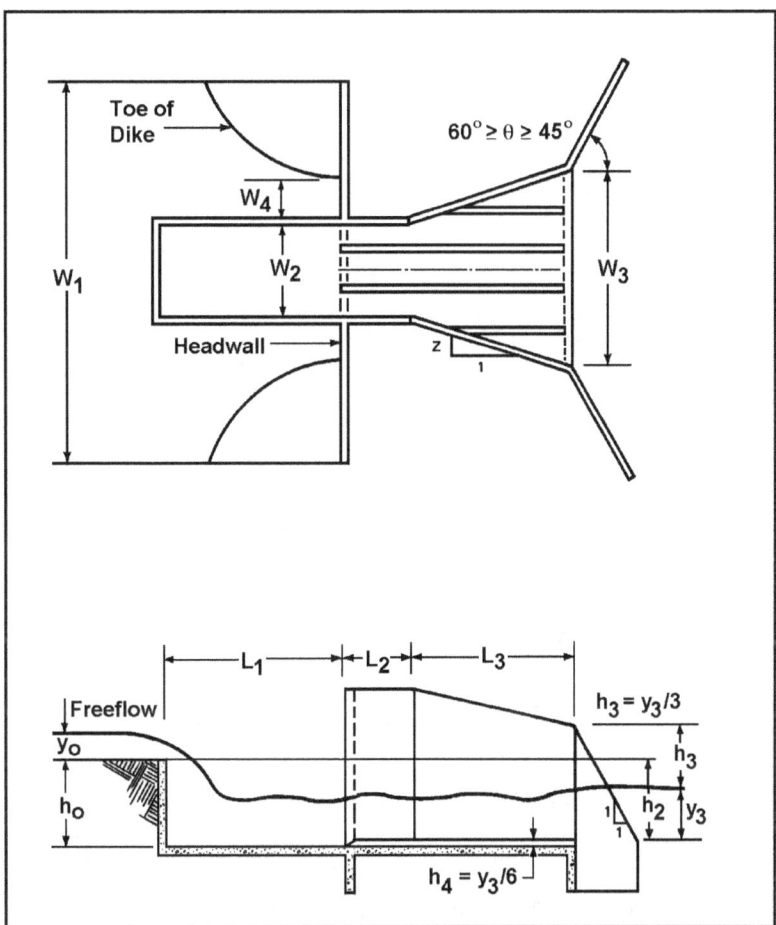

Figure 11.4. Box Inlet Drop Structure

One of two different sections will control the flow: the crest of the box inlet or the opening in the headwall. The flow at which the control changes from one point to the other is dependent upon a number of factors, the principal factors being the box inlet depth and its length. The design of the box inlet drop structure involves determining which section (crest or headwall opening) controls at the design flow.

First, assume crest control and calculate the head, y_o, at the crest of the box inlet drop structure required to pass the design discharge. The general equation relating discharge to head for a rectangular weir is:

$$Q = C_w \sqrt{2g} \, Lh^{3/2} \tag{11.17}$$

where,

C_w = dimensionless weir coefficient = 0.43
L = weir length, m (ft)
h = head on the weir crest, m (ft)

Calling the weir length, L_c, and the head, y_o, and solving for head yields the following relationship.

$$y_o = \left(\frac{Q}{C_w \sqrt{2gL_c}} \right)^{\frac{2}{3}}$$ (11.18)

where,

y_o = required head on the weir crest to pass the design flow, m (ft)
L_c = length of box inlet crest, m (ft)
Q = design discharge, m^3/s (ft^3/s)

By inspection of Figure 11.4, it is apparent that the weir crest length is:

$$L_c = W_2 + 2L_1$$ (11.19)

where,

L_c = length of box inlet crest, m (ft)
W_2 = width of box inlet, m (ft)
L_1 = length of box inlet, m (ft)

Various lengths of crest, L_c, are evaluated in Equation 11.18 to obtain a head consistent with the hydraulic conditions in the approach channel.

Several corrections to the weir coefficient used in Equation 11.18 are appropriate if the crest does control the structure hydraulics. However, for determining crest versus headwall control it is not necessary to make these corrections. Four multiplicative corrections are considered:

1. Correction for low relative head given in Figure 11.5.

2. Correction for box inlet shape given in Figure 11.6. This correction is only applicable for $W_1/L_c \geq 3$.

3. Correction for approach channel width given in Figure 11.7. This correction is only applicable for $W_1/L_c < 3$.

4. Correction for dike proximity to the box inlet crest given in Table 11.1. These values have a low precision.

The precision of the correction curves is within 7 percent when there is no dike present and within 15 percent when dikes are used.

Second, assume headwall control and calculate the head, y_o, to determine if this head is greater than that obtained for the box inlet crest control. The general equation relating discharge and head for a rectangular weir was described in Equation 11.17. For this case, the weir length is W_2, the head is $y_o + C_H$, and the weir coefficient is C_2. Solving for head yields the following relationship.

$$y_o = \left(\frac{Q}{C_2\sqrt{2g}W_2} \right)^{\frac{2}{3}} - C_H \qquad (11.20)$$

where,

y_o = required head on the headwall crest to pass the design flow, m (ft)

W_2 = width of the box inlet, m (ft)

Q = design discharge, m³/s (ft³/s)

C_2 = dimensionless weir coefficient (discharge coefficient)

C_H = head correction, m (ft)

The discharge coefficient, C_2, is obtained from Figure 11.8. The head correction, C_H, is given in Figure 11.9. If h_o/W_2 is between 1/4 and 1, C_H may be more readily determined from Figure 11.10. The precision of the design curves for headwall control is estimated to be within 10 percent.

When the box inlet drop structure operates under submerged conditions, reference should be made to Blaisdell and Donnelly (1956) to determine the submerged design. However, this is not a desirable design condition.

The outlet for a box inlet drop structure should be designed as follows. Critical depth in the straight section is:

$$y_c = \left(\left(\frac{Q}{W_2} \right)^2 \bigg/ g \right)^{\frac{1}{3}} \qquad (11.21)$$

Similarly, critical depth at the exit of the stilling basin is:

$$y_{c3} = \left(\left(\frac{Q}{W_3} \right)^2 \bigg/ g \right)^{\frac{1}{3}} \qquad (11.22)$$

The minimum length of the straight section for values of L_1/W_2 equal to or greater than 0.25 is:

$$L_2 = y_c \left(\frac{0.2W_2}{L_1} + 1 \right) \qquad (11.23)$$

As shown in Figure 11.4, the sidewalls of the stilling basin may flare from z=0 (parallel extensions of the section walls) to z=0.5.

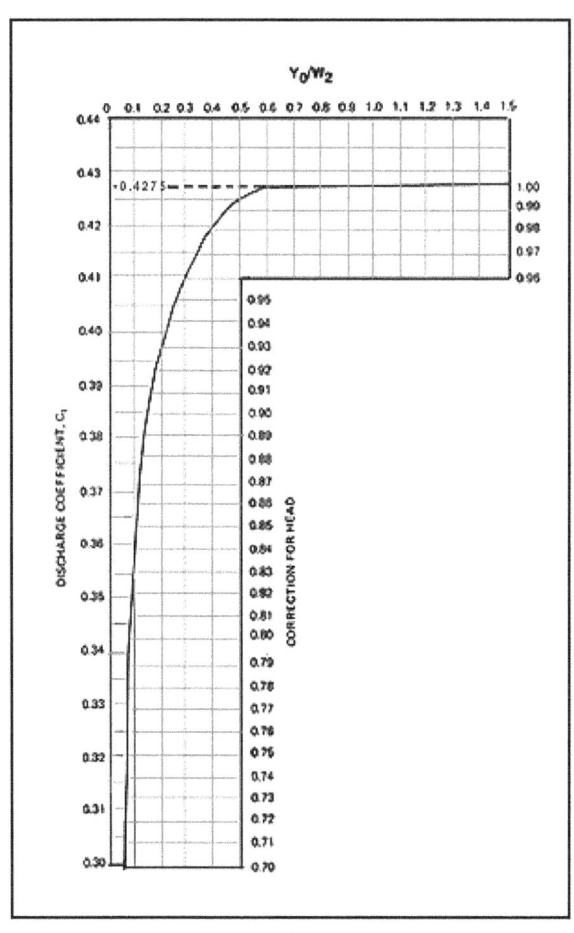

Figure 11.5. Discharge Coefficients/Correction for Head with Control at Box Inlet Crest

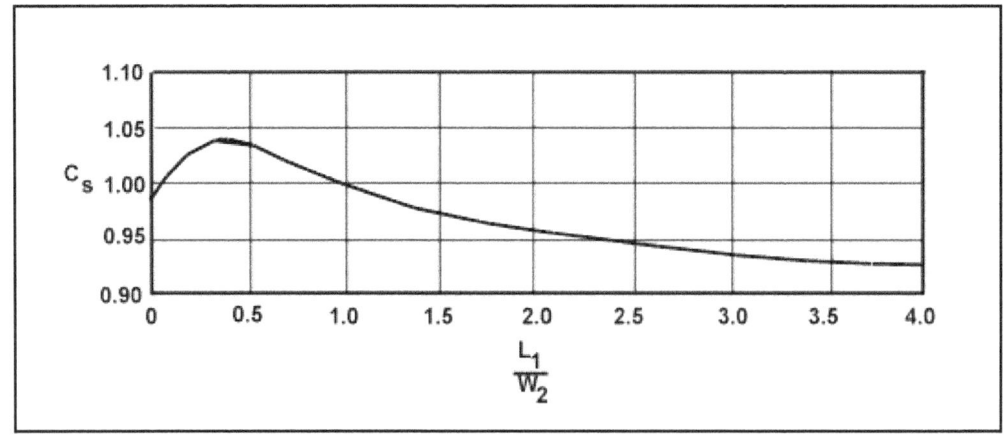

Figure 11.6. Correction for Box Inlet Shape with Control at Box Inlet Crest

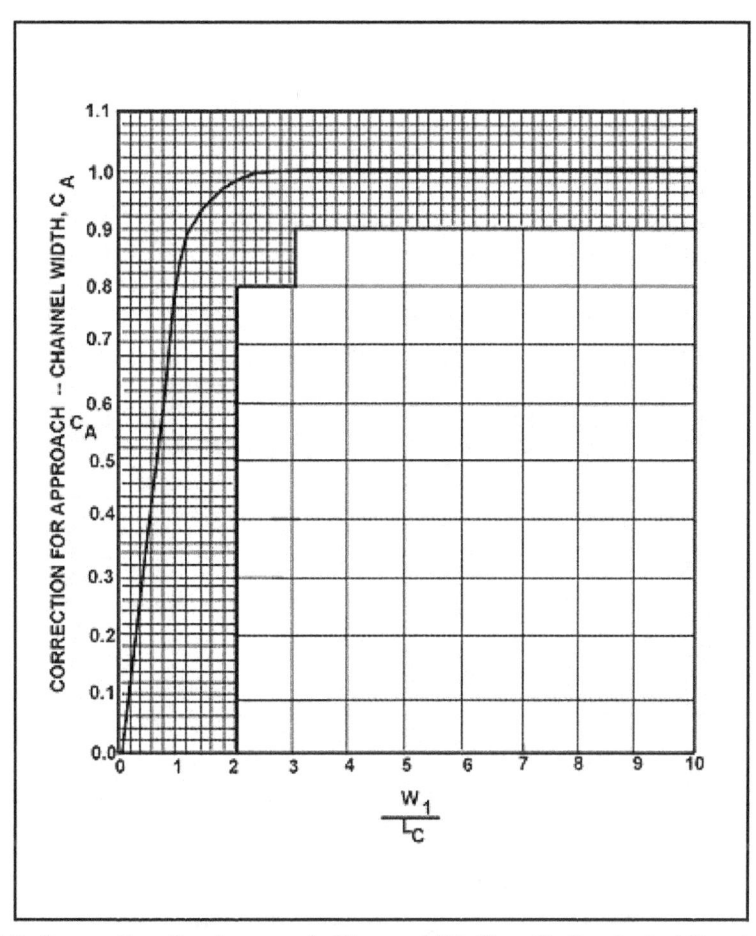

Figure 11.7. Correction for Approach Channel Width with Control at Box Inlet Crest

Table 11.1. Correction for Dike Effect, C_E, with Control at Box Inlet Crest

L_1/W_2	W_4/W_2						
	0.0	0.1	0.2	0.3	0.4	0.5	0.6
0.5	0.90	0.96	1.00	1.02	1.04	1.05	1.05
1.0	0.80	0.88	0.93	0.96	0.98	1.00	1.01
1.5	0.76	0.83	0.88	0.92	0.94	0.96	0.97
2.0	0.76	0.83	0.88	0.92	0.94	0.96	0.97

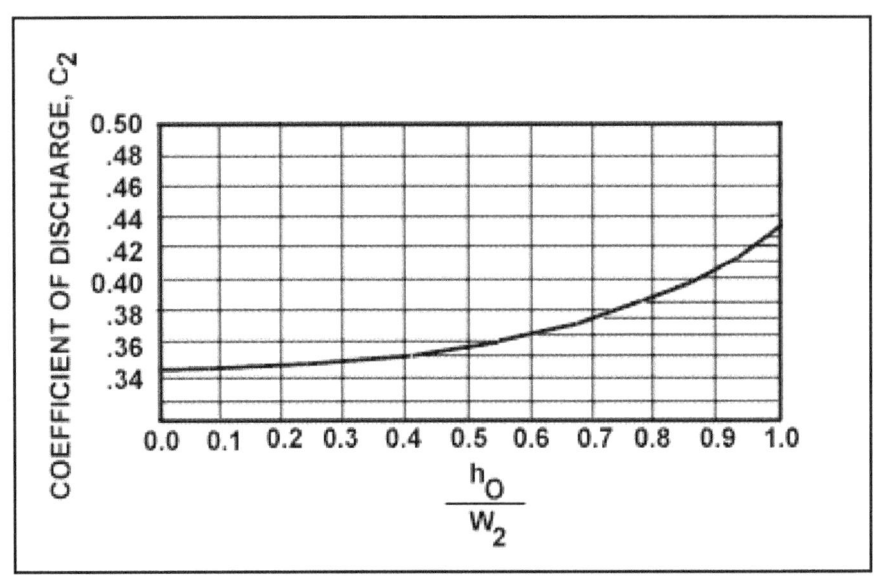

Figure 11.8. Coefficient of Discharge with Control at Headwall Opening

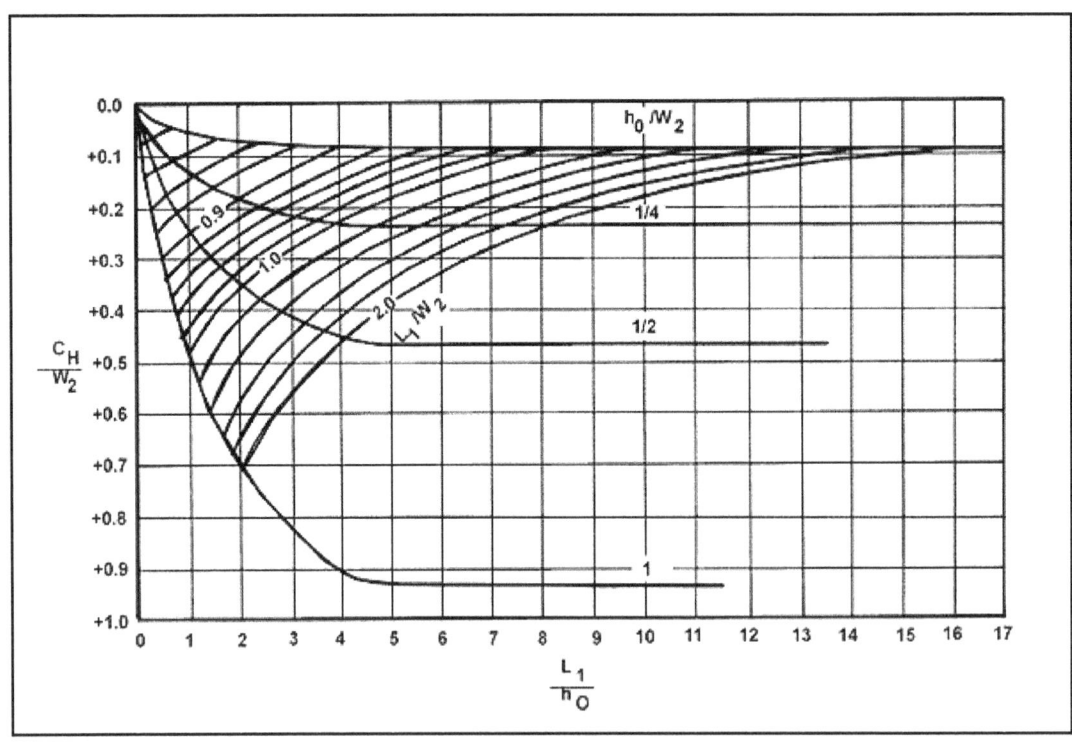

Figure 11.9. Relative Head Correction with Control at Headwall Opening

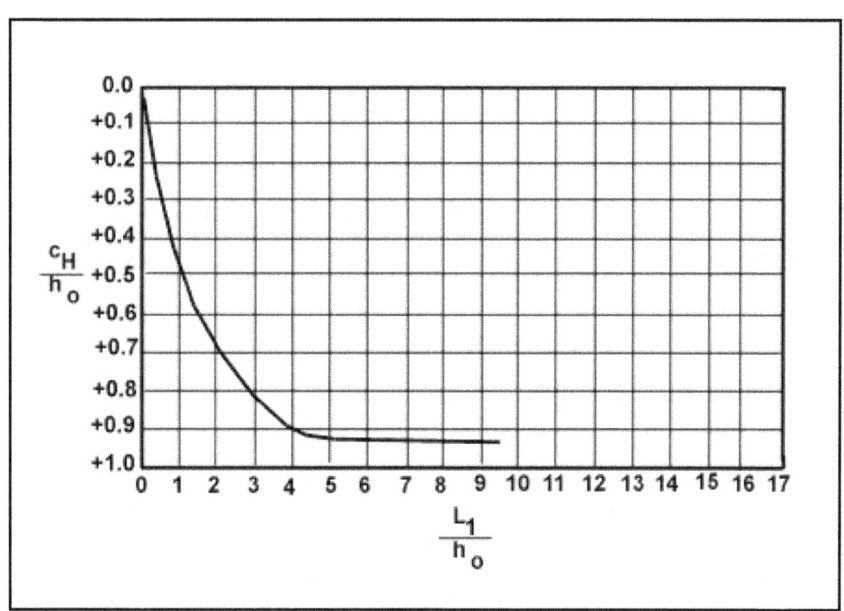

Figure 11.10. Relative Head Correction for $h_o/W_2 \geq 1/4$ with Control at Headwall Opening

The minimum length of the final, potentially flared portion of the stilling basin is taken as the larger of the following two equations. However, Equation 11.24b is only valid for L_1/W_2 values equal to or greater than 0.25.

$$L_3 = \frac{L_c W_2}{2L_1} \tag{11.24a}$$

or

$$L_3 = \frac{W_3 - W_2}{2z} \tag{11.24b}$$

Frequently, it is desirable to design the stilling basin outlet width to equal the width of the tailwater channel. When the stilling basin width at the exit is less than $11.5y_{c3}$, the minimum tailwater depth over the basin floor is:

$$y_3 = 1.6y_{c3} \tag{11.25}$$

When the stilling basin width at the exit is greater than $11.5\ y_{c3}$, the minimum tailwater depth over the basin floor is calculated from Equation 11.26. However, such a stilling basin may make inefficient use of the outlet.

$$y_3 = y_{c3} + 0.52W_3 \tag{11.26}$$

The height of the end sill is:

$$h_4 = \frac{y_3}{6} \qquad\qquad (11.27)$$

Longitudinal sills will improve the flow distribution in the outlet. Considerations for their use are:

- When the stilling basin sidewalls are parallel (z=0), the longitudinal sills may be omitted.
- The center pair of longitudinal sills should start at the exit of the box inlet and extend through the straight section and stilling basin to the end sill.
- When W_3 is less than $2.5W_2$, only two sills are needed. These sills should be located at a distance W_5, each side of the centerline.
- When W_3 exceeds $2.5W_2$ two additional sills are required. These sills should be located parallel to the outlet centerline and midway between the center sills and the sidewalls at the exit of the stilling basin.
- The height of the longitudinal sills should be the same as the height of the end sill.

The minimum height of the sidewalls above the water surface at the exit of the stilling basin is calculated from Equation 11.28. The sidewalls should extend above the tailwater surface under all conditions.

$$h_3 = \frac{y_3}{3} \qquad\qquad (11.28)$$

The wingwalls should be triangular in elevation and have top slope of 45° with the horizontal. Top slopes as flat as 30° are permissible. The wingwalls should flare in plan at an angle of 60° with the outlet centerline. Flare angles as small as 45° are permissible; however, wingwalls parallel to the outlet centerline are not recommended.

The design procedure for the box inlet drop structure may be summarized in the following steps.

Step 1. Select the initial box inlet trial dimensions, h_o, L_1, and W_2.

Step 2. Assume crest control and estimate the crest control head using Equation 11.18.

Step 3. Assume headwall control and estimate the headwall control head using Equation 11.20.

Step 4. Select the largest head from steps 2 and 3. If the largest head is crest control, adjust the crest control head with the correction factors from Figures 11.5, 11.6, and 11.7 as well as Table 11.1.

Step 5. Calculate critical depths in both the straight and flared basin sections using Equations 11.21 and 11.22, respectively.

Step 6. Determine the basin length from Equations 11.23 and 11.24.

Step 7. Calculate the outlet depth from Equations 11.25 and 11.26 and compare this depth with the tailwater depth.

Step 8. Calculate the sill height using Equation 11.27 and determine the need for longitudinal sills.

Step 9. Determine the height of the sidewalls using Equation 11.28. Lay out the wingwalls.

Design Example: Box Inlet Drop Structure (SI)

Find the dimensions for a box inlet drop structure used to reduce channel slope. Given:

Q $\quad$ = 7.1 m³/s
h_o $\quad$ = 1.20 m
W_2 $\quad$ = 1.20 m
L_1 $\quad$ = 1.20 m

Upstream and downstream channel (trapezoidal)

B $\quad$ = 6.0 m
Z $\quad$ = 1V:3H
S_o $\quad$ = 0.002 m/m
n $\quad$ = 0.030

Solution

Step 1. The initial box inlet trial dimensions, h_o, L_1, and W_2 were given.

Step 2. Assume crest control and estimate the crest control head using Equation 11.18. Equation 11.19 gives us the crest length for Equation 11.18.

$$L_c = W_2 + 2L_1 \ = 1.2 + 2(1.2) = 3.6 \text{ m}$$

$$y_o = \left(\frac{Q}{C_w\sqrt{2gL_c}} \right)^{\frac{2}{3}} = \left(\frac{7.1}{0.43\sqrt{2(9.81)3.6}} \right)^{\frac{2}{3}} = 1.024 \text{ m}$$

Step 3. Assume headwall control and estimate the headwall control head using Equation 11.20. First we need to determine C_2 and C_H.

From Figure 11.8 for a value of $h_o/W_2 = 1.2/1.2 = 1.0$ we determine $C_2 = 0.43$.

From Figure 11.10 for a value of $L_1/h_o = 1.2/1.2 = 1.0$ we determine $C_H/h_o = 0.49$. Therefore, $C_H = 0.588$ m

$$y_o = \left(\frac{Q}{C_2\sqrt{2gW_2}} \right)^{\frac{2}{3}} - C_H = \left(\frac{7.1}{0.43\sqrt{2(9.81)1.2}} \right)^{\frac{2}{3}} - 0.588 = 1.541 \text{ m}$$

Step 4. Select the largest head from steps 2 and 3. In this case, the headwall controls and $y_o = 1.541$ m

Step 5. Calculate critical depths in the straight section using Equation 11.21.

$$y_c = \left(\frac{\left(\frac{Q}{W_2} \right)^2}{g} \right)^{\frac{1}{3}} = \left(\frac{\left(\frac{7.1}{1.2} \right)^2}{9.81} \right)^{\frac{1}{3}} = 1.528 \text{ m}$$

Calculate critical depth at the exit using Equation 11.22 and taking the structure width equal to the channel width.

$$y_{c3} = \left[\frac{\left(\frac{Q}{W_3}\right)^2}{g} \right]^{\frac{1}{3}} = \left[\frac{\left(\frac{7.1}{6.0}\right)^2}{9.81} \right]^{\frac{1}{3}} = 0.523 \text{ m}$$

Step 6. Determine the length of the straight section beyond the inlet from Equation 11.23.

$$L_2 = y_c \left(\frac{0.2W_2}{L_1} + 1 \right) = 1.528 \left(\frac{0.2(1.2)}{1.2} + 1 \right) = 1.834 \text{ m} \quad \text{(round to 1.8 m)}$$

The length of the flared section is determined by the maximum of Equations 11.24a and 11.24b.

$$L_3 = \frac{L_c W_2}{2L_1} = \frac{3.6(1.2)}{2(1.2)} = 1.8 \text{ m}$$

$$L_3 = \frac{W_3 - W_2}{2z} = \frac{6.0 - 1.2}{2(0.5)} = 4.8 \text{ m}$$

Therefore, L_3 = 4.8 m

Step 7. Calculate the outlet depth from Equations 11.25 or 11.26 depending on whether or not W_3 is less than $11.5y_{c3}$. $11.5y_{c3} = 11.5(0.523) = 6.01$ m. Therefore, use Equation 11.25.

$y_3 = 1.6y_{c3} = 1.6(0.523) = 0.84$ m

Normal depth in the tailwater channel is 0.80 m. Since y_3 is slightly greater than the tailwater channel depth, some form of channel protection at the exit may be advisable to protect against erosion from the accelerating flow.

Step 8. Calculate the sill height using Equation 11.27.

$$h_4 = \frac{y_3}{6} = \frac{0.84}{6} = 0.14 \text{ m}$$

We determine if longitudinal sills are necessary by comparing W_3 to $2.5W_2$. Since $2.5W_2=2.5(1.2)=3.0$ m and this is less than W_3, 4 sills are needed.

Step 9. Determine the height of the sidewalls above the water surface elevation using Equation 11.28.

$$h_3 = \frac{y_3}{3} = \frac{0.84}{3} = 0.28 \text{ m}$$

Design Example: Box Inlet Drop Structure (CU)

Find the dimensions for a box inlet drop structure used to reduce channel slope. Given:

Q = 250 ft³/s
h_o = 4.0 ft
W_2 = 4.0 ft
L_1 = 4.0 ft

Upstream and downstream channel (trapezoidal)

B = 20.0 ft
Z = 1V:3H
S_o = 0.002 ft/ft
n = 0.030

Solution

Step 1. The initial box inlet trial dimensions, h_o, L_1, and W_2 were given.

Step 2. Assume crest control and estimate the crest control head using Equation 11.18. Equation 11.19 gives us the crest length for Equation 11.18.

$$L_c = W_2 + 2L_1 = 4.0 + 2(4.0) = 12.0 \text{ ft}$$

$$y_o = \left(\frac{Q}{C_w\sqrt{2gL_c}} \right)^{\frac{2}{3}} = \left(\frac{250}{0.43\sqrt{2(32.2)12.0}} \right)^{\frac{2}{3}} = 3.32 \text{ ft}$$

Step 3. Assume headwall control and estimate the headwall control head using Equation 11.20. First we need to determine C_2 and C_H.

From Figure 11.8 for a value of $h_o/W_2 = 4.0/4.0 = 1.0$ we determine $C_2 = 0.43$.

From Figure 11.10 for a value of $L_1/h_o = 4.0/4.0 = 1.0$ we determine $C_H/h_o = 0.49$. Therefore, $C_H = 1.96$ ft

$$y_o = \left(\frac{Q}{C_2\sqrt{2gW_2}} \right)^{\frac{2}{3}} - C_H = \left(\frac{250}{0.43\sqrt{2(32.2)12.0}} \right)^{\frac{2}{3}} - 1.96 = 4.94 \text{ ft}$$

Step 4. Select the largest head from steps 2 and 3. In this case, the headwall controls and $y_o = 4.94$ ft

Step 5. Calculate critical depths in the straight section using Equation 11.21.

$$y_c = \left(\frac{\left(\frac{Q}{W_2} \right)^2}{g} \right)^{\frac{1}{3}} = \left(\frac{\left(\frac{250}{4.0} \right)^2}{32.2} \right)^{\frac{1}{3}} = 4.95 \text{ ft}$$

Calculate critical depth at the exit using Equation 11.22 and taking the structure width equal to the channel width.

$$y_{c3} = \left[\frac{\left(\frac{Q}{W_3}\right)^2}{g} \right]^{\frac{1}{3}} = \left[\frac{\left(\frac{250}{20.0}\right)^2}{32.2} \right]^{\frac{1}{3}} = 1.69 \text{ ft}$$

Step 6. Determine the length of the straight section beyond the inlet from Equation 11.23.

$$L_2 = y_c \left(\frac{0.2W_2}{L_1} + 1 \right) = 4.95 \left(\frac{0.2(4.0)}{4.0} + 1 \right) = 5.94 \text{ ft} \quad \text{(round to 5.9 ft)}$$

The length of the flared section is determined by the maximum of Equations 11.24a and 11.24b.

$$L_3 = \frac{L_c W_2}{2L_1} = \frac{12(4.0)}{2(4.0)} = 6.0 \text{ ft}$$

$$L_3 = \frac{W_3 - W_2}{2z} = \frac{20.0 - 4.0}{2(0.5)} = 16.0 \text{ ft}$$

Therefore, $L_3 = 16.0$ ft

Step 7. Calculate the outlet depth from Equations 11.25 or 11.26 depending on whether or not W_3 is less than $11.5y_{c3}$. $11.5y_{c3} = 11.5(1.69) = 19.4$ ft. Therefore, use Equation 11.25.

$y_3 = 1.6y_{c3} = 1.6(1.69) = 2.7$ ft

Normal depth in the tailwater channel is 2.6 ft. Since y_3 is slightly greater than the tailwater channel depth, some form of channel protection at the exit may be advisable to protect against erosion from the accelerating flow.

Step 8. Calculate the sill height using Equation 11.27.

$$h_4 = \frac{y_3}{6} = \frac{2.7}{6} = 0.45 \text{ ft}$$

We determine if longitudinal sills are necessary by comparing W_3 to $2.5W_2$. Since $2.5W_2 = 2.5(4.0) = 10$ ft and this is less than W_3, 4 sills are needed.

Step 9. Determine the height of the sidewalls above the water surface elevation using Equation 11.28.

$$h_3 = \frac{y_3}{3} = \frac{2.7}{3} = 0.90 \text{ ft}$$

CHAPTER 12: STILLING WELLS

The design of the US Army Corps of Engineers' stilling well energy dissipator is based on model tests conducted by the US Army Corps of Engineers (USACE, 1963; Grace and Pickering, 1971). It is illustrated in Figure 12.1. The stilling well can be used in channels with moderate to high concentrations of sand or silt and where debris is not a serious problem. The stilling well should not be used in areas where large floating or rolling debris is expected unless suitable debris-control structures are used. The highway uses of stilling wells are at the outfalls of storm drains, median, and pipe down drains where little debris is expected.

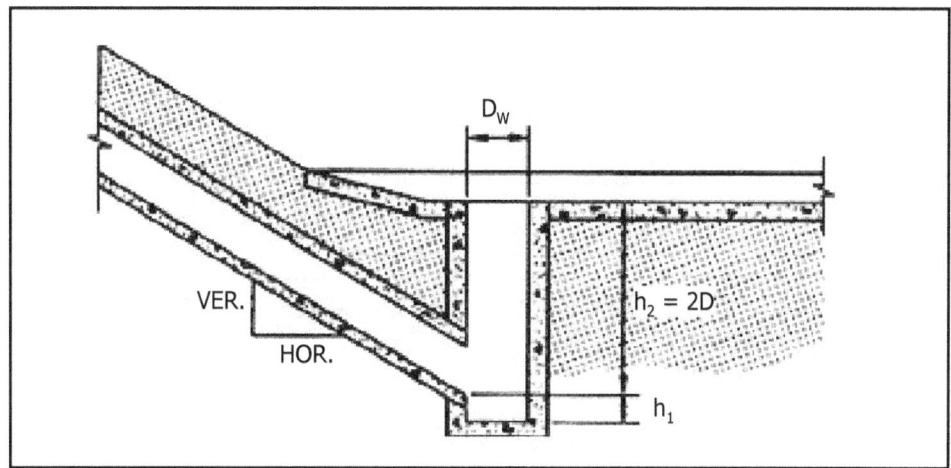

Figure 12.1. US Army Corps of Engineers' Stilling Well (USACE, 1963)

The design of the stilling well is initiated after the size and discharge of the incoming pipe are determined. Figure 12.2 is used to select the stilling well diameter, D_W. The model tests indicated that satisfactory performance can be maintained for $K_u Q/D^{5/2}$ ratios as large as 10, with stilling well diameters from 1 to 5 times that of the incoming conduits. (K_u is a unit conversion constant equal to 1.811 in SI and 1.0 in CU.) These ratios were used to define the curves shown in Figure 12.2.

The optimum depth of stilling well below the invert of the incoming pipe is determined by entering Figure 12.3 with the slope of the incoming pipe and using the stilling well diameter, D_W, previously obtained from Figure 12.2. The height of the stilling well above the invert is fixed at twice the diameter of the incoming pipe, 2D. This dimension results in satisfactory operation and is practical from a cost standpoint; however, if increased, greater efficiency will result.

Tailwater also increases the efficiency of the stilling well. Whenever possible, it should be located in a sump or depressed area.

Riprap or other types of channel protection should be provided around the stilling well outlet and for a distance of at least $3D_W$ downstream.

The outlet may also be covered with a screen or grate for safety. However, the screen or grate should have a clear opening area of at least 75 percent of the total stilling well area and be capable of passing small floating debris such as cans and bottles.

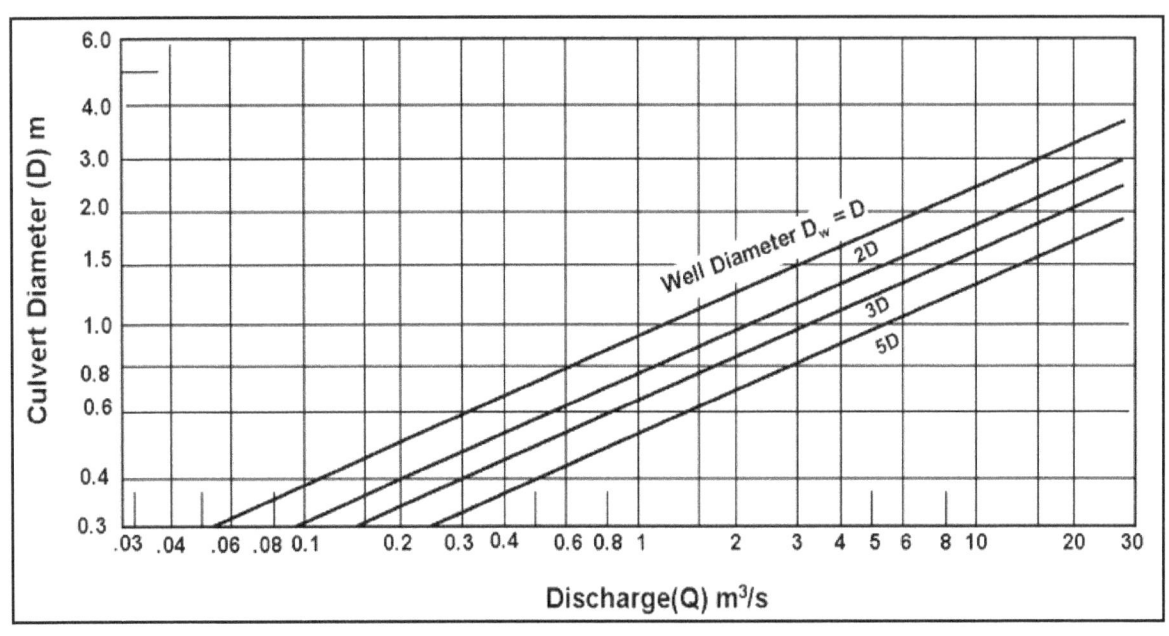

Figure 12.2 (SI). Stilling Well Diameter, D$_W$ (USACE, 1963)

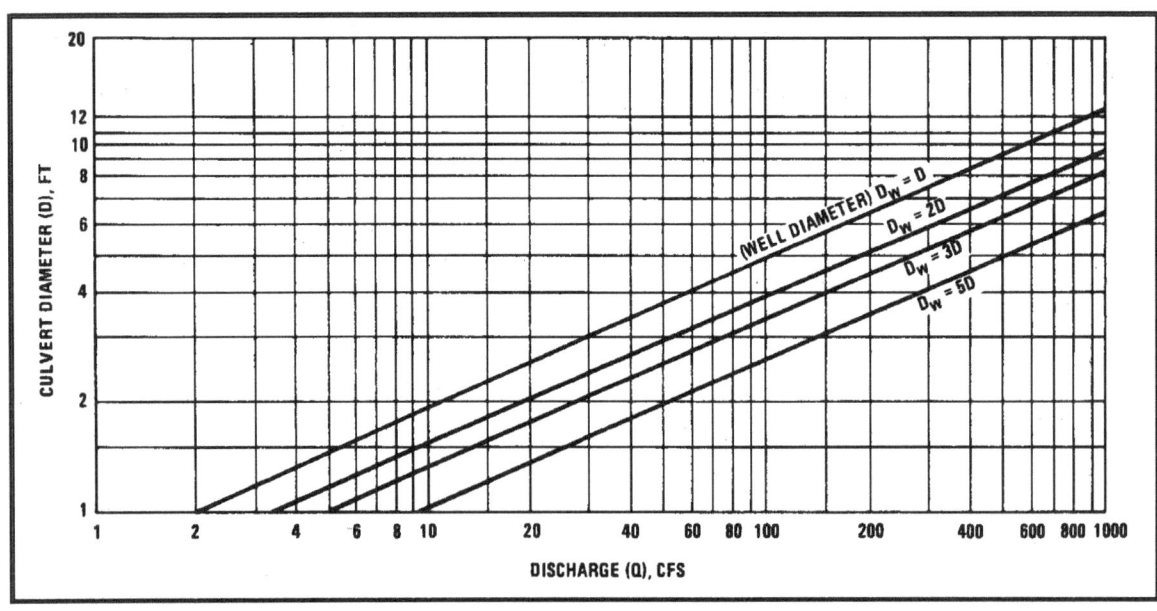

Figure 12.2 (CU). Stilling Well Diameter, D$_W$ (USACE, 1963)

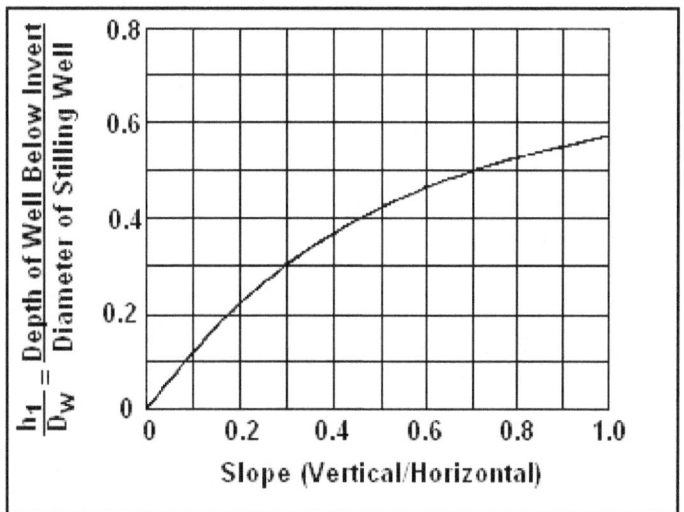

Figure 12.3. Depth of Stilling Well Below Invert (USACE, 1963)

The design procedure is summarized as follows:

Step 1.   Select approach pipe diameter, D, and discharge, Q.

Step 2.   Obtain well diameter, D_W from Figure 12.2.

Step 3.   Calculate the culvert slope. The depth of the well below the culvert invert, h_1 is determined from Figure 12.3.

Step 4.   The depth of the well above the culvert invert, h_2, is equal to 2D as a minimum, but may be greater if the site permits.

Step 5.   The total height of the well, $h_W = h_1 + h_2$.

Design Example: U.S. Army Corps of Engineers' Stilling Well (SI)

Determine the stilling well dimensions. Given:

D = 600 mm CMP

S = 0.5 V/H

Q = 0.424 m³/s

Solution

Step 1.   Select approach pipe diameter and discharge.

D = 0.600 m

Q = 0.424 m³/s

Step 2.   Obtain well diameter (D_W) from Figure 12.2

D_W =1.5 D = 1.5 (0.600 m) = 0.90 m

Step 3.   Calculate the culvert slope. The depth of the well below the culvert invert is determined from Figure 12.3.

Slope = 0.5

12-3

$$h_1/D_W = 0.42$$

$$h_1 = 0.42(0.90) = 0.378 \text{ m, use } h_1 = 0.38 \text{ m}$$

Step 4. The depth of the well above the culvert invert is equal to 2D as a minimum, but may be greater if the site permits.

$$h_2 = 2(D) = 2(0.600) = 1.20 \text{ m}$$

Step 5. The total height of the well.

$$h_W = h_1 + h_2 = 0.38 + 1.20 = 1.58 \text{ m}$$

Design Example: U.S. Army Corps of Engineers' Stilling Well (CU)

Determine the stilling well dimensions. Given:

D = 24 in CMP

S = 0.5 V/H

Q = 15 ft³/s

Solution

Step 1. Select approach pipe diameter and discharge.

D = 2 ft

Q = 15 ft³/s

Step 2. Obtain well diameter from Figure 12.2

$$D_W = 1.5 D = 1.5 (2 \text{ ft}) = 3 \text{ ft}$$

Step 3. Calculate the culvert slope. The depth of the well below the culvert invert is determined from Figure 12.3.

Slope = 0.5

$$h_1/D_W = 0.42$$

$$h_1 = 0.42(3) = 1.26 \text{ ft, use } h_1 = 1.3 \text{ ft}$$

Step 4. The depth of the well above the culvert invert is equal to 2D as a minimum, but may be greater if the site permits.

$$h_2 = 2(D) = 2(2 \text{ ft}) = 4 \text{ ft}$$

Step 5. The total height of the well.

$$h_W = h_1 + h_2 = 1.3 + 4 = 5.3 \text{ ft}$$

APPENDIX A: METRIC SYSTEM, CONVERSION FACTORS, AND WATER PROPERTIES

The following information is summarized from the Federal Highway Administration, National Highway Institute (NHI) Course No. 12301, "Metric (SI) Training for Highway Agencies." For additional information, refer to the Participant Notebook for NHI Course No. 12301.

In SI there are seven base units, many derived units and two supplemental units (Table A.1). Base units uniquely describe a property requiring measurement. One of the most common units in civil engineering is length, with a base unit of meters in SI. Decimal multiples of meter include the kilometer (1000 m), the centimeter (1m/100) and the millimeter (1 m/1000). The second base unit relevant to highway applications is the kilogram, a measure of mass that is the inertia of an object. There is a subtle difference between mass and weight. In SI, mass is a base unit, while weight is a derived quantity related to mass and the acceleration of gravity, sometimes referred to as the force of gravity. In SI the unit of mass is the kilogram and the unit of weight/force is the newton. Table A.2 illustrates the relationship of mass and weight. The unit of time is the same in SI as in the Customary (English) system (seconds). The measurement of temperature is Centigrade. The following equation converts Fahrenheit temperatures to Centigrade, $°C = 5/9 (°F - 32)$.

Derived units are formed by combining base units to express other characteristics. Common derived units in highway drainage engineering include area, volume, velocity, and density. Some derived units have special names (Table A.3).

Table A.4 provides useful conversion factors from Customary to SI units. The symbols used in this table for metric (SI) units, including the use of upper and lower case (e.g., kilometer is "km" and a newton is "N") are the standards that should be followed. Table A.5 provides the standard SI prefixes and their definitions.

Table A.6 provides physical properties of water at atmospheric pressure in SI units. Table A.7 gives the sediment grade scale and Table A.8 gives some common equivalent hydraulic units.

Table A.1. Overview of SI Units

	Units	Symbol
Base units		
length	meter	m
mass	kilogram	kg
time	second	s
temperature*	kelvin	K
electrical current	ampere	A
luminous intensity	candela	cd
amount of material	mole	mol
Derived units		
Supplementary units		
angles in the plane	radian	rad
solid angles	steradian	sr
*Use degrees Celsius (°C), which has a more common usage than kelvin.		

Table A.2. Relationship of Mass and Weight

	Mass	Weight or Force of Gravity	Force
Customary	slug pound-mass	pound pound-force	pound pound-force
Metric	kilogram	newton	newton

Table A.3. Derived Units With Special Names

Quantity	Name	Symbol	Expression
Frequency	hertz	Hz	s^{-1}
Force	newton	N	$kg \cdot m/s^2$
Pressure, stress	pascal	Pa	N/m^2
Energy, work, quantity of heat	joule	J	$N \cdot m$
Power, radiant flux	watt	W	J/s
Electric charge, quantity	coulomb	C	$A \cdot s$
Electric potential	volt	V	W/A
Capacitance	farad	F	C/V
Electric resistance	ohm	Ω	V/A
Electric conductance	siemens	S	A/V
Magnetic flux	weber	Wb	$V \cdot s$
Magnetic flux density	tesla	T	Wb/m^2
Inductance	henry	H	Wb/A
Luminous flux	lumen	lm	$cd \cdot sr$
Illuminance	lux	lx	lm/m^2

Table A.4. Useful Conversion Factors

Quantity	From Customary Units	To Metric Units	Multiplied By*
Length	mile	Km	1.609
	yard	m	0.9144
	foot	m	<u>0.3048</u>
	inch	mm	<u>25.40</u>
Area	square mile	km^2	2.590
	acre	m^2	4047
	acre	hectare	0.4047
	square yard	m^2	0.8361
	square foot	m^2	0.09290
	square inch	mm^2	645.2
Volume	acre foot	m^3	1233
	cubic yard	m^3	0.7646
	cubic foot	m^3	0.02832
	cubic foot	L (1000 cm^3)	28.32
	100 board feet	m^3	0.2360
	gallon	L (1000 cm^3)	3.785
	cubic inch	cm^3	16.39
Mass	lb	kg	0.4536
	kip (1000 lb)	metric ton (1000 kg)	0.4536
Mass/unit length	plf	kg/m	1.488
Mass/unit area	psf	kg/m^2	4.882
Mass density	pcf	kg/m^3	16.02
Force	lb	N	4.448
	kip	kN	4.448
Force/unit length	plf	N/m	14.59
	klf	kN/m	14.59
Pressure, stress, modulus of elasticity	psf	Pa	47.88
	ksf	kPa	47.88
	psi	kPa	6.895
	ksi	MPa	6.895
Bending moment, torque, moment of force	ft-lb	N ∧ m	1.356
	ft-kip	kN ∧ m	1.356
Moment of mass	lb•ft	m	0.1383
Moment of inertia	lb•ft^2	kg•m^2	0.04214
Second moment of area	In^4	mm^4	416200
Section modulus	in^3	mm^3	16390
Power	ton (refrig)	kW	3.517
	Btu/s	kW	1.054
	hp (electric)	W	745.7
	Btu/h	W	0.2931
Volume rate of flow	ft^3/s	m^3/s	0.02832
	cfm	m^3/s	0.0004719
	cfm	L/s	0.4719
	mgd	m^3/s	0.0438
Velocity, speed	ft/s	M/s	<u>0.3048</u>
Acceleration	F/s^2	m/s^2	<u>0.3048</u>
Momentum	lb•ft/sec	kg•m/s	0.1383
Angular momentum	lb•ft^2/s	kg•m^2/s	0.04214
Plane angle	degree	rad	0.01745
		mrad	17.45
*4 significant figures; underline denotes exact conversion			

Table A.5. Prefixes

Submultiples			Multiples		
Deci	10^{-1}	d	deka	10^{1}	da
Centi	10^{-2}	c	hecto	10^{2}	h
Milli	10^{-3}	m	kilo	10^{3}	k
Micro	10^{-6}	μ	mega	10^{6}	M
Nano	10^{-9}	n	giga	10^{9}	G
Pica	10^{-12}	p	tera	10^{12}	T
Femto	10^{-15}	f	peta	10^{15}	P
Atto	10^{-18}	a	exa	10^{18}	E
Zepto	10^{-21}	z	zetta	10^{21}	Z
Yocto	10^{-24}	y	yotto	10^{24}	Y

Table A.6. Physical Properties of Water at Atmospheric Pressure in SI Units.

Temperature		Density	Specific Weight	Dynamic Viscosity	Kinematic Viscosity	Vapor Pressure	Surface Tension[1]	Bulk Modulus
Centigrade	Fahrenheit	kg/m^3	N/m^3	N · s/m^2	m^2/s	N/m^2 abs.	N/m	GN/m^2
0°	32°	1,000	9,810	1.79×10^{-3}	1.79×10^{-6}	611	0.0756	1.99
5°	41°	1,000	9,810	1.51×10^{-3}	1.51×10^{-6}	872	0.0749	2.05
10°	50°	1,000	9,810	1.31×10^{-3}	1.31×10^{-6}	1,230	0.0742	2.11
15°	59°	999	9,800	1.14×10^{-3}	1.14×10^{-6}	1,700	0.0735	2.16
20°	68°	998	9,790	1.00×10^{-3}	1.00×10^{-6}	2,340	0.0728	2.20
25°	77°	997	9,781	8.91×10^{-4}	8.94×10^{-7}	3,170	0.0720	2.23
30°	86°	996	9,771	7.97×10^{-4}	8.00×10^{-7}	4,250	0.0712	2.25
35°	95°	994	9,751	7.20×10^{-4}	7.24×10^{-7}	5,630	0.0704	2.27
40°	104°	992	9,732	6.53×10^{-4}	6.58×10^{-7}	7,380	0.0696	2.28
50°	122°	988	9,693	5.47×10^{-4}	5.53×10^{-7}	12,300	0.0679	
60°	140°	983	9,643	4.66×10^{-4}	4.74×10^{-7}	20,000	0.0662	
70°	158°	978	9,594	4.04×10^{-4}	4.13×10^{-7}	31,200	0.0644	
80°	176°	972	9,535	3.54×10^{-4}	3.64×10^{-7}	47,400	0.0626	
90°	194°	965	9,467	3.15×10^{-4}	3.26×10^{-7}	70,100	0.0607	
100°	212°	958	9,398	2.82×10^{-4}	2.94×10^{-7}	101,300	0.0589	

[1]Surface tension of water in contact with air

Table A.7. Physical Properties of Water at Atmospheric Pressure in English Units.

Temperature		Density	Specific Weight	Dynamic Viscosity	Kinematic Viscosity	Vapor Pressure	Surface Tension[1]	Bulk Modulus
Fahrenheit	Centigrade	Slugs/ft³	Weight lb/ft³	lb-sec/ft²	ft²/sec	lb/in²	lb/ft	lb/in²
32	0	1.940	62.416	0.374×10^{-4}	1.93×10^{-5}	0.09	0.00518	287,000
39.2	4.0	1.940	62.424					
40	4.4	1.940	62.423	0.323	1.67	0.12	.00514	296,000
50	10.0	1.940	62.408	0.273	1.41	0.18	.00508	305,000
60	15.6	1.939	62.366	0.235	1.21	0.26	.00504	313,000
70	21.1	1.936	62.300	0.205	1.06	0.36	.00497	319,000
80	26.7	1.934	62.217	0.180	0.929	0.51	.00492	325,000
90	32.2	1.931	62.118	0.160	0.828	0.70	.00486	329,000
100	37.8	1.927	61.998	0.143	0.741	0.95	.00479	331,000
120	48.9	1.918	61.719	0.117	0.610	1.69	.00466	332,000
140	60.0	1.908	61.386	0.0979	0.513	2.89		
160	71.1	1.896	61.006	0.0835	0.440	4.74		
180	82.2	1.883	60.586	0.0726	0.385	7.51		
200	93.3	1.869	60.135	0.0637	0.341	11.52		
212	100	1.847	59.843	0.0593	0.319	14.70		

[1] Surface tension of water in contact with air

Table A.8. Sediment Particles Grade Scale.

Size				Approximate Sieve Mesh Openings Per Inch		Class
Millimeters	Microns	Inches		Tyler	U.S. Standard	
4000-2000	-----	160-80		-----	-----	Very large boulders
2000-1000	-----	80-40		-----	-----	Large boulders
1000-500	-----	40-20		-----	-----	Medium boulders
500-250	-----	20-10		-----	-----	Small boulders
250-130	-----	10-5		-----	-----	Large cobbles
130-64	-----	5-2.5		-----	-----	Small cobbles
64-32	-----	2.5-1.3		-----	-----	Very coarse gravel
32-16	-----	1.3-0.6		-----	-----	Coarse gravel
16-8	-----	0.6-0.3		2 1/2	-----	Medium gravel
8-4	-----	0.3-0.16		5	5	Fine gravel
4-2	-----	0.16-0.08		9	10	Very fine gravel
2-1	2000-1000	-----		16	18	Very coarse sand
1-1/2	1000-500	-----		32	35	Coarse sand
1/2-1/4	500-250	-----		60	60	Medium sand
1/4-1/8	250-125	-----		115	120	Fine sand
1/8-1/16	125-62	-----		250	230	Very fine sand
1/16-1/32	62-31	-----		-----	-----	Coarse silt
1/32-1/64	31-16	-----		-----	-----	Medium silt
1/64-1/128	16-8	-----		-----	-----	Fine silt
1/128-1/256	8-4	-----		-----	-----	Very fine silt
1/256-1/512	4-2	-----		-----	-----	Coarse clay
1/512-1/1024	2-1	-----		-----	-----	Medium clay
1/1024-1/2048	1-0.5	-----		-----	-----	Fine clay
1/2048-1/4096	0.5-0.24	-----		-----	-----	Very fine clay

Millimeters column (lower set):
4000-2000 ... ----- ; 2000-1000 ... ----- ; 1000-500 ... ----- ; 500-250 ... ----- ; 250-130 ... ----- ; 130-64 ... ----- ; 64-32 ... ----- ; 32-16 ... ----- ; 16-8 ... ----- ; 8-4 ... ----- ; 4-2 ... ----- ; 2.00-1.00 ; 1.00-0.50 ; 0.50-0.25 ; 0.25-0.125 ; 0.125-0.062 ; 0.062-0.031 ; 0.031-0.016 ; 0.016-0.008 ; 0.008-0.004 ; 0.004-0.0020 ; 0.0020-0.0010 ; 0.0010-0.0005 ; 0.0005-0.0002

Table A.9. Common Equivalent Hydraulic Units.

Volume

Unit	Equivalent							
	cubic inch	liter	u.s. gallon	cubic foot	cubic yard	cubic meter	acre-foot	sec-foot-day
liter	61.02	1	0.264 2	0.035 31	0.001 308	0.001	810.6 E - 9	408.7 E - 9
U.S. gallon	231.0	3.785	1	0.133 7	0.004 951	0.003 785	3.068 E - 6	1.547 E - 6
cubic foot	1728	28.32	7.481	1	0.037 04	0.028 32	22.96 E - 6	11.57 E - 6
cubic yard	46 660	764.6	202.0	27	1	0.746 6	619.8 E - 6	312.5 E - 6
meter^3	61 020	1000	264.2	35.31	1.308	1	810.6 E - 6	408.7 E - 6
acre-foot	75.27 E + 6	1 233 000	325 900	43 560	1 613	1 233	1	0.504 2
sec-foot-day	149.3 E + 6	2 447 000	646 400	86 400	3 200	2 447	1.983	1

Discharge (Flow Rate, Volume/Time)

Unit	Equivalent					
	gallon/min	liter/sec	acre-foot/day	foot^3/sec	million gal/day	meter^3/sec
gallon/minute	1	0.063 09	0.004 419	0.002 228	0.001 440	63.09 E - 6
liter/second	15.85	1	0.070 05	0.035 31	0.022 82	0.001
acre-foot/day	226.3	14.28	1	0.504 2	0.325 9	0.014 28
feet^3/second	448.8	28.32	1.983	1	0.646 3	0.028 32
million gal/day	694.4	43.81	3.068	1.547	1	0.043 82
meter^3/second	15 850	1000	70.04	35.31	22.82	1

This page intentionally left blank.

APPENDIX B: CRITICAL DEPTH AND UNIFORM FLOW FOR VARIOUS CULVERT AND CHANNEL SHAPES

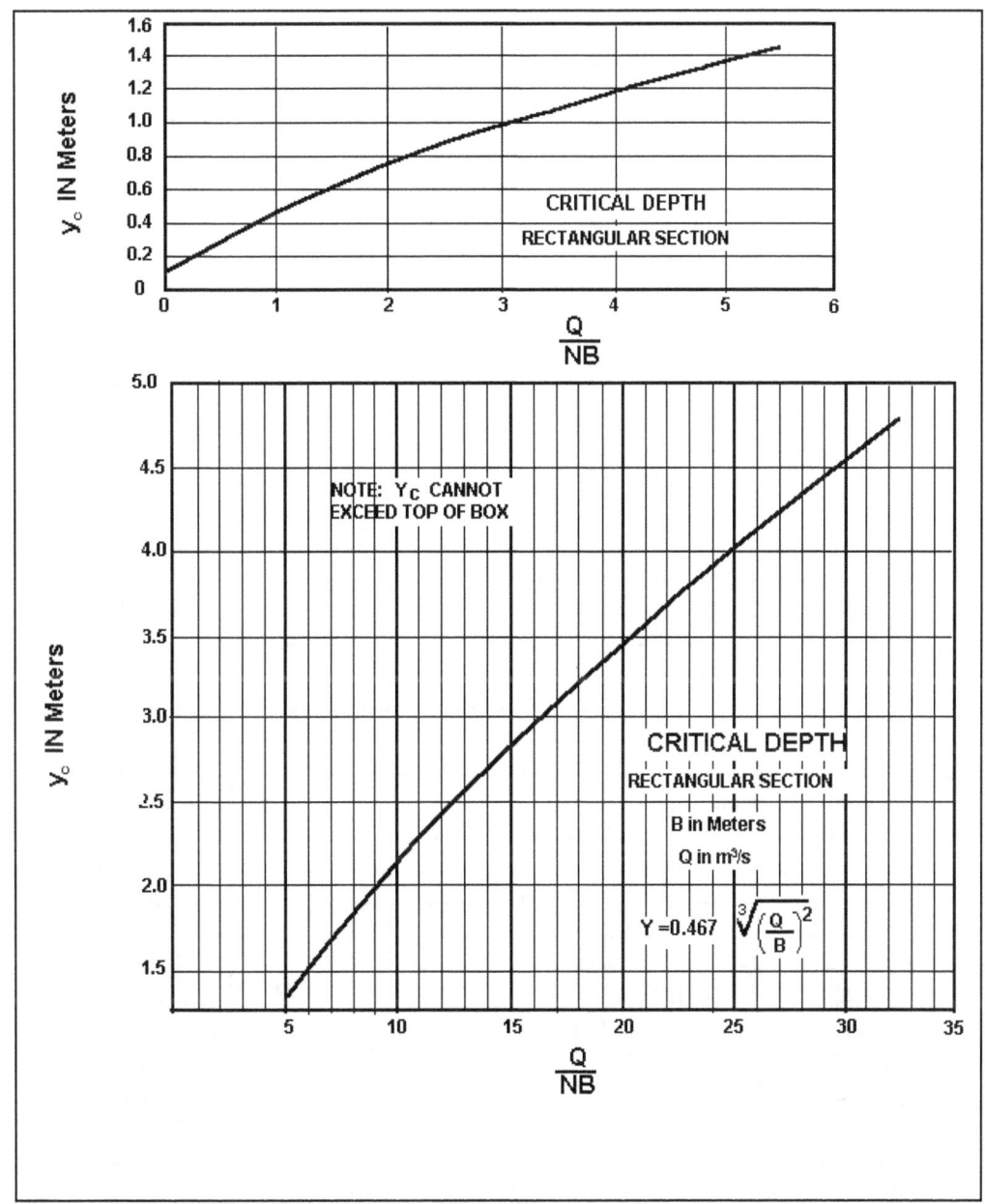

Figure B.1 (SI). Critical Depth Rectangular Section (Normann, et al., 2001)

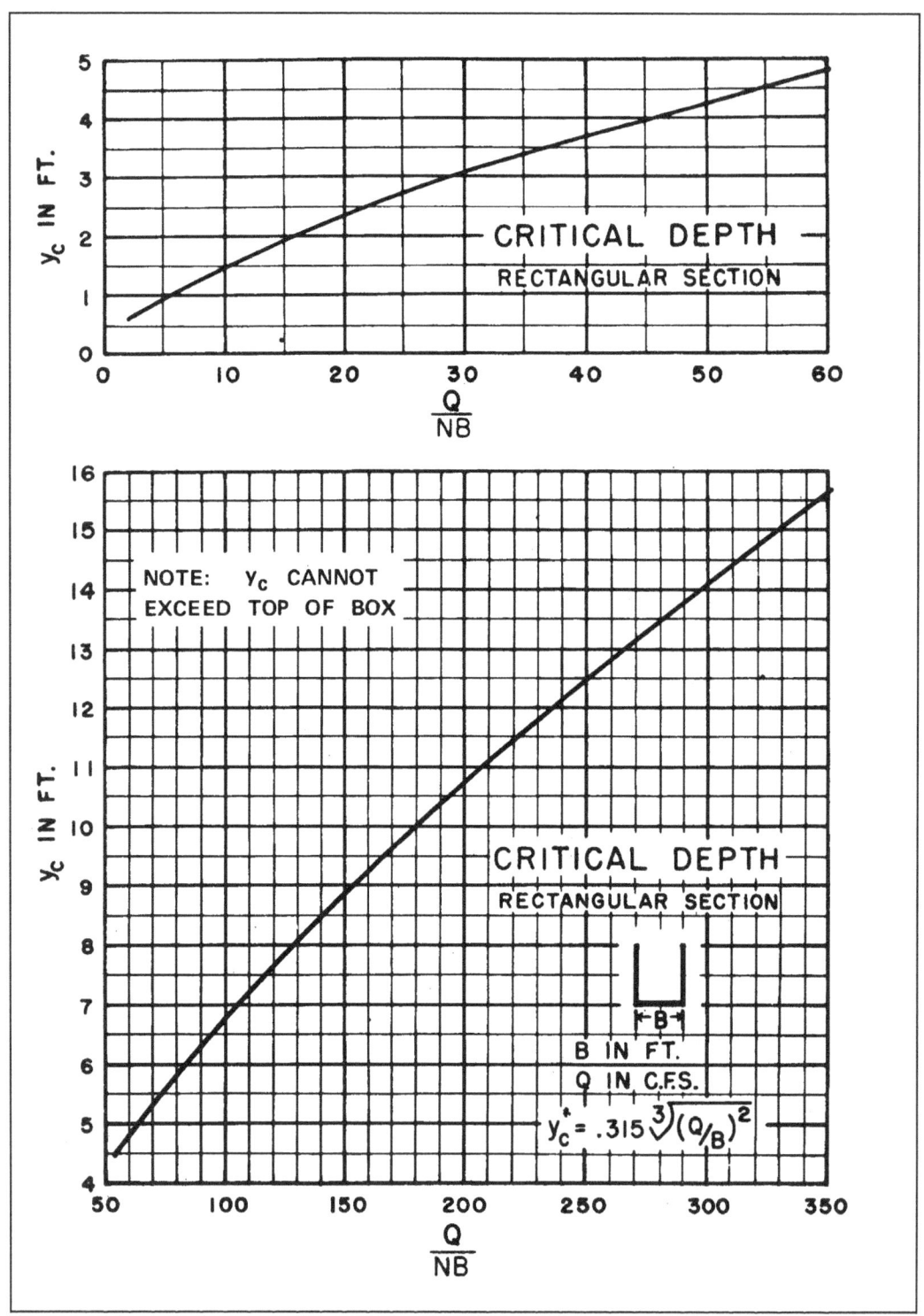

Figure B.1 (CU). Critical Depth Rectangular Section (Normann, et al., 2001)

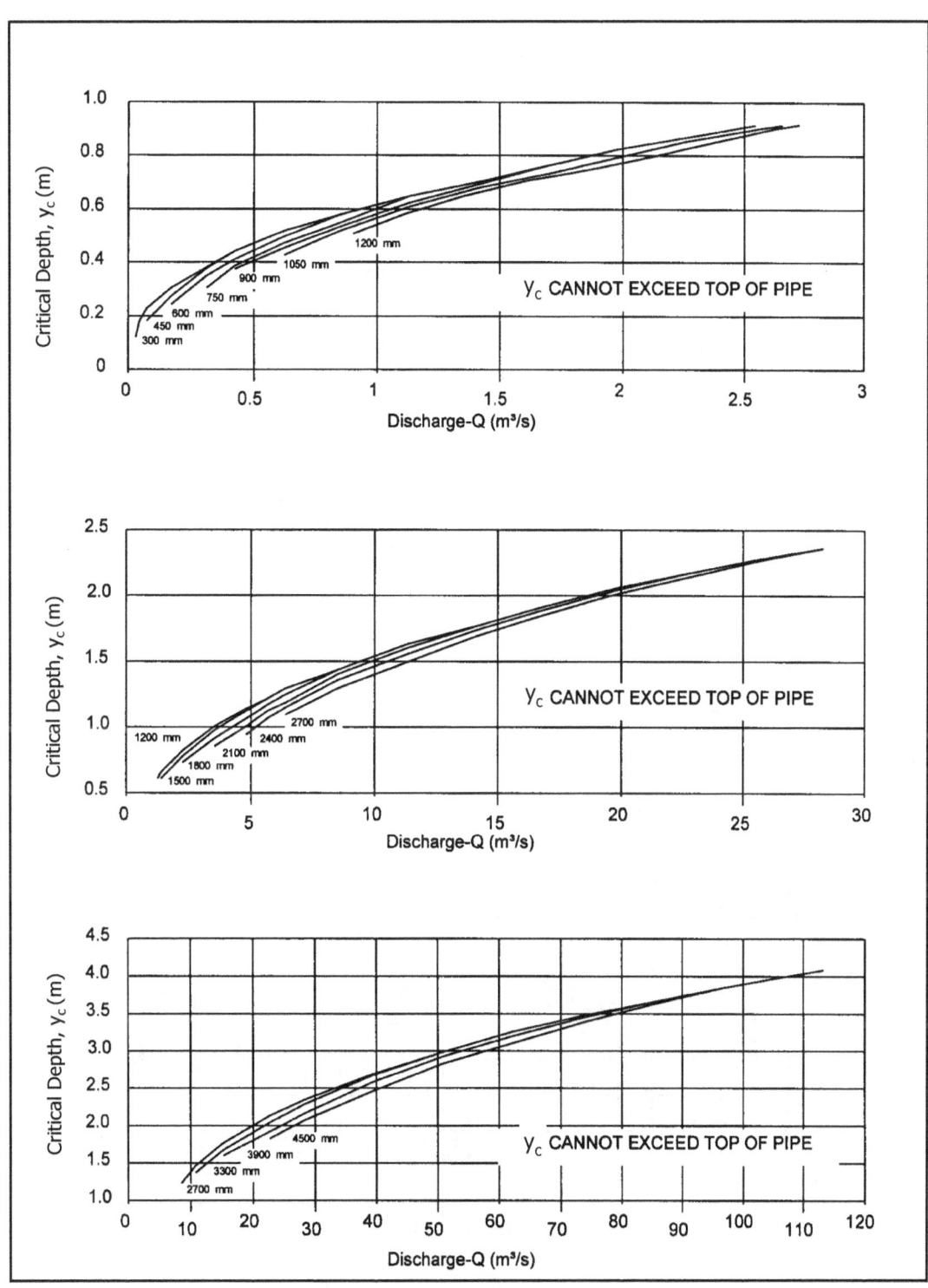

Figure B.2 (SI). Critical Depth of Circular Pipe

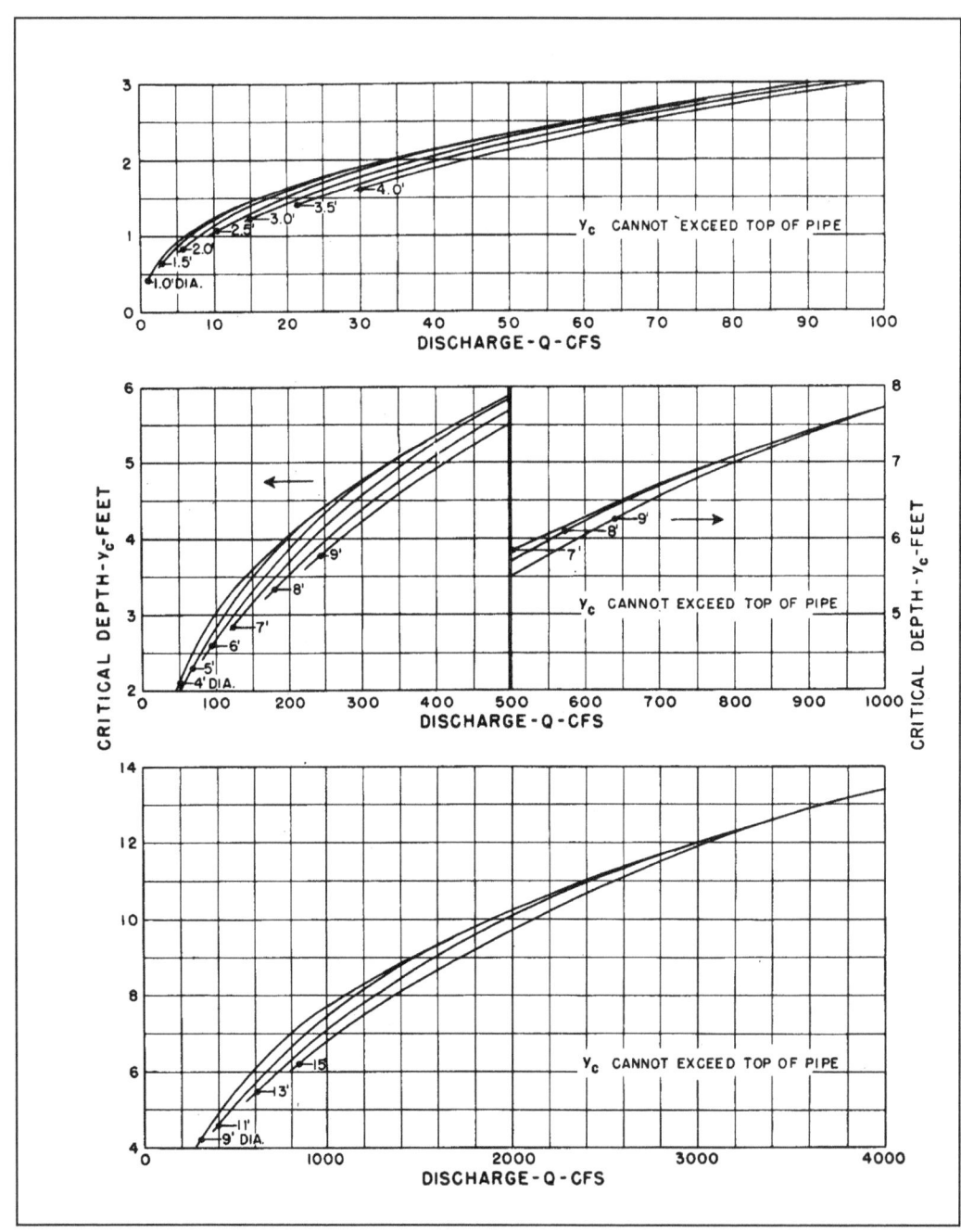

Figure B.2 (CU). Critical Depth of Circular Pipe

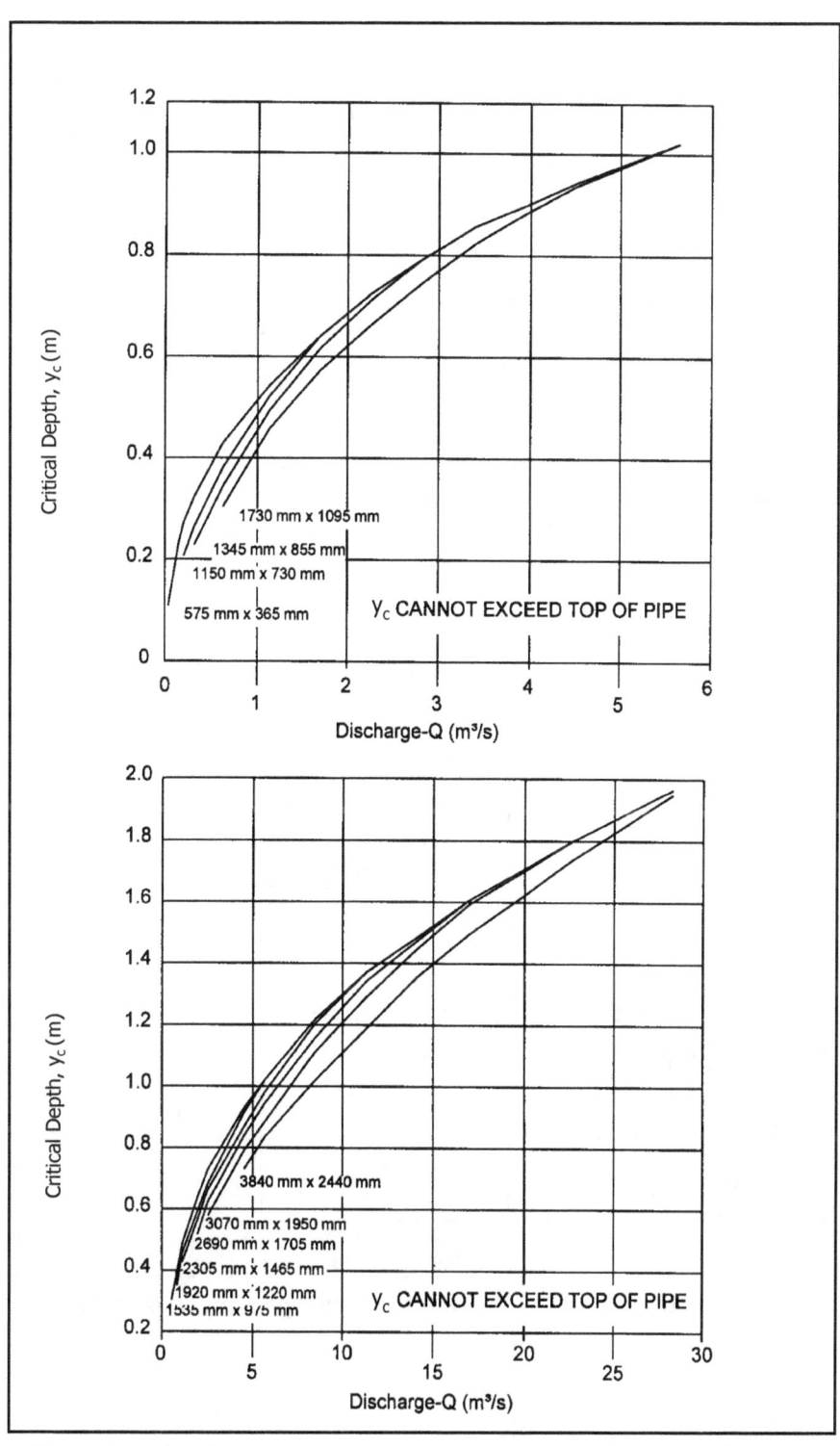

Figure B.3 (SI). Critical Depth Oval Concrete Pipe Long Axis Horizontal

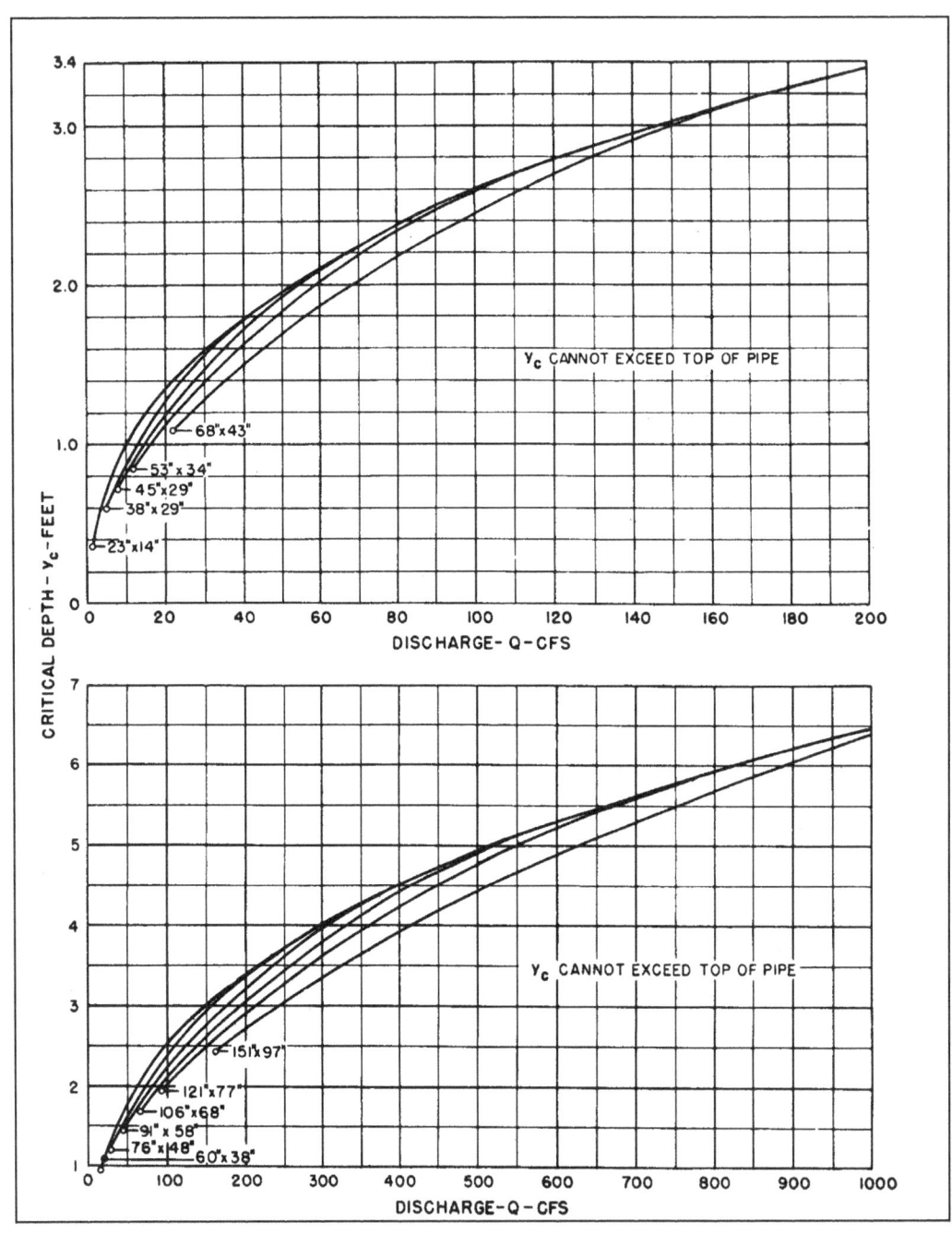

Figure B.3 (CU). Critical Depth Oval Concrete Pipe Long Axis Horizontal

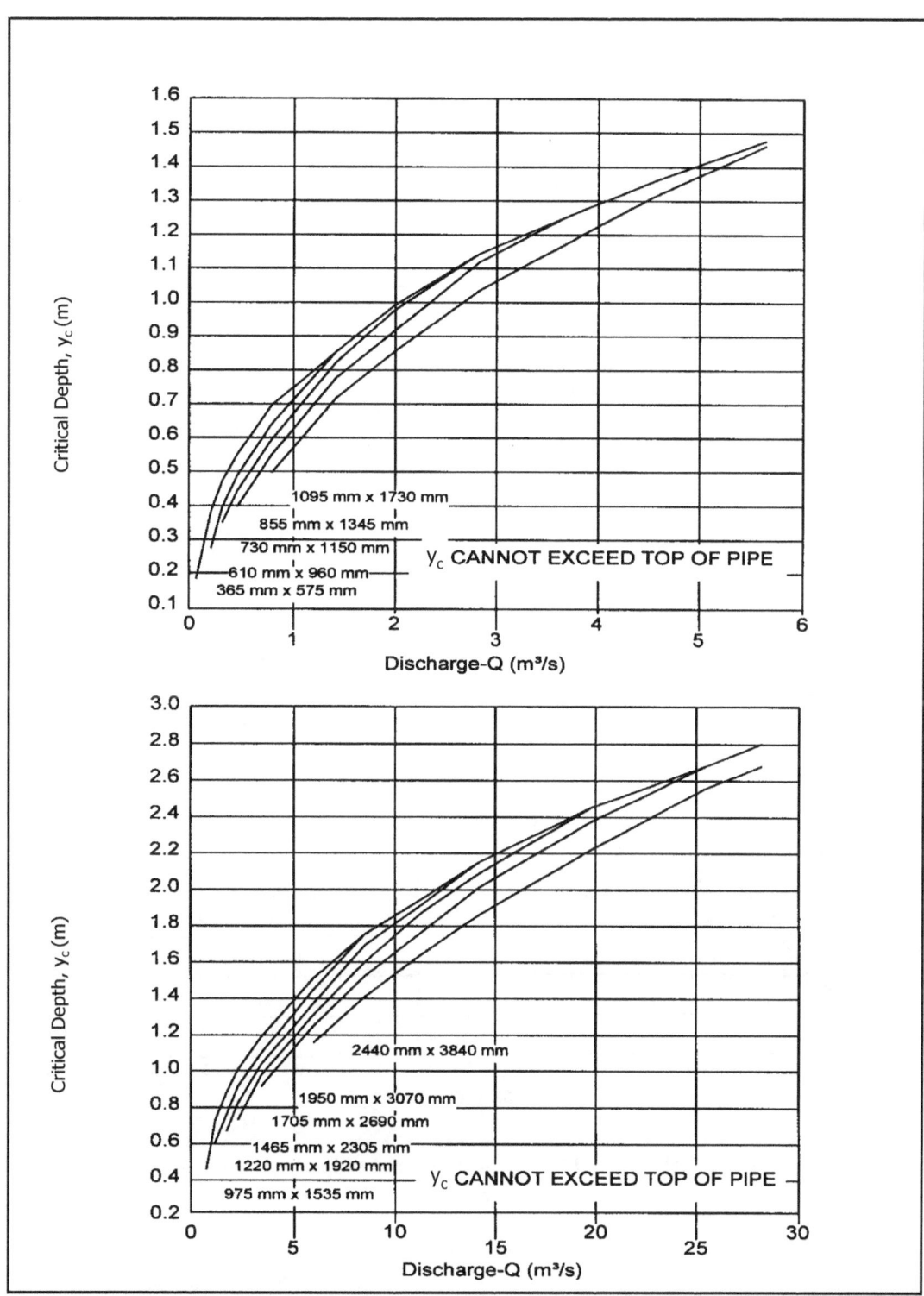

Figure B.4 (SI). Critical Depth Oval Concrete Pipe Long Axis Vertical

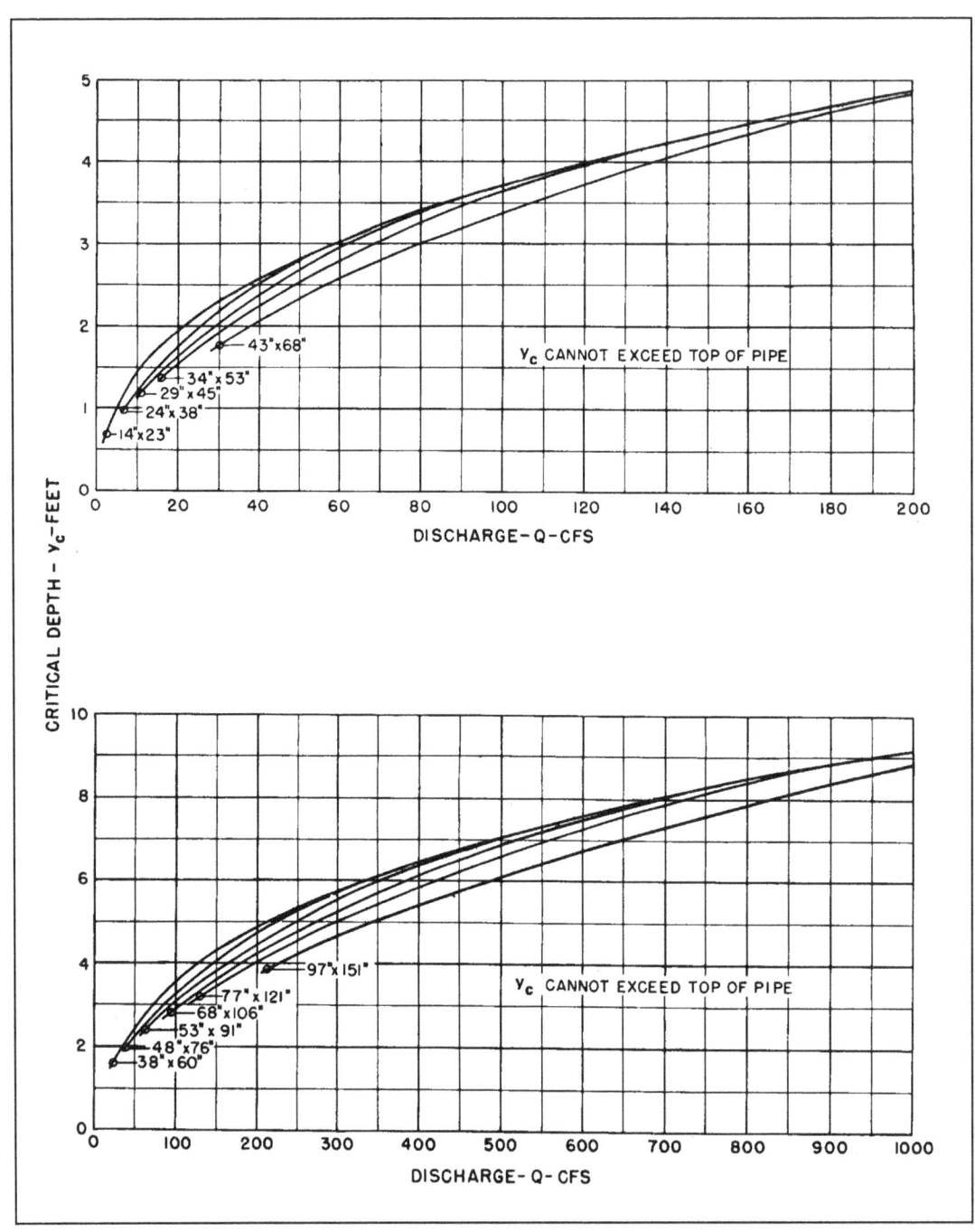

Figure B.4 (CU). Critical Depth Oval Concrete Pipe Long Axis Vertical

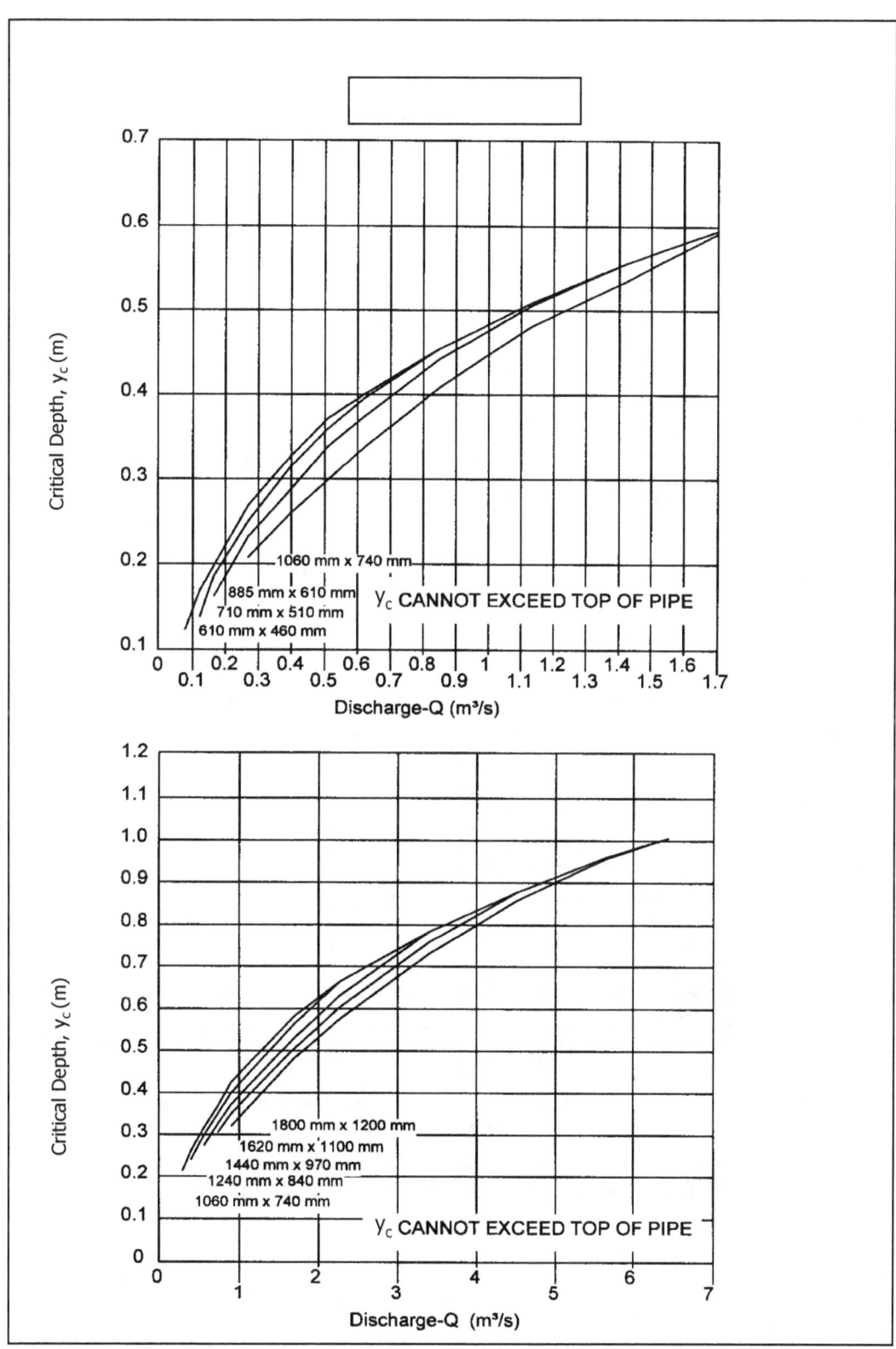

Figure B.5 (SI). Critical Depth Standard C.M. Pipe-Arch

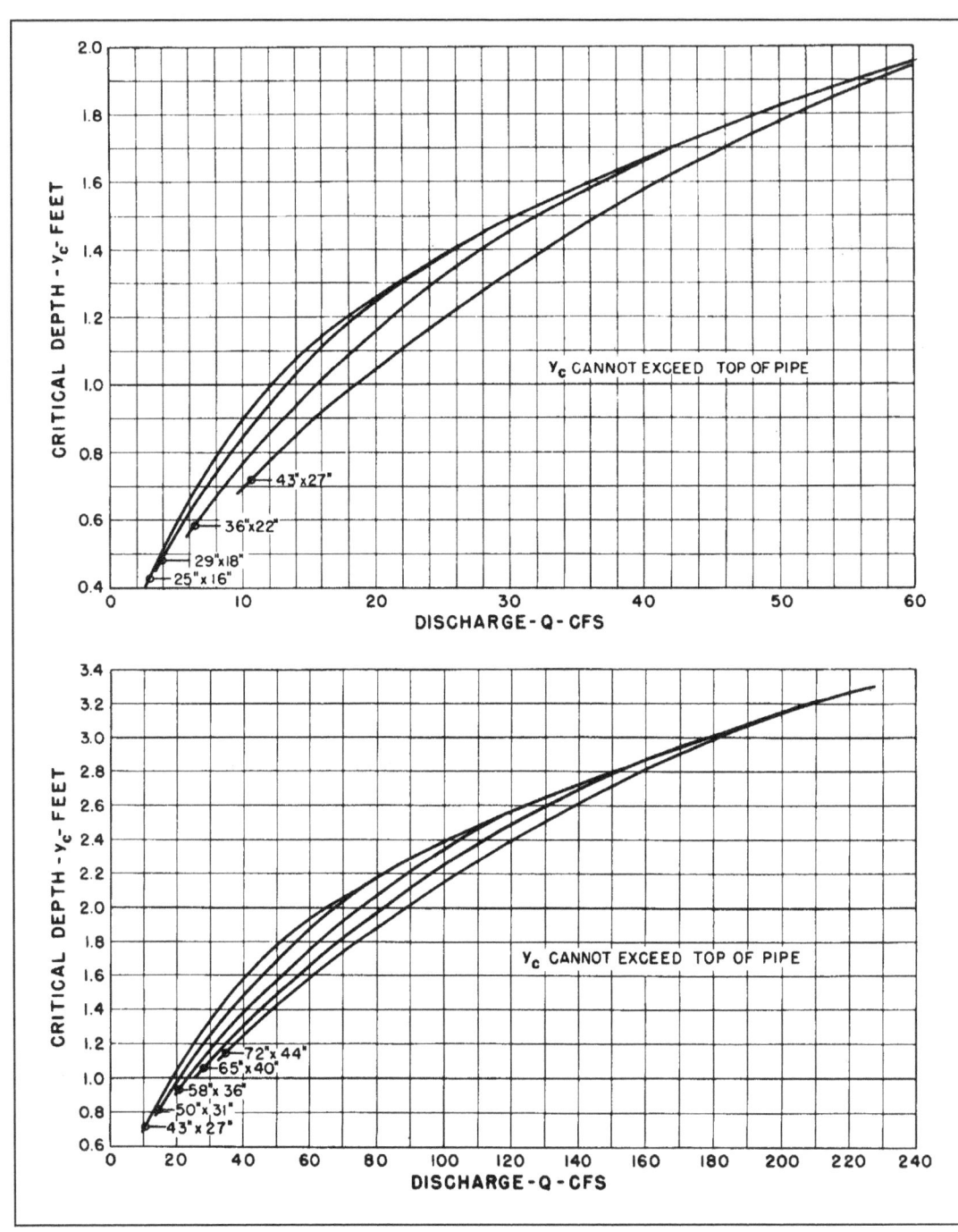

Figure B.5 (CU). Critical Depth Standard C.M. Pipe-Arch

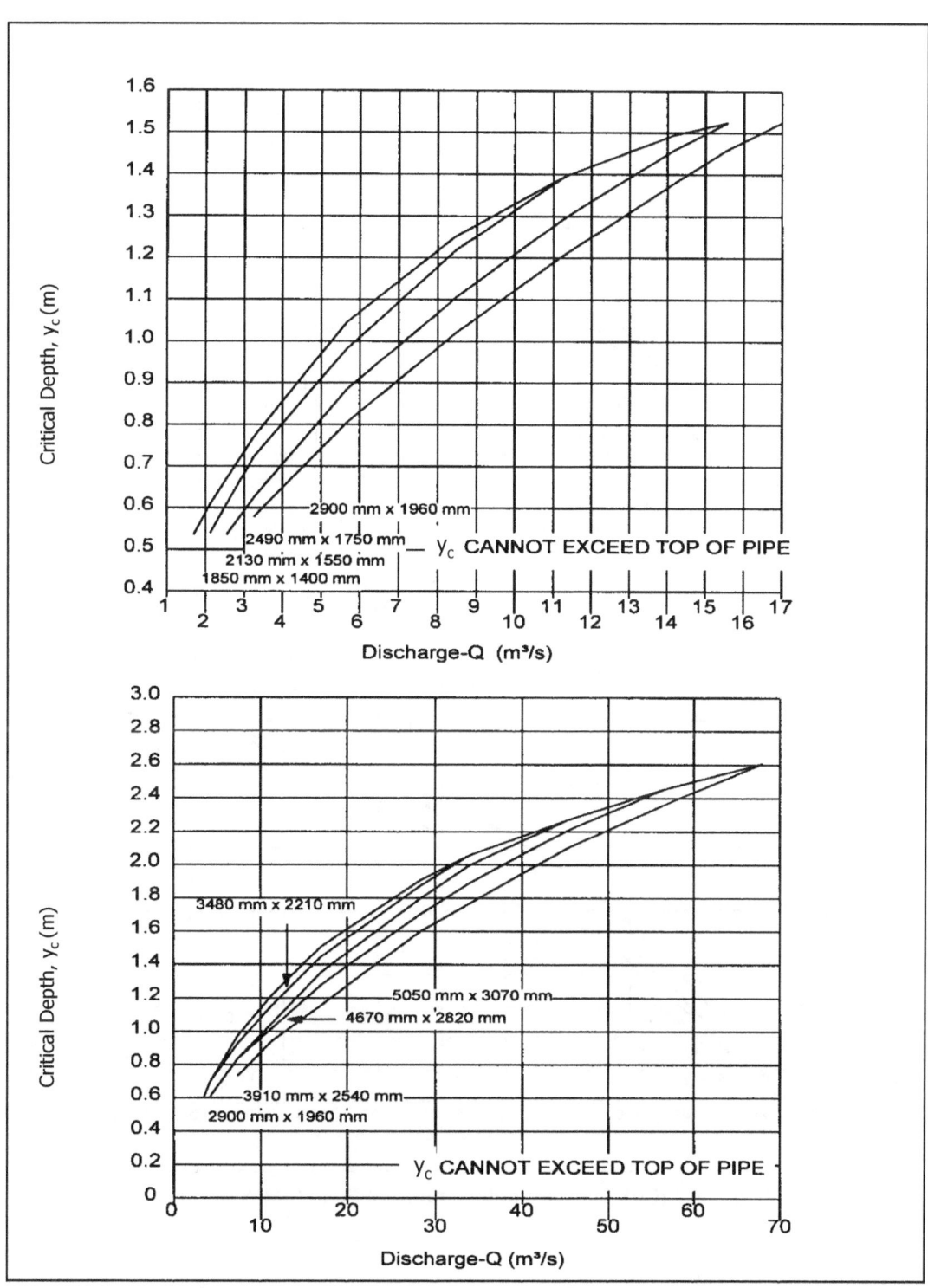

Figure B.6 (SI). Critical Depth Structural Plate C.M. Pipe-Arch

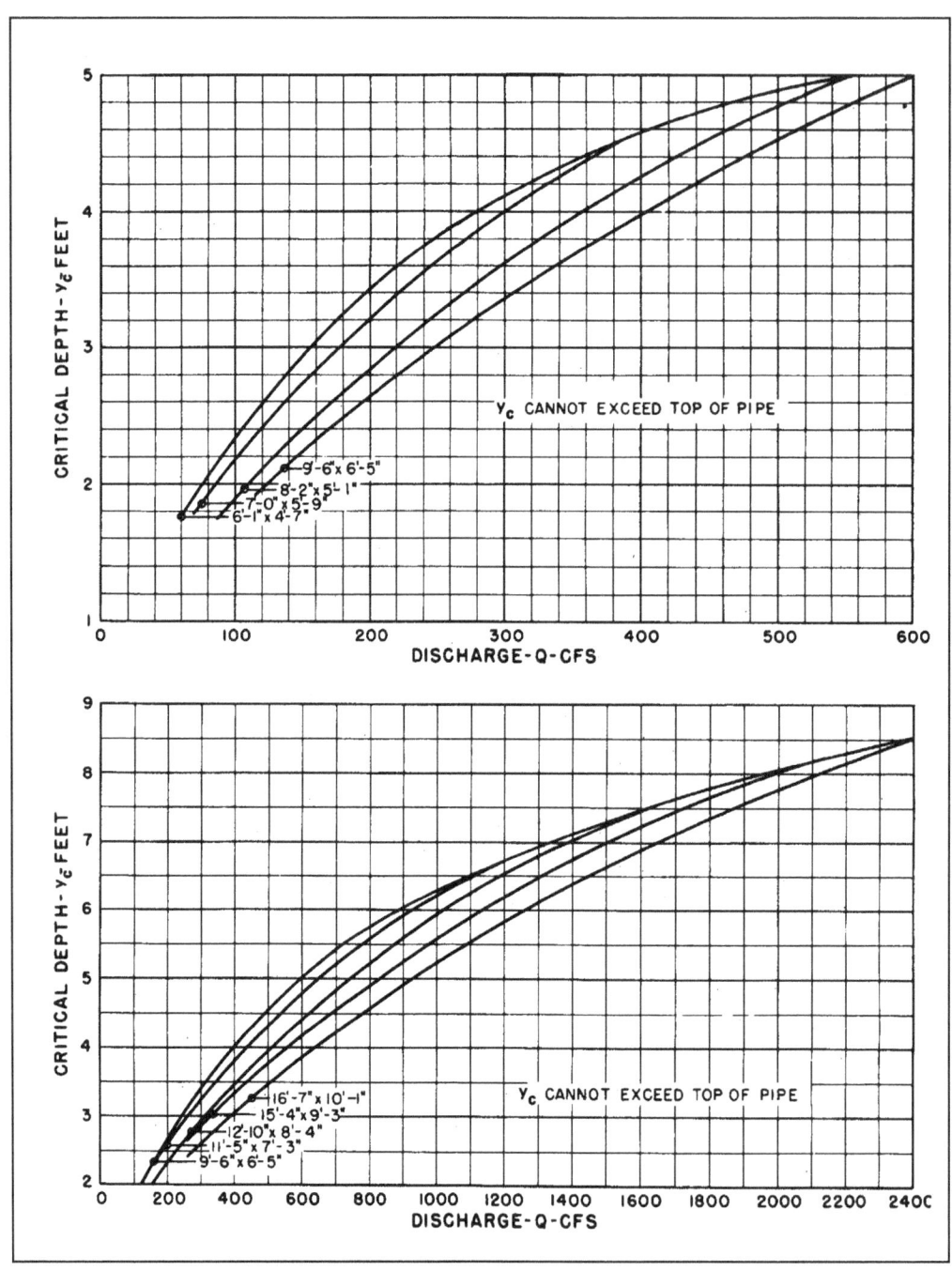

Figure B.6 (CU). Critical Depth Structural Plate C.M. Pipe-Arch

Table B.1. Uniform Flow in Trapezoidal Channels by Manning's Formula

| | | | | | Values of $(\alpha Qn)/(b^{8/3}S^{1/2})$ | | | | | |
y/b^1	z = 0	z = 0.25	z = 0.5	z = 0.75	z = 1	z = 1.25	z = 1.5	z = 1.75	z = 2	z = 3
0.02	0.00213	0.00215	0.00216	0.00217	0.00218	0.00219	0.00220	0.00220	0.00221	0.00223
0.03	0.00414	0.00419	0.00423	0.00426	0.00429	0.00429	0.00433	0.00434	0.00437	0.00443
0.04	0.00661	0.00670	0.00679	0.00685	0.00690	0.00690	0.00700	0.00704	0.00707	0.00722
0.05	0.00947	0.00964	0.00980	0.00991	0.01000	0.01010	0.01020	0.01030	0.01030	0.01060
0.06	0.0127	0.0130	0.0132	0.0134	0.0136	0.0137	0.0138	0.0140	0.0141	0.0145
0.07	0.0162	0.0166	0.0170	0.0173	0.0176	0.0177	0.0180	0.0182	0.0183	0.0190
0.08	0.0200	0.0206	0.0211	0.0215	0.0219	0.0222	0.0225	0.0228	0.0231	0.0240
0.09	0.0240	0.0249	0.0256	0.0262	0.0267	0.0271	0.0275	0.0279	0.0282	0.0296
0.10	0.0283	0.0294	0.0305	0.0311	0.0318	0.0324	0.0329	0.0334	0.0339	0.0358
0.11	0.0329	0.0342	0.0354	0.0364	0.0373	0.0380	0.0387	0.0394	0.0400	0.0424
0.12	0.0376	0.0393	0.0408	0.0420	0.0431	0.0441	0.0450	0.0458	0.0466	0.0497
0.13	0.0425	0.0446	0.0464	0.0480	0.0493	0.0505	0.0516	0.0527	0.0537	0.0575
0.14	0.0476	0.0501	0.0524	0.0542	0.0559	0.0573	0.0587	0.0599	0.0312	0.0659
0.15	0.0528	0.0559	0.0585	0.0608	0.0628	0.0645	0.0662	0.0677	0.0692	0.0749
0.16	0.0582	0.0619	0.0650	0.0676	0.0699	0.0720	0.0740	0.0759	0.0776	0.0845
0.17	0.0638	0.0680	0.0717	0.0748	0.0775	0.0800	0.0823	0.0845	0.0867	0.0947
0.18	0.0695	0.0744	0.0786	0.0822	0.0854	0.0883	0.0910	0.0936	0.0961	0.1050
0.19	0.0753	0.0809	0.0857	0.0900	0.0936	0.0970	0.1000	0.1030	0.1060	0.1170
0.20	0.0813	0.0875	0.0932	0.0979	0.1020	0.1060	0.1100	0.1130	0.1160	0.1290
0.21	0.0873	0.0944	0.1010	0.1060	0.1110	0.1150	0.1200	0.1230	0.1270	0.1420
0.22	0.0935	0.1010	0.1090	0.1150	0.1200	0.1250	0.1300	0.1340	0.1390	0.1550
0.23	0.0997	0.1090	0.1170	0.1240	0.1300	0.1350	0.1410	0.1460	0.1510	0.1690
0.24	0.106	0.116	0.125	0.133	0.139	0.146	0.152	0.157	0.163	0.184
0.25	0.113	0.124	0.133	0.142	0.150	0.157	0.163	0.170	0.176	0.199
0.26	0.119	0.131	0.142	0.152	0.160	0.168	0.175	0.182	0.189	0.215
0.27	0.126	0.139	0.151	0.162	0.171	0.180	0.188	0.195	0.203	0.232
0.28	0.133	0.147	0.160	0.172	0.182	0.192	0.201	0.209	0.217	0.249
0.29	0.139	0.155	0.170	0.182	0.193	0.204	0.214	0.223	0.232	0.267
0.30	0.146	0.163	0.179	0.193	0.205	0.217	0.227	0.238	0.248	0.286
0.31	0.153	0.172	0.189	0.204	0.217	0.230	0.242	0.253	0.264	0.306
0.32	0.160	0.180	0.199	0.215	0.230	0.243	0.256	0.269	0.281	0.327
0.33	0.167	0.189	0.209	0.227	0.243	0.257	0.271	0.285	0.298	0.348
0.34	0.174	0.198	0.219	0.238	0.256	0.272	0.287	0.301	0.315	0.369
0.35	0.181	0.207	0.230	0.251	0.270	0.287	0.303	0.318	0.334	0.392
0.36	0.190	0.216	0.241	0.263	0.283	0.302	0.319	0.336	0.353	0.416
0.37	0.196	0.225	0.251	0.275	0.297	0.317	0.336	0.354	0.372	0.440
0.38	0.203	0.234	0.263	0.289	0.311	0.333	0.354	0.373	0.392	0.465
0.39	0.210	0.244	0.274	0.301	0.326	0.349	0.371	0.392	0.412	0.491
0.40	0.218	0.254	0.286	0.314	0.341	0.366	0.389	0.412	0.433	0.518
0.41	0.225	0.263	0.297	0.328	0.357	0.383	0.408	0.432	0.455	0.545

				Values of $(\alpha Qn)/(b^{8/3}S^{1/2})$						
y/b^1	z = 0	z = 0.25	z = 0.5	z = 0.75	z = 1	z = 1.25	z = 1.5	z = 1.75	z = 2	z = 3
0.42	0.233	0.279	0.310	0.342	0.373	0.401	0.427	0.453	0.478	0.574
0.43	0.241	0.282	0.321	0.356	0.389	0.418	0.447	0.474	0.501	0.604
0.44	0.249	0.292	0.334	0.371	0.405	0.437	0.467	0.496	0.524	0.634
0.45	0.256	0.303	0.346	0.385	0.442	0.455	0.487	0.519	0.548	0.665
0.46	0.263	0.313	0.359	0.401	0.439	0.475	0.509	0.541	0.547	0.696
0.47	0.271	0.323	0.371	0.417	0.457	0.494	0.530	0.565	0.600	0.729
0.48	0.279	0.333	0.384	0.432	0.475	514.000	0.552	0.589	0.626	0.763
0.49	0.287	0.345	0.398	0.448	0.492	0.534	0.575	0.614	0.652	0.797
0.50	0.295	0.356	0.411	0.463	0.512	0.556	0.599	0.639	0.679	0.833
0.52	0.310	0.377	0.438	0.496	0.548	0.599	0.646	0.692	0.735	0.906
0.54	0.327	0.398	0.468	0.530	0.590	0.644	0.696	0.746	0.795	0.984
0.56	0.343	0.421	0.496	0.567	0.631	0.690	0.748	0.803	0.856	1.070
0.58	0.359	0.444	0.526	0.601	0.671	0.739	0.802	0.863	0.922	1.150
0.60	0.375	0.468	0.556	0.640	0.717	0.789	0.858	0.924	0.988	1.240
0.62	0.391	0.492	0.590	0.679	0.763	0.841	0.917	0.989	1.060	1.330
0.64	0.408	0.516	0.620	0.718	0.809	0.894	0.976	1.050	1.130	1.430
0.66	0.424	0.541	0.653	0.759	0.858	0.951	1.040	1.130	1.210	1.530
0.68	0.441	0.566	0.687	0.801	0.908	1.010	1.100	1.200	1.290	1.640
0.70	0.457	0.591	0.722	0.842	0.958	1.070	1.170	1.270	1.370	1.750
0.72	0.474	0.617	0.757	0.887	1.010	1.130	1.240	1.350	1.450	1.870
0.74	0.491	0.644	0.793	0.932	1.070	1.190	1.310	1.430	1.550	1.980
0.76	0.508	0.670	0.830	0.981	1.120	1.260	1.390	1.510	1.640	2.110
0.78	0.525	0.698	0.868	1.030	1.180	1.320	1.460	1.600	1.730	2.240
0.80	0.542	0.725	0.906	1.080	1.240	1.400	1.540	1.690	1.830	2.370
0.82	0.559	0.753	0.945	1.130	1.300	1.470	1.630	1.780	1.930	2.510
0.84	0.576	0.782	0.985	1.180	1.360	1.540	1.710	1.870	2.030	2.650
0.86	0.593	0.810	1.030	1.230	1.430	1.610	1.790	1.970	2.140	2.800
0.88	0.610	0.839	1.070	1.290	1.490	1.690	1.880	2.070	2.250	2.950
0.90	0.627	0.871	1.110	1.340	1.560	1.770	1.980	2.170	2.360	3.110
0.92	0.645	0.898	1.150	1.400	1.630	1.860	2.070	2.280	2.480	3.270
0.94	0.662	0.928	1.200	1.460	1.700	1.940	2.160	2.380	2.600	3.430
0.96	0.680	0.960	1.250	1.520	1.780	2.030	2.270	2.500	2.730	3.610
0.98	0.697	0.991	1.290	1.580	1.850	2.110	2.370	2.610	2.850	3.790
										3.970
										4.450
										4.960
										5.520
										6.110
										6.730

y/b[1]	z = 0	z = 0.25	z = 0.5	z = 0.75	z = 1	z = 1.25	z = 1.5	z = 1.75	z = 2	z = 3
										7.390
										8.10
										8.83
										9.62
										10.40
										11.30
										12.20
										13.20
										14.20
										15.20
										16.30
										17.40
										18.70
										19.90
										21.10
										23.90
										26.80
										30.00
										33.40
										37.00
										40.80
										44.80
										49.10
										53.70
										58.40
										68.90
										80.20
										92.80
										107
										122
										164
										216

Values of $(\alpha Qn)/(b^{8/3}S^{1/2})$

for y /b less than 0.04, use of the assumption R = y is more convenient and more accurate than interpolation in the table.

y = depth of flow, m (ft)

Q = discharge by Manning's Equation, m^3/s (ft^3/s)

n = Manning's coefficient

S = channel bottom and water surface slope

α = units conversion = 1.49 for SI, 1 for CU

z = side slope, 1:z (V:H)

b = bottom width

Source: USBR (1974)

Table B.2. Uniform Flow in Circular Sections Flowing Partly Full

y/D	A/D^2	R/D	$\frac{(\alpha Qn)}{(D^{8/3} S^{1/2})}$	$\frac{(\alpha Qn)}{(y^{8/3} S^{1/2})}$	y/D	A/D^2	R/D	$\frac{(\alpha Qn)}{(D^{8/3} S^{1/2})}$	$\frac{(\alpha Qn)}{(y^{8/3} S^{1/2})}$
0.01	0.0013	0.0066	0.00007	15.04	0.51	0.4027	0.2531	0.239	1.442
0.02	0.0037	0.0132	0.00031	10.57	0.52	0.4127	0.2562	0.247	0.415
0.03	0.0069	0.0197	0.00074	8.56	0.53	0.4227	0.2592	0.255	1.388
0.04	0.0105	0.0262	0.00138	7.38	0.54	0.4327	0.2621	0.263	1.362
0.05	0.0147	0.0325	0.00222	6.55	0.55	0.4426	0.2649	0.271	1.336
0.06	0.0192	0.0389	0.00328	5.95	0.56	0.4526	0.2676	0.279	1.311
0.07	0.0294	0.0451	0.00455	5.47	0.57	0.1626	0.2703	0.287	1.286
0.08	0.0350	0.0513	0.00604	5.09	0.58	0.4724	0.2728	0.295	1.262
0.09	0.0378	0.0575	0.00775	4.76	0.59	0.4822	0.2753	0.303	1.238
0.10	0.0409	0.0635	0.0097	4.49	0.60	0.4920	0.2776	0.311	1.215
0.11	0.0470	0.0695	0.0118	4.25	0.61	0.5018	0.2799	0.319	1.192
0.12	0.0534	0.0755	0.0142	4.04	0.62	0.5115	0.2821	0.327	1.170
0.13	0.0600	0.0813	0.0167	3.86	0.63	0.5212	0.2842	0.335	1.148
0.14	0.0668	0.0871	0.0195	3.69	0.64	0.5308	0.2862	0.343	1.126
0.15	0.0739	0.0929	0.0225	3.54	0.65	0.5405	0.2988	0.350	1.105
0.16	0.0811	0.0985	0.0257	3.41	0.66	0.5499	0.2900	0.358	1.084
0.17	0.0885	0.1042	0.0291	3.28	0.67	0.5594	0.2917	0.366	1.064
0.18	0.0961	0.1097	0.0327	3.17	0.68	0.5687	0.2933	0.373	1.044
0.19	0.0139	0.1152	0.0365	3.06	0.69	0.5780	0.2948	0.380	1.024
0.20	0.1118	0.1206	0.0406	2.96	0.70	0.5872	0.2962	0.388	1.004
0.21	0.1199	0.1259	0.0448	2.87	0.71	0.5964	0.2975	0.395	0.985
0.22	0.1281	0.1312	0.0492	2.79	0.72	0.6054	0.2987	0.402	0.965
0.23	0.1365	0.1364	0.0537	2.71	0.73	0.6143	0.2998	0.409	0.947
0.24	0.1449	0.1416	0.0585	2.63	0.74	0.6231	0.3008	0.416	0.928
0.25	0.1535	0.1466	0.0634	2.56	0.75	0.6319	0.3042	0.422	0.910
0.26	0.1623	0.1516	0.0686	2.49	0.76	0.6405	0.3043	0.429	0.891
0.27	0.1711	0.1566	0.0739	2.42	0.77	0.6489	0.3043	0.435	0.873
0.28	0.1800	0.1614	0.0793	2.36	0.78	0.6573	0.3041	0.441	0.856
0.29	0.1890	0.1662	0.0849	2.30	0.79	0.6655	0.3039	0.447	0.838
0.30	0.1982	0.1709	0.0907	2.25	0.80	0.6736	0.3042	0.453	0.821
0.31	0.2074	0.1756	0.0966	2.20	0.81	0.6815	0.3043	0.458	0.804
0.32	0.2167	0.1802	0.1027	2.14	0.82	0.6893	0.3043	0.463	0.787
0.33	0.2260	0.1847	0.1089	2.09	0.83	0.6969	0.3041	0.468	0.770
0.34	0.2355	0.1891	0.1153	2.05	0.84	0.7043	0.3038	0.473	0.753
0.35	0.2450	0.1935	0.1218	2.00	0.85	0.7115	0.3033	0.453	0.736
0.36	0.2546	0.1978	0.1284	1.958	0.86	0.7186	0.3026	0.458	0.720

y/D	A/D^2	R/D	$\dfrac{(\alpha Qn)}{(D^{8/3} S^{1/2})}$	$\dfrac{(\alpha Qn)}{(y^{8/3} S^{1/2})}$	y/D	A/D^2	R/D	$\dfrac{(\alpha Qn)}{(D^{8/3} S^{1/2})}$	$\dfrac{(\alpha Qn)}{(y^{8/3} S^{1/2})}$
0.37	0.2642	0.2020	0.1351	1.915	0.87	0.7254	0.3018	0.485	0.703
0.38	0.2739	0.2062	0.1420	1.875	0.88	0.7320	0.3007	0.488	0.687
0.39	0.2836	0.2102	0.1490	1.835	0.89	0.7384	0.2995	0.491	0.670
0.40	0.2934	0.2142	0.1561	1.797	0.90	0.7445	0.2980	0.494	0.654
0.41	0.3032	0.2182	0.1633	1.760	0.91	0.7504	0.2963	0.496	0.637
0.42	0.3130	0.2220	0.1705	1.724	0.92	0.7560	0.2944	0.497	0.621
0.43	0.3229	0.2258	0.1779	1.689	0.93	0.7612	0.2921	0.498	0.604
0.44	0.3328	0.2295	0.1854	1.655	0.94	0.7662	0.2895	0.498	0.588
0.45	0.3428	0.2331	0.1929	1.622	0.95	0.7707	0.2865	0.498	0.571
0.46	0.3527	0.2366	0.201	1.590	0.96	0.7749	0.2829	0.496	0.553
0.47	0.3627	0.2401	0.208	1.559	0.97	0.7785	0.2787	0.494	0.535
0.48	0.3727	0.2435	0.216	1.530	0.98	0.7817	0.2735	0.489	0.517
0.49	0.3827	0.2468	0.224	1.500	0.99	0.7841	0.2666	0.483	0.496
0.50	0.3927	0.2500	0.232	1.471	1.00	0.7854	0.2500	0.463	0.463

y = depth of flow, m (ft)
D = diameter of pipe, m (ft)
A = area of flow, m^2 (ft^2)
R = hydraulic radius, m (ft)
Source: USBR (1974)

Q = discharge by Manning's Equation, m^3/s (ft^3/s)
n = Manning's coefficient
S = channel bottom and water surface slope
α = units conversion = 1.49 for SI, 1 for CU

This page intentionally left blank.

APPENDIX C: STRUCTURAL CONSIDERATIONS FOR ROUGHNESS ELEMENTS

Blocks, sills, and other roughness elements are used to impose exaggerated resistance to flow and to force and stabilize the hydraulic jump. They may be employed inside the culvert barrel, at the culvert exit or in open channels. Roughness elements must be anchored sufficiently to withstand the drag forces on the elements. The fluid dynamic drag equation is:

$$F_D = C_D \, A_F \, \rho \, V_a^2 \, /2 \qquad \text{(C.1)}$$

where,

C_D = coefficient of drag (The maximum C_D for a structural angle or a rectangular block is 1.98 (Horner, 1965).)

ρ = density of water, 1000 kg/m³ (1.94 slugs/ft³)

V_a = approach velocity acting on roughness element, m/s (ft/s)

The roughness elements in the CSU rigid boundary basin, the USBR basins, the SAF basin, and internal dissipators must be able to satisfactorily resist the drag force over the lifetime of the structure. The drag force may be assumed to act at the center of the roughness element as shown in Figure C.1.

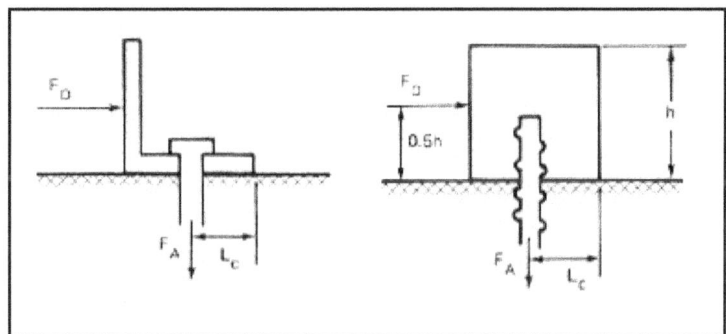

Figure C.1. Forces Acting on a Roughness Element

The anchor forces necessary to resist overturning can be computed as follows:

$$F_A = h F_D / 2 L_c = 0.5 \, (h/L_c) \, A_F \rho V_a^2 \qquad \text{(C.2)}$$

where,

F_A = total force on anchors

F_D = drag force on roughness element

h = height of roughness, m (ft)

L_c = distance from downstream edge of roughness element to the centroid of the anchors, m (ft)

A_F = frontal area of roughness element, m² (ft²)

V_a = approach velocity acting on roughness element, m/s (ft/s)

ρ = density of water, 1000 kg/m³ (1.94 slugs/ft³)

The approach velocity, V_a, should be selected as a worst case using the approach velocity at the first row for V_a. In cases of tumbling flow or increased resistance on steep slopes, use the normal velocity of the culvert without roughness elements for V_a.

This page intentionally left blank.

APPENDIX D: RIPRAP APRON SIZING EQUATIONS

A variety of relationships for sizing riprap aprons have been developed. Six are summarized and compared in this appendix. The first is from the Urban Drainage and Flood Control District in Denver Colorado (UD&FCD, 2004). These equations consider tailwater in addition to a measure of flow intensity.

$$D_{50} = 0.023D\left(\frac{Q}{\alpha D^{2.5}}\right)\left(\frac{D}{TW}\right)^{1.2}$$ (D.1a)

$$D_{50} = 0.014D\left(\frac{Q}{\alpha BD^{1.5}}\right)\left(\frac{D}{TW}\right)$$ (D.1b)

where,

D_{50} = riprap size, m (ft)

Q = design discharge, m^3/s (ft^3/s)

D = culvert diameter (circular) or culvert rise (rectangular), m (ft)

B = culvert span (rectangular), m (ft)

TW = tailwater depth, m (ft)

α = unit conversion constant, 1.811 (SI) and 1.0 (CU)

An equation in Berry (1948) and Peterka (1978) has been used for apron riprap sizing. It is only based on velocity.

$$D_{50} = \alpha V^2$$ (D.2)

where,

V = culvert exit velocity, m/s (ft/s)

α = unit conversion constant, 0.0413 (SI) and 0.0126 (CU)

A relationship used in the previous edition of HEC 14 from Searcy (1967) and also found in HEC 11 (Brown and Clyde, 1989) for sizing riprap protection for piers is based on velocity.

$$D_{50} = \frac{0.692}{S-1}\left(\frac{V^2}{2g}\right)$$ (D.3)

where,

S = riprap specific gravity

Bohan (1970) developed two relationships based on laboratory testing that considered, among other factors, whether the culvert was subjected to "minimum" tailwater (TW/D < 0.5) or "maximum" tailwater (TW/D > 0.5). The equations for minimum and maximum tailwater, respectively, are as follows:

$$D_{50} = 0.25DFr_o$$ (D.4a)

$$D_{50} = D(0.25Fr_o - 0.15)$$ (D.4b)

where,

  Fr_o = Froude number at the outlet defined as $V_o/(gD)^{0.5}$

Fletcher and Grace (1972) used the laboratory data from Bohan and other sources to develop a similar equation to Equation D.1.

$$D_{50} = 0.020D\left(\frac{Q}{\alpha D^{2.5}}\right)^{4/3}\left(\frac{D}{TW}\right)$$

(D.5)

where,

  α = unit conversion constant, 0.55 (SI) and 1.0 (CU)

Finally, the USDA/SCS has a series of charts for sizing riprap for aprons. These charts appear to be based on Bohan (Equation D.4a and D.4b).

Equation D.2 (Berry) and Equation D.3 (Searcy) are similar in their exclusive reliance on velocity as the predictor variable and differ only in terms of their coefficient. Equation D.1 (UD&FCD), Equation D.4 (Bohan), and Equation D.5 (Fletcher and Grace) incorporate some sort of flow intensity parameter, i.e. relative discharge or Froude number, as well as relative tailwater depth. (Bohan incorporates tailwater by having separate minimum and maximum tailwater equations.) UD&FCD and Fletcher and Grace have identical forms but differ in their coefficient and exponents.

These equations and the USDA charts were compared based on a series of hypothetical situations. A total of 10 scenarios were run with HY8 to generate outlet velocity conditions for testing the equations. The 10 scenarios included the following variations:

- Two culvert sizes, 760 and 1200 mm (30 to 48 in) metal pipe culverts

- Discharges ranging from (1.1 to 4.2 m^3/s) (40 to 150 ft^3/s)

- Slope and tailwater changes resulting in 5 inlet control and 5 outlet control cases

Figures D.1, D.2, and D.3 compare the recommended riprap size, D_{50}, relative to the outlet velocity, V, discharge intensity, $Q/D^{2.5}$, and relative tailwater depth, TW/D. The recommended D50 varies widely, but it is clear that the Berry equation (Equation D.2) results in the highest values for the range of conditions evaluated.

Equations D.2 and D.3 are not recommended because they do not consider tailwater effects. Equation D.4 is not further considered because it treats tailwater only as two separate conditions, minimum and maximum. Equations D.1 and D.5 are similar in their approach and are based on laboratory data. Both would probably both generate reasonable designs. For the ten hypothetical cases evaluated Equation D.1 produced the higher recommendation 3 times and the lower recommendation 7 times. Therefore, Equation D.5 is included in Chapter 10 of this manual.

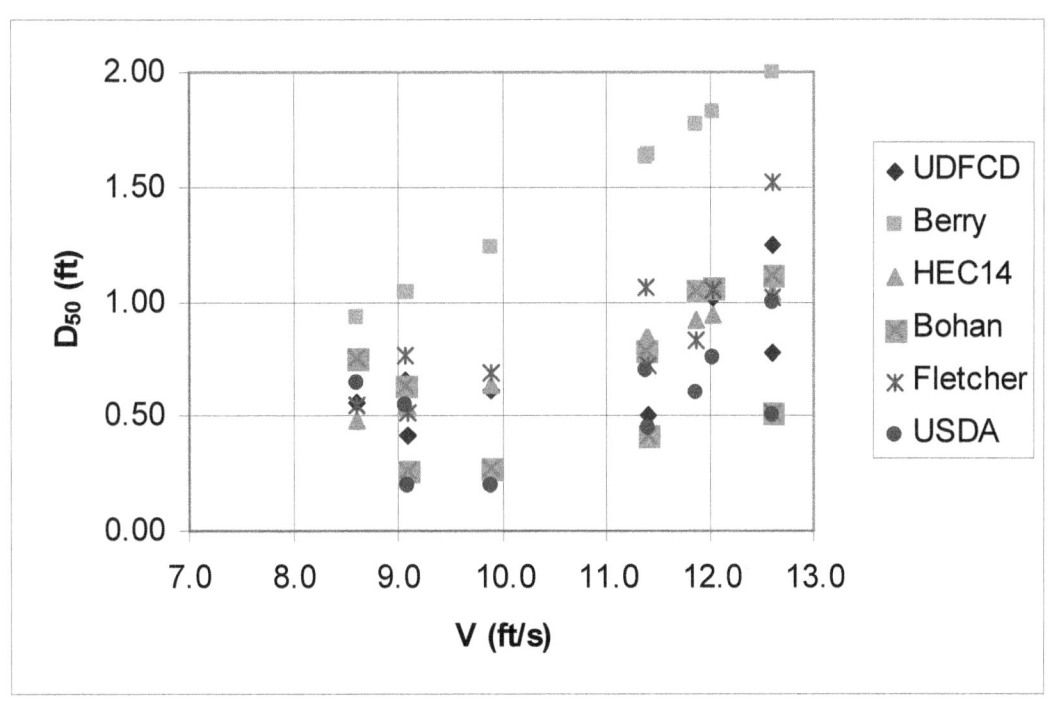

Figure D.1. D_{50} versus Outlet Velocity

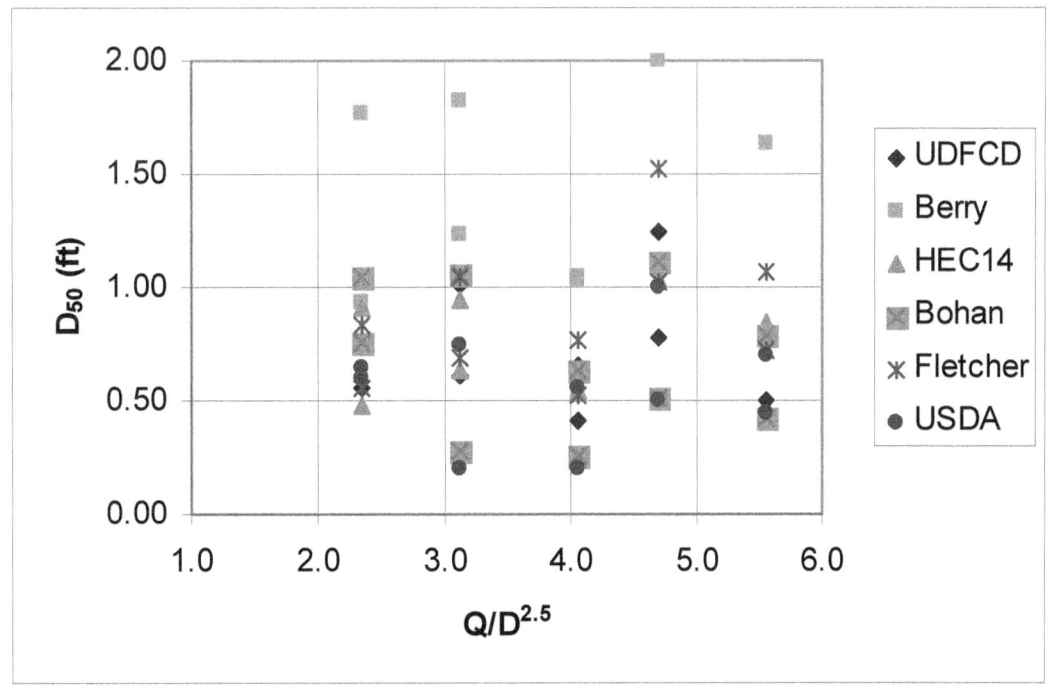

Figure D.2. D_{50} versus Discharge Intensity

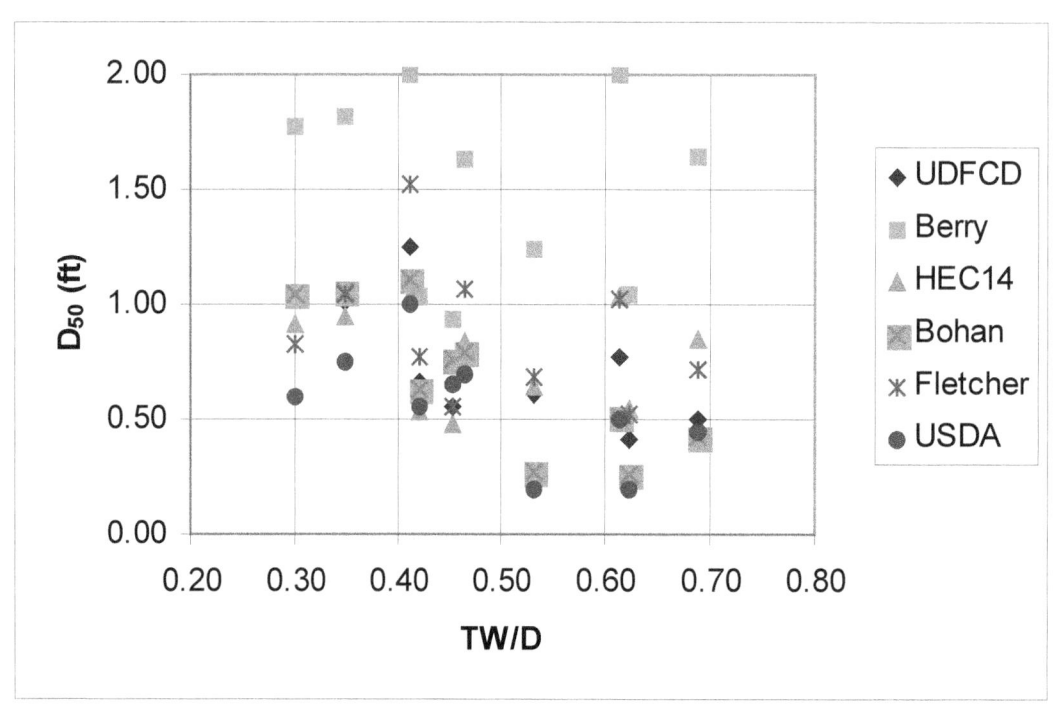

Figure D.3. D_{50} versus Relative Tailwater Depth

REFERENCES

AASHTO, 1999. Model Drainage Manual, Task Force on Hydrology and Hydraulics.

AASHTO, 2005. Model Drainage Manual, Task Force on Hydrology and Hydraulics.

Abt, S. R., Donnell, C. A., Ruff, J. F., and Doehring, F. K., 1985a. "Culvert Slope and Shape Effects on Outlet Scour," Transportation Research Record 1017, pp. 24-30.

Abt, S. R., Ruff, J. F., and Doehring, F. K., 1985b. "Culvert Slope Effects on Outlet Scour," Journal of Hydraulic Engineering, ASCE, 111(10), pp. 1363-1367, October.

Abt, S. R., Ruff, J. F., Doehring, F. K., and Donnell, C.A., 1987. "Influence of Culvert Shape on Outlet Scour," Journal of Hydraulic Engineering, ASCE, (113)(3), p. 393-400, March.

Abt, S. R., P. L. Thompson, and T. M. Lewis, 1996. "Enhancement of the Culvert Outlet Scour Estimation Equations," Transportation Research Record 1523.

ACPA, 1972. "Culvert Velocity Reduction By Internal Energy Dissipators," Concrete Pipe News, pp. 87-94, American Concrete Pipe Association, Arlington, VA, Oct..

ASCE, 1957. "Hydraulic Design of Stilling Basins," Journal of the Hydraulics Division, ASCE, paper 1406, October.

Berry, N. K., 1948. "The Start of Bedload Movement," Thesis, University of Colorado.

Blaisdell, Fred W. and Charles A. Donnelly, 1956. "The Box Inlet Drop Spillway and Its Outlet," Transactions, ASCE, Vol. 121, pp. 955-986.

Blaisdell, Fred W., 1959. "The SAF Stilling Basin," U.S. Government Printing Office.

Bohan, J. P., 1970. "Erosion and Riprap Requirements at Culvert and Storm-Drain Outlets," Research Report H-70-2, U.S. Army Engineer Waterways Experiment Station, Vicksburg, Mississippi.

Bradley, J. B., D. L. Richards, and C. D. Bahner, 2005. "Debris Control Structures Evaluation and Countermeasures," Hydraulic Engineering Circular No. 9 (HEC 9), Federal Highway Administration, FHWA-IF-04-16.

Bradley, J. N. and Peterka, A. J., 1961. "Stilling Basins with Sloping Apron," Symposium on Stilling Basins and Energy Dissipators, ASCE Series No. 5., June, pp. 1405-1.

Brown, Scott A. and Eric S. Clyde, 1989. "Design of Riprap Revetment," Hydraulic Engineering Circular No. 11 (HEC 11), FHWA-IP-89-016, Washington, D.C.

Chow, Ven Te, 1959. Open-Channel Hydraulics, McGraw-Hill Book Company, Inc., New York.

Colgate, D., 1971. "Hydraulic Model Studies of Corrugated Metal Pipe Underdrain Energy Dissipators," USBR, REC-ERC-71-10, January.

Corry, M. L., P. L. Thompson, F. J. Watts, J. S. Jones, and D. L. Richards, 1983. "The Hydraulic Design of Energy Dissipators for Culverts and Channels," Hydraulic Engineering Circular No. 14 (HEC 14), Federal Highway Administration, FHWA-EPD-86-110.

Doehring F. K. and S. R. Abt, 1994. "Drop Height Influence on Outlet Scour," Journal of Hydraulic Engineering, ASCE, 120(12), December.

Donnell, C. A. and S. R. Abt, 1983. "Culvert Shape Effects on Localized Scour," Hydrographic Engineering Publications CER82-83CAD-SRA42, Colorado State University, Fort Collins, CO.

Donnelly, Charles A., and Fred W. Blaisdell, 1954. "Straight Drop Spillway Stilling Basin," University of Minnesota, St. Anthony Falls Hydraulic Laboratory, Technical Paper 15, Series B. November.

Dunn, I. S.,1959. "Tractive Resistance of Cohesive Channels," Journal of the Soil Mechanics and Foundations, Vol. 85, No. SM3, June.

Federal Highway Administration, 1961. "Design Charts for Open-Channel Flow," Hydraulic Design Series No. 3 (HDS 3), U.S. Government Printing Office Washington, D.C., 105 pp.

Federal Highway Administration, 2003. "Standard Specifications for Construction of Roads and Bridges on Federal Highway Projects," Federal Lands Highway Division, FP-03, Metric Units.

Fletcher, B. P. and J. L. Grace, 1972. "Practical Guidance for Estimating and Controlling Erosion at Culvert Outlets," Misc. Paper H-72-5, U.S. Army Waterways Experiment Station, Vicksburg, Mississippi.

Forster, J. W. and R. A. Skrinde, 1950. "Control of Hydraulic Jump by Sills," Transactions, ASCE, Vol.115, pp. 973-987.

French, Richard H., 1985. Open-Channel Hydraulics, New York, New York, McGraw-Hill Book Co.

Grace, J. L., and G. A. Pickering, 1971. "Evaluation of Three Energy Dissipators for Storm Drain Outlets," U.S. Army WES, HRB, Washington, D.C.

Hinds, J., 1928. "The Hydraulic Design of Flume and Siphon Transitions," Transactions, ASCE, vol. 92, pp. 1423-1459.

Horner, S. F., 1965. "Fluid Dynamic Drag," published by author, 2 King Lane, Greenbriar, Bricktown, N.J. 08723.

Hotchkiss, R. H., P. J. Flanagan, and K. Donahoo, 2004. "Hydraulics of Broken-back Culverts," Transportation Research Record 1851, p. 35 – 44.

Hotchkiss, Rollin H. and Emily A. Larson, 2004. "Energy Dissipation in Culverts by Forcing a Hydraulic Jump at the Outlet," Draft Report NDOR Research Project SPR-p1(04) P566, Nebraska Department of Roads.

Hotchkiss, Rollin H., Emily A. Larson, and David M. Admiraal, 2005. "Energy Dissipation in Culverts by Forced Hydraulic Jump Within a Barrel. Transportation Research Record, Transportation Research Board, Issue 1904.

Ippen, A. T., 1951. "Mechanics of Supercritical Flow," ASCE Transactions Volume 116, pp. 268-295.

Jones, J. S., 1975. Unpublished FHWA in-house research incorporated into the first edition of HEC 14.

Keim, R. S., 1962. "The Contra Costa Energy Dissipator," Journal of Hydraulic Division, ASCE Paper 3077, March, p. 109.

Kilgore, R. T. and G. K. Cotton, 2005. "Design of Stable Channels with Flexible Linings, 3rd Edition," Hydraulic Engineering Circular No. 15 (HEC 15), Federal Highway Administration, FHWA-NHI-05-114, August.

Lagasse, P. F., J. D. Schall, and E. V. Richardson, 2001. "Stream Stability at Highway Structures," Hydraulic Engineering Circular No 20 (HEC 20), Federal Highway Administration, FHWA-NHI-01-002, March.

MacDonald, T. C., 1967. "Model Studies of Energy Dissipators for Large Culverts," Hydraulic Engineering Laboratory Study HEL-13-5, University of California, Berkeley, November.

Meshgin, K. and W. L. Moore, 1970. "Design Aspects and Performance Characteristics of Radical Flow Energy Dissipators," University of Texas at Austin, Research Report 116-2F, August.

Mohanty, P. K., 1959. "The Dynamics of Turbulent Flow in Steep, Rough, Open Channels," Ph.D. Dissertation, Utah State University, Logan, Utah.

Morris, H. M., 1963. Applied Hydraulics In Engineering, The Ronald Press Company, New York, 1963.

Morris, H. M., 1968. "Hydraulics of Energy Dissipation in Steep Rough Channels," Virginia Polytechnical Institute Bulletin 19, VPI & SU, Blacksburg, VA, November.

Morris, H. M., 1969. "Design of Roughness Elements for Energy Dissipation in Highway Drainage Chutes," HRR #261, pp. 25-37, TRB, Washington, D.C.

Normann, Jerome M., 1974. "Hydraulic Aspects of Fish Ladder Baffles in Box Culverts," Preliminary FHWA Hydraulics Engineering Circular, Jan.

Normann, Jerome M., Robert J. Houghtalen, and William J. Johnston, 2001. "Hydraulic Design of Highway Culverts," Hydraulic Design Series No 5 (HDS 5), Second Edition, FHWA-NHI-01-020, September.

Peterka, A. J., 1978. "Hydraulic Design of Stilling Basins and Energy Dissipators," USBR, Engineering Monograph No. 25, January.

Peterson, D. F. and P. K. Mohanty, 1960. "Flume Studies of Flow in Steep Rough Channels," ASCE Hydraulics Journal, HY-9, Nov.

Powell, R. W., 1946. "Flow in a Channel of Definite Roughness," ASCE Transactions, Figure 10, p. 544.

Rand, Walter, 1955. "Flow Geometry at Straight Drop Spillways," Paper 791, Proceedings, ASCE, Volume 81, pp. 1-13, September.

Ruff, J. F., S. R. Abt, C. Mendoza, A. Shaikh, and R. Kloberdanz, 1982. "Scour at Culvert Outlets in Mixed Bed Materials," FHWA-RD-82-011, Offices of Research and Development, Washington D.C. 20590, September.

Sarikelle, S. and Andrew L. Simon, 1980. "Field and Laboratory Evaluation of Energy Dissipators for Culvert and Storm Drain Outlets, Volume I, Modular Energy Dissipators, Internal Energy Dissipators, Rock Channel Protection," Book viii, 164 p. Akron, Ohio, University of Akron, Dept. of Civil Engineering, OHIO-DOT-03-79, CEHY80-2, FHWA-OH-79-03

Schall, James D., Everitt V. Richardson, and Johnny L. Morris, 2001. "Introduction to Highway Hydraulics," Hydraulic Design Series Number 4 (HDS 4), Federal Highway Administration, FHWA-NHI-01-019, August.

Searcy, James K., 1967. "Use of Riprap for Bank Protection," Federal Highway Administration, Washington, D.C., pp. 43.

Shoemaker, R. F., 1956. "Hydraulics of Box Culverts With Fish Ladder Baffles," Proceedings of the 35th Annual Meeting of the HRB, Washington, D.C., pp.196-209.

Simons, D. B., M. A. Stevens, and F. J. Watts, 1970. "Flood Protection at Culvert Outlets," Colorado State University Fort Collins, Colorado, CER 69-70 DBS-MAS-FJW4.

Stevens. M. A. and D. B. Simons, 1971. "Experimental Programs and Basic Data for Studies of Scour in Riprap at Culvert Outfalls," Colorado State University, Fort Collins, Colorado, CER 70-71-MAS-DBS-57.

Stevens, M. A., D. B. Simons, and F.J. Watts, 1971. "Riprapped Basins for Culvert Outfalls," Highway Research Record No. 373, Highway Research Board, Washington, DC.

Sylvester, R., 1964. "Hydraulic Jump in All Shapes of Horizontal Channels," Journal Hydraulics Division ASCE, Vol. 90, HY1. Jan, pp. 23-55.

U.S. Army Corps of Engineers, 1963. "Impact Type Energy Dissipators for Storm-Drainage Outfalls Stilling Well Design," Technical Report No. 2-620 March, WES, Vicksburg, Mississippi.

U.S. Army Corps of Engineers, 1994. "Hydraulic Design of Flood Control Channels," Engineering and Design Manual, EM 1110-2-1601, July 1991, Change 1 (June 1994).

U.S. Army Corps of Engineers, 2002. "HEC-RAS River Analysis System, User's Manual," The Hydrologic Engineering Center, Davis, CA, Version 3.1, November.

U.S. Bureau of Reclamation, 1974. "Design of Small Canal Structures," pp. 127-130, U.S. Department of the Interior.

U.S. Bureau of Reclamation, 1987. "Design of Small Dams, 3rd Edition," U.S. Department of the Interior.

Urban Drainage and Flood Control District, 2004. "Drainage Criteria Manual: Major Drainage," Denver, Colorado.

Watts, F. J., 1968. "Hydraulics of Rigid Boundary Basins," Ph.D. Dissertation, Colorado State University, Fort Collins, Colorado, August.

Wiggert, J. M. and P. D. Erfle, 1971. "Roughness Elements as Energy Dissipators of Free Surface Flow in Circular Pipes," HPR #373, pp. 64-73, TRB, Washington, D.C.